AF353103

In Vivo Nuclear Molecular Imaging in Drug Development and Pharmacological Research

In Vivo Nuclear Molecular Imaging in Drug Development and Pharmacological Research

Editor

Xuyi Yue

MDPI • Basel • Beijing • Wuhan • Barcelona • Belgrade • Manchester • Tokyo • Cluj • Tianjin

Editor
Xuyi Yue
Radiology
Nemours Children's Health
Wilmington
United States

Editorial Office
MDPI
St. Alban-Anlage 66
4052 Basel, Switzerland

This is a reprint of articles from the Special Issue published online in the open access journal *Pharmaceuticals* (ISSN 1424-8247) (available at: www.mdpi.com/journal/pharmaceuticals/special_issues/Molecular_Imaging_Drug_Development).

For citation purposes, cite each article independently as indicated on the article page online and as indicated below:

LastName, A.A.; LastName, B.B.; LastName, C.C. Article Title. *Journal Name* **Year**, *Volume Number*, Page Range.

ISBN 978-3-0365-7391-5 (Hbk)
ISBN 978-3-0365-7390-8 (PDF)

© 2023 by the authors. Articles in this book are Open Access and distributed under the Creative Commons Attribution (CC BY) license, which allows users to download, copy and build upon published articles, as long as the author and publisher are properly credited, which ensures maximum dissemination and a wider impact of our publications.
The book as a whole is distributed by MDPI under the terms and conditions of the Creative Commons license CC BY-NC-ND.

Contents

About the Editor

Xuyi Yue

Dr. Yue is a research scientist and lab head at the Department of Radiology, Nemours Children's Hospital, Delaware, and a research assistant professor of pediatrics at Thomas Jefferson University in Philadelphia. He also holds an affiliate assistant professor position at the Department of Psychological and Brain Sciences at the University of Delaware. Dr. Yue obtained his Ph.D. degree from the Shanghai Institute of Organic Chemistry, Chinese Academy of Sciences. Dr. Yue received his molecular imaging training, particularly PET imaging, from the NIH and Washington University in St. Louis. Dr. Yue's lab focuses on developing radioactive imaging agents for preclinical research and clinical translation for brain disorders. He has successfully translated several imaging agents into clinical investigations. Dr. Yue received the Ursula Mary Kocemba-Slosky, Ph.D., Best ACNM Nuclear Medicine Research Abstract Award in 2020 and the NIH/NIBIB Trailblazer Award in 2021. His research has also been supported by the Delaware CTR ACCEL Pilot Award, Delaware INBRE Faculty Start-Up Award, Delaware INBRE Pilot Project Award and The Nemours Foundation.

Preface to "In Vivo Nuclear Molecular Imaging in Drug Development and Pharmacological Research"

Molecular imaging is a valuable tool for the real-time visualization of biological processes at the molecular level in intact living subjects. Nuclear molecular imaging has been receiving increasing attention due to its high sensitivity and critical role in precision medicine, as well as its use for personalized care in patient management. This Special Issue of *Pharmaceuticals*, "In Vivo Nuclear Molecular Imaging in Drug Development and Pharmacological Research", features original and review articles related to in vivo molecular imaging, including topics of PET imaging modality, in drug development and pharmacological research. The Special Issue published one editorial, eight articles and four reviews from approximately one hundred authors. It aimed to provide cutting-edge research on pre-clinical development and the clinical translation of radiopharmaceuticals within the molecular imaging community. The Special Issue covered radioligand development, existing radiotracer optimization, imaging agent evaluation in animal models, the clinical production of radiopharmaceuticals, and investigative research on the use of molecular imaging probes in human subjects. We appreciate all the authors' significant contributions to this Special Issue and hope the readers will enjoy the content.

Xuyi Yue
Editor

Editorial

Special Issue "In Vivo Nuclear Molecular Imaging in Drug Development and Pharmacological Research"

Xuyi Yue [1,2]

1 Diagnostic & Research PET/MR Center, Nemours Children's Health, Wilmington, DE 19803, USA; xuyi.yue@nemours.org
2 Department of Radiology, Nemours Children's Health, Wilmington, DE 19803, USA

Citation: Yue, X. Special Issue "In Vivo Nuclear Molecular Imaging in Drug Development and Pharmacological Research". *Pharmaceuticals* **2023**, *16*, 459. https://doi.org/10.3390/ph16030459

Received: 24 February 2023
Revised: 15 March 2023
Accepted: 16 March 2023
Published: 20 March 2023

Copyright: © 2023 by the author. Licensee MDPI, Basel, Switzerland. This article is an open access article distributed under the terms and conditions of the Creative Commons Attribution (CC BY) license (https://creativecommons.org/licenses/by/4.0/).

Nuclear molecular imaging is increasingly important in aiding diagnosis, monitoring disease progression, and assessing response to treatment. It is an essential tool in drug development and pharmacological research to study mechanisms of action, identify targets, evaluate receptor occupancy, determine dose regimens, and investigate pharmacokinetic and pharmacodynamic properties. In recent years, the United States Food and Drug Administration (FDA) approved several radioactive diagnostic agents, including piflufolastat F-18 ([18F]DCFPyL), gallium 68 PSMA-11, gallium Ga 68 gozetotide, and Fluorodopa F18 injection, for prostate cancer and suspected Parkinsonian syndromes, highlighting the critical and complementary role of nuclear molecular imaging in addition to traditional imaging modalities.

Our call for papers for this Special Issue, In Vivo Nuclear Molecular Imaging in Drug Development and Pharmacological Research, received great interest from a broad range of researchers from different fields. We have published 12 papers, involving approximately 100 authors from nine countries. We are delighted to see that the research includes a range of fields, from basic science studies to clinical translation investigations. This reprint describes cutting-edge research from a diverse community.

N-methyl-D-aspartate receptors (NMDAR) play a pivotal function in neurodegenerative diseases. However, the therapeutics targeting NMDA receptor subunits GluN1/2B need to be improved due to the lack of a selective radioligand for drug screening. To circumvent the limitations of the most commonly used but unselective [3H]ifenprodil for GluN1/2B competitive binding assay, Ahmed et al. developed a tritiated version of OF-NB1 [1]. This research is a continuation of the team's previous exciting work [2–5]. [3H]OF-NB1 showed good selectivity over the sigma 1 receptor. Furthermore, in vitro binding assay of the known GluN1/2B antagonist and sigma one compound with [3H]OF-NB1 and [3H]ifenprodil and in vivo receptor occupancy study in rats validate the favorable profile of [3H]OF-NB1.

Prostate-specific membrane antigen (PSMA) is expressed in more than 90% of prostate cancer patients [6]. Unfortunately, several reported PSMA radiotracers either show unfavorable kinetics or high uptake in non-target organs. Basuli et al. developed a series of fluorine-18-labeled oxime radiotracers based on the reported Lys-Urea-Glu scaffold by modulating the lipophilicity. The most lipophilic radiotracer maintained good in vitro binding affinity, a high tumor-to-non-target ratio in vivo, and comparable tumor uptake when compared with the FDA-approved [18F]DCFPyL. Furthermore, the simple structural modification significantly lowered the kidney uptake, which may provide a strategy to reduce nephrotoxicity [7].

Marie et al. used FDA-approved ^{99m}Tc-mebrofenin to evaluate hepatocyte transporter function. The dysregulation of hepatocyte transporters is closely associated with impaired liver function and hepatotoxicity. This study found that lipopolysaccharide (LPS)-treated rats showed a dramatic downregulation of hepatocyte transporters, including multidrug resistance-associated proteins 2 and 3. Interestingly, the antituberculosis drug rifampicin, a potent inhibitor of hepatocyte transporters, showed very different effects on the hepatocyte transporters in both control and LPS-treated rats. ^{99m}Tc-mebrofenin imaging may show potential in precision medicine with optimized dose selection for various drugs [8].

Microvascular disease is frequently associated with major pathologies, including atherosclerosis, diabetes, dementia, and stroke. It can occur in several vital organs, such as the brain, heart, and kidneys [9]. Existing strategies to non-invasively detect microvascular disease are limited. Wang et al. successfully used ^{18}F-fluorodeoxyglucose (^{18}F-FDG)-labeled rat red blood cells (^{18}F-FDG RBC) to map brain total vascular volume and intramyocardial vascular volume changes in rats challenged by a coronary artery vasodilator. In addition, ^{18}F-FDG-labeled erythrocytes can localize infarcted myocardium in a myocardial infarction rat model. The results correlate with metabolic ^{18}F-FDG positron emission tomography (PET) imaging and were further validated by tissue staining. Furthermore, ^{18}F-FDG RBC PET can map drug-induced intra-myocardial vasodilation in diabetic rats and normal controls [10]. This technique is operationally simple and may be promising in the non-invasive detection of whole-body microvascular pathologies and evaluation of treatment response with therapeutics targeting microvascular diseases.

Mesenchymal stem cell-derived extracellular vesicles (MSC-EV) therapy is promising as a treatment for type 1 and type 2 diabetes due to its efficiency in transferring serial biological molecules to modulate immune responses and metabolic functions. Therefore, the safe delivery and tracking of MSC-EVs are critical for diabetes therapies. Li et al. developed iodine-124-labeled umbilical cord MSC-EV, which showed over 95% purity over 4 h. The researchers used two administration routes (intra-arterial vs. intravenous) to conduct a pilot study in non-diabetic Lewis rats to guide iodine-124-labeled umbilical cord MSC-EV delivery. The results show that the two strategies display similar delivery efficacies, except in the spleen and liver. However, the intravenous administration method is preferred since it is much less invasive and operationally simple compared with the invasive and challenging intra-arterial delivery [11].

Chen et al. reported a heterobivalent peptide modified with thin layer-protected gold nanoparticles for multiple imaging of esophageal cancer in a human xenograft model. The nanoprobe features good stability and biocompatibility, dual targeting of epidermal growth factor receptors and erb-b2 receptor tyrosine kinase 2, multimodal imaging with photoacoustic and computed tomography, and favorable in vivo kinetics. In addition, the dual targeting strategy shows promise for detecting cancers in the early stages due to improved sensitivity [12].

Alzheimer's disease (AD) is the leading cause of dementia. The recent failure of crenezumab, an investigational anti-amyloid drug, is the latest setback in effective AD treatment [13]. Therefore, preclinical animal models with various PET probes are critical to research mechanisms of action and develop potential therapeutics. Ni contributed to a comprehensive review of AD imaging in several animal models with PET modality. The author reviewed well-studied biomarkers, including amyloid, brain glucose metabolism, and synaptic and neurotransmitter receptors, and discussed new biomarkers in AD, such as microtubule and mitochondria imaging. The author also addressed the challenges of translating the rodent AD model to a clinical investigation and proposed models close to human AD pathology [14].

On 23 March 2022, the FDA approved gallium Ga 68 gozetotide injection, a peptide conjugate, for the diagnosis of PSMA-positive lesions in males with prostate cancer. On the same day, the FDA approved the amino acid-based Lutetium (^{177}Lu) vipivotide tetraxetan for treating patients with castration-resistant prostate cancer [15]. Radiometal-based agents have received increasing attention due to favorable half-life, easily adaptable clinical production, and radiotheranostics implementation. New radionuclides that can easily be distributed to satellite sites, have favorable positron emission energies, and are operationally simple for a therapeutic match to meet personalized medicine requirements are still attractive. Fonseca et al. reported two routes to produce clinical doses of ^{61}Cu-based radiopharmaceuticals with fully automated Good Manufacturing Practice

(GMP)-compliant procedures. The purity of the two targets significantly affects the yield of copper-61. The utilization of the produced copper-61 was demonstrated by a fully automated GMP-compliant production of three radiopharmaceuticals labeled with gallium-68 in clinical practice. Copper-61 may serve as an alternative radionuclide to the widely used gallium-68 [16].

Fibroblast activation protein (FAP) is a novel target for the molecular imaging of oncology and cardiovascular disease [17,18]. The research by Diekmann et al. using ^{68}Ga-fibroblast-activation protein-46 (^{68}Ga-FAP-46) PET/CT to predict myocardial infarction was selected as the Society of Nuclear Medicine and Molecular Imaging (SNMMI) Image of the Year 2022 [19,20]. In this Special Issue of *Pharmaceuticals*, Vallejo-Armenta et al. reported the findings of a boronic acid derivative, $[^{99m}$Tc]Tc-{(R)-1-[(6-hydrazinylnicotinoyl)-D-alanyl]pyrrolidin-2-yl}-labeled boronic acid ($[^{99m}$Tc]Tc-iFAP), as the radioligand targeting FAP in 32 patients with six different cancer entities. The results show that $[^{99m}$Tc]Tc-iFAP can effectively detect high-grade World Health Organization (WHO) III–IV gliomas with a 100% sensitivity for primary tumors, while it is inferior to ^{18}F-FDG in lymph node metastases and distant metastases cases. However, patients with peritoneal carcinomatosis lesions in recurrent colorectal cancer show only $[^{99m}$Tc]Tc-iFAP uptake, demonstrating its valuable complementary role for prognostic evaluation [21].

Son et al. reviewed PET/MR hybrid systems and their applications in psychiatric disorders. The authors discussed the advancements of PET, MRI, and fusion PET-MRI technology in clinical settings. While there are many carbon-11 and fluorine-18-labeled tracers targeting serotonin receptors and transporters, glucose, dopamine receptors, and phosphodiesterase 10A, etc., for clinical investigations, the authors stated that improving the PET spatial resolution to match MRI is crucial for reliable and quantitative analysis. The team achieved a 1.56 mm full width at the half-maximum transaxial resolution, a resolution even higher than the high-resolution research tomograph (2.47 mm) [22].

Netufo et al. reviewed intraoperative fluorescence imaging agents for guiding glioblastoma surgery in preclinical research and clinical practice. 5-Aminolevulinic acid (5-ALA) is the only FDA-approved intra-operative fluorescence imaging agent for glioblastoma patients. However, 5-ALA has several limitations, including challenges in identifying critical neurological components under dark-field conditions, photobleaching, and 2D-only images. Therefore, the authors emphasize the importance of using a targeting strategy and a combination of multimodal imaging, such as PET-guide surgical planning with intra-operative fluorescence imaging agents, to determine the extent of resection and improve overall survival [23].

Lung ventilation–perfusion scintigraphy is a critical technique to assess regional ventilation and perfusion function. Currently, most radiopharmaceuticals for lung function are technetium-99m (^{99m}Tc)-based single-photon emission computerized tomography (SPECT) agents. However, PET has a higher sensitivity, resolution, and better quantitative capacity than SPECT. Therefore, Blanc-Béguin et al. reviewed the chemical, technical, and pharmacological aspects of ^{99m}Tc- and ^{68}Ga-based lung ventilation and perfusion imaging agents and discussed the advantages and challenges of transition from ^{99m}Tc- to ^{68}Ga-labeled agents for optimal clinical use. The authors concluded that minimal pharmacological property changes and simplified and GMP-compliant automated procedures are essential for switching from ^{99m}Tc- to ^{68}Ga-labeled nuclear imaging agents for lung functions [24].

In summary, this Special Issue highlights the opportunities and challenges in nuclear molecular imaging from preclinical research to clinical translation and covers a broad overview of the field. I sincerely thank all the authors for their valuable contributions to this Special Issue. I also thank all the reviewers and editors for their tremendous support. I hope the articles and reviews in this Special Issue meet readers' expectations in the field and further promote nuclear molecular imaging research in the community.

Funding: The author received financial support from a Delaware INBRE Pilot Project Award from the National Institute of General Medical Sciences of the National Institutes of Health under grant number P20GM103446 (PI: Duncan), an NIH R21 grant EB032025 from the National Institute of Biomedical Imaging and Bioengineering (NIBIB), and The Nemours Foundation.

Institutional Review Board Statement: Not applicable.

Informed Consent Statement: Not applicable.

Data Availability Statement: Data is contained within the article.

Conflicts of Interest: The author declares no conflict of interest.

References

1. Ahmed, H.; Gisler, L.; Elghazawy, N.H.; Keller, C.; Sippl, W.; Liang, S.H.; Haider, A.; Ametamey, S.M. Development and validation of [3H]OF-NB1 for preclinical assessment of GluN1/2B candidate drugs. *Pharmaceuticals* **2022**, *15*, 960. [CrossRef] [PubMed]
2. Haider, A.; Iten, I.; Ahmed, H.; Herde, A.M.; Gruber, S.; Krämer, S.D.; Keller, C.; Schibli, R.; Wünsch, B.; Mu, L.; et al. Identification and preclinical evaluation of a radiofluorinated benzazepine derivative for imaging the GluN2B subunit of the ionotropic NMDA receptor. *J. Nucl. Med.* **2019**, *60*, 259–266. [CrossRef]
3. Ahmed, H.; Wallimann, R.; Haider, A.; Hosseini, V.; Gruber, S.; Robledo, M.; Nguyen, T.A.N.; Herde, A.M.; Iten, I.; Keller, C.; et al. Preclinical development of 18F-OF-NB1 for imaging GluN2B-containing N-Methyl-d-Aspartate receptors and its utility as a biomarker for amyotrophic lateral sclerosis. *J. Nucl. Med.* **2021**, *62*, 259–265. [CrossRef] [PubMed]
4. Ahmed, H.; Zheng, M.Q.; Smart, K.; Fang, H.; Zhang, L.; Emery, P.R.; Gao, H.; Ropchan, J.; Haider, A.; Tamagnan, G.; et al. Evaluation of (rac)-, (R)-, and (S)-18F-OF-NB1 for Imaging GluN2B subunit–containing N-Methyl-d-Aspartate receptors in nonhuman primates. *J. Nucl. Med.* **2022**, *63*, 1912–1918. [CrossRef] [PubMed]
5. Rischka, L.; Vraka, C.; Pichler, V.; Rasul, S.; Nics, L.; Gryglewski, G.; Handschuh, P.; Murgaš, M.; Godbersen, G.M.; Silberbauer, L.R.; et al. First-in-humans brain PET imaging of the GluN2B-containing N-methyl-d-aspartate receptor with (R)-11C-Me-NB1. *J. Nucl. Med.* **2022**, *63*, 936–941. [CrossRef] [PubMed]
6. Calais, J.; Czernin, J. PSMA expression assessed by PET imaging is a required biomarker for selecting patients for any PSMA-targeted therapy. *J. Nucl. Med.* **2021**, *62*, 1489. [CrossRef] [PubMed]
7. Basuli, F.; Phelps, T.E.; Zhang, X.; Woodroofe, C.C.; Roy, J.; Choyke, P.L.; Swenson, R.E.; Jagoda, E.M. Fluorine-18 labeled urea-based ligands targeting Prostate-Specific Membrane Antigen (PSMA) with increased tumor and decreased renal uptake. *Pharmaceuticals* **2022**, *15*, 597. [CrossRef]
8. Marie, S.; Hernández-Lozano, I.; Le Vée, M.; Breuil, L.; Saba, W.; Goislard, M.; Goutal, S.; Truillet, C.; Langer, O.; Fardel, O.; et al. Pharmacokinetic imaging Using99m Tc-Mebrofenin to untangle the pattern of hepatocyte transporter disruptions induced by endotoxemia in rats. *Pharmaceuticals* **2022**, *15*, 392. [CrossRef]
9. Berry, C.; Sidik, N.; Pereira, A.C.; Ford, T.J.; Touyz, R.M.; Kaski, J.C.; Hainsworth, A.H. Small-vessel disease in the heart and brain: Current knowledge, unmet therapeutic need, and future directions. *J. Am. Heart Assoc.* **2019**, *8*, 11104. [CrossRef]
10. Wang, S.; Budzevich, M.; Abdalah, M.A.; Balagurunathan, Y.; Choi, J.W. In vivo imaging of rat vascularity with FDG-labeled erythrocytes. *Pharmaceuticals* **2022**, *15*, 292. [CrossRef]
11. Li, J.; Komatsu, H.; Poku, E.K.; Olafsen, T.; Huang, K.X.; Huang, L.A.; Chea, J.; Bowles, N.; Chang, B.; Rawson, J.; et al. Biodistribution of intra-arterial and intravenous delivery of human umbilical cord mesenchymal stem cell-derived extracellular vesicles in a rat model to guide delivery strategies for diabetes therapies. *Pharmaceuticals* **2022**, *15*, 595. [CrossRef] [PubMed]
12. Chen, J.; Nguyen, V.P.; Jaiswal, S.; Kang, X.; Lee, M.; Paulus, Y.M.; Wang, T.D. Thin layer-protected gold nanoparticles for targeted multimodal imaging with photoacoustic and CT. *Pharmaceuticals* **2021**, *14*, 1075. [CrossRef] [PubMed]
13. NIA Statement on Crenezumab Trial Results: Anti-Amyloid Drug Did Not Demonstrate a Statistically Significant Clinical Benefit in People with Inherited form of Alzheimer's Disease | National Institute on Aging. Available online: https://www.nia.nih.gov/news/nia-statement-crenezumab-trial-results-anti-amyloid-drug-did-not-demonstrate-statistically (accessed on 10 February 2023).
14. Ni, R. Positron emission tomography in animal models of Alzheimer's Disease amyloidosis: Translational implications. *Pharmaceuticals* **2021**, *14*, 1179. [CrossRef] [PubMed]
15. FDA Approves Pluvicto for Metastatic Castration-Resistant Prostate Cancer | FDA. Available online: https://www.fda.gov/drugs/resources-information-approved-drugs/fda-approves-pluvicto-metastatic-castration-resistant-prostate-cancer (accessed on 10 February 2023).
16. Fonseca, A.I.; Alves, V.H.; Do Carmo, S.J.C.; Silva, M.; Hrynchak, I.; Alves, F.; Falcão, A.; Abrunhosa, A.J. Production of GMP-compliant clinical amounts of Copper-61 radiopharmaceuticals from liquid targets. *Pharmaceuticals* **2022**, *15*, 723. [CrossRef]
17. Kuyumcu, S.; Sanli, Y.; Subramaniam, R.M. Fibroblast-activated protein inhibitor PET/CT: Cancer diagnosis and management. *Front. Oncol.* **2021**, *11*, 4605. [CrossRef]
18. Heckmann, M.B.; Reinhardt, F.; Finke, D.; Katus, H.A.; Haberkorn, U.; Leuschner, F.; Lehmann, L.H. Relationship between cardiac fibroblast activation protein activity by positron emission tomography and cardiovascular disease. *Circ. Cardiovasc. Imaging* **2020**, *13*, 10628. [CrossRef]

19. SNMMI Image of the Year 2022: PET/CT biomarker predicts post-MI cardiac remodeling. *J. Nucl. Med.* **2022**, *63*, 16N.
20. Diekmann, J.; Koenig, T.; Thackeray, J.T.; Derlin, T.; Czerner, C.; Neuser, J.; Ross, T.L.; Schäfer, A.; Tillmanns, J.; Bauersachs, J.; et al. Cardiac fibroblast activation in patients early after acute myocardial infarction: Integration with MR tissue characterization and subsequent functional outcome. *J. Nucl. Med.* **2022**, *63*, 1415–1423. [CrossRef]
21. Vallejo-Armenta, P.; Ferro-Flores, G.; Santos-Cuevas, C.; García-Pérez, F.O.; Casanova-Triviño, P.; Sandoval-Bonilla, B.; Ocampo-García, B.; Azorín-Vega, E.; Luna-Gutiérrez, M. [99mTc]Tc-iFAP/SPECT tumor stroma imaging: Acquisition and analysis of clinical images in six different cancer entities. *Pharmaceuticals* **2022**, *15*, 729. [CrossRef]
22. Son, Y.-D.; Kim, Y.-B.; Kim, J.-H.; Kim, J.-H.; Kwon, D.-H.; Lee, H.; Cho, Z.-H. Future prospects of positron emission tomography–magnetic resonance imaging hybrid systems and applications in psychiatric disorders. *Pharmaceuticals* **2022**, *15*, 583. [CrossRef]
23. Netufo, O.; Connor, K.; Shiels, L.P.; Sweeney, K.J.; Wu, D.; O'shea, D.F.; Byrne, A.T.; Miller, I.S. Refining glioblastoma surgery through the use of intra-operative fluorescence imaging agents. *Pharmaceuticals* **2022**, *15*, 550. [CrossRef] [PubMed]
24. Blanc-Béguin, F.; Hennebicq, S.; Robin, P.; Tripier, R.; Salaün, P.Y.; Le Roux, P.Y. Radiopharmaceutical labelling for lung ventilation/perfusion PET/CT imaging: A review of production and optimization processes for clinical use. *Pharmaceuticals* **2022**, *15*, 518. [CrossRef] [PubMed]

Disclaimer/Publisher's Note: The statements, opinions and data contained in all publications are solely those of the individual author(s) and contributor(s) and not of MDPI and/or the editor(s). MDPI and/or the editor(s) disclaim responsibility for any injury to people or property resulting from any ideas, methods, instructions or products referred to in the content.

pharmaceuticals

MDPI

Article

Development and Validation of [^{3}H]OF-NB1 for Preclinical Assessment of GluN1/2B Candidate Drugs

Hazem Ahmed [1,†], Livio Gisler [1,†], Nehal H. Elghazawy [2], Claudia Keller [1], Wolfgang Sippl [2], Steven H. Liang [3], Ahmed Haider [3,*] and Simon M. Ametamey [1,*]

1   Center for Radiopharmaceutical Sciences ETH-PSI-USZ, Institute of Pharmaceutical Sciences ETH, Vladimir-Prelog-Weg 4, 8093 Zurich, Switzerland; hazem.farouk20@gmail.com (H.A.); liviogisler@hotmail.com (L.G.); claudia.keller@pharma.ethz.ch (C.K.)
2   Institute of Pharmacy, Department of Medicinal Chemistry, Martin-Luther-University Halle-Wittenberg, W.-Langenbeck-Str. 4, 06120 Halle, Germany; nehal.elghazawy@gmail.com (N.H.E.); wolfgang.sippl@pharmazie.uni-halle.de (W.S.)
3   Division of Nuclear Medicine and Molecular Imaging, Massachusetts General Hospital & Department of Radiology, Harvard Medical School, Boston, MA 02114, USA; liang.steven@mgh.harvard.edu
*   Correspondence: ahmed.haider@usz.ch (A.H.); simon.ametamey@pharma.ethz.ch (S.M.A.)
†   These authors contributed equally to this work.

Abstract: GluN2B-enriched *N*-methyl-D-aspartate receptors (NMDARs) are implicated in several neurodegenerative and psychiatric diseases, such as Alzheimer's disease. No clinically valid GluN1/2B therapeutic exists due to a lack of selective GluN2B imaging tools, and the state-of-the-art [^{3}H]ifenprodil shows poor selectivity in drug screening. To this end, we developed a tritium-labeled form of OF-NB1, a recently reported selective GluN1/2B positron emission tomography imaging (PET) agent, with a molar activity of 1.79 GBq/µmol. The performance of [^{3}H]OF-NB1 and [^{3}H]ifenprodil was compared through head-to-head competitive binding experiments, using the GluN1/2B ligand CP-101,606 and the sigma-1 receptor (σ1R) ligand SA-4503. Contrary to [^{3}H]ifenprodil, the usage of [^{3}H]OF-NB1 differentiated between GluN1/2B and σ1R binding components. These results were corroborated by observations from PET imaging experiments in Wistar rats using the σ1R radioligand [^{18}F]fluspidine. To unravel the binding modes of OF-NB1 and ifenprodil in GluN1/2B and σ1Rs, we performed a retrospective in silico study using a molecular operating environment. OF-NB1 maintained similar interactions to GluN1/2B as ifenprodil, but only ifenprodil successfully fitted in the σ1R pocket, thereby explaining the high GluN1/2B selectivity of OF-NB1 compared to ifenprodil. We successfully showed in a proof-of-concept study the superiority of [^{3}H]OF-NB1 over the gold standard [^{3}H]ifenprodil in the screening of potential GluN1/2B drug candidates.

Keywords: GluN1/2B receptors; NMDA; [^{3}H]ifenprodil; σ1 and σ2 receptors; receptor occupancy; PET imaging; drug development; neurodegenerative diseases

Citation: Ahmed, H.; Gisler, L.; Elghazawy, N.H.; Keller, C.; Sippl, W.; Liang, S.H.; Haider, A.; Ametamey, S.M. Development and Validation of [^{3}H]OF-NB1 for Preclinical Assessment of GluN1/2B Candidate Drugs. *Pharmaceuticals* **2022**, *15*, 960. https://doi.org/10.3390/ph15080960

Academic Editors: Xuyi Yue and Gerald Reischl

Received: 31 May 2022
Accepted: 27 July 2022
Published: 2 August 2022

Publisher's Note: MDPI stays neutral with regard to jurisdictional claims in published maps and institutional affiliations.

Copyright: © 2022 by the authors. Licensee MDPI, Basel, Switzerland. This article is an open access article distributed under the terms and conditions of the Creative Commons Attribution (CC BY) license (https://creativecommons.org/licenses/by/4.0/).

1. Introduction

N-methyl-D-aspartate receptors (NMDARs) are ionotropic excitatory neurotransmission mediators with pivotal functions in the central nervous system (CNS), particularly in learning and memory [1,2]. NMDARs are heterotetrameric constructs of GluN1/2/3 subunits, which are encoded by one, four and two genes, respectively [1,2]. The type of GluN2 subunit (A–D) dictates the function of the receptor and the unique spatial and temporal distribution across the brain [1,3–6]. In the adult brain, for instance, GluN1/2A is ubiquitously expressed, whereas the GluN1/2B is localized in the forebrain area [6–8]. Furthermore, the composition of NMDAR subunits varies depending on the localization within the neurons. While GluN1/2A is abundant at synaptic junctions, extrasynaptic

NMDARs are particularly enriched with GluN1/2B [9]. Finally, the localization and function are closely linked; synaptic NMDARs promote neuronal survival upon stimulation, while extrasynaptic NMDARs are pro-apoptotic [9]. Notwithstanding the advances in our understanding of NMDARs, much remains to be elucidated with regards to the role of distinct subtypes in neuropathologies [4,10].

NMDARs are activated by the binding of the most abundant CNS excitatory neurotransmitter, glutamate, to the GluN2 subunit, while co-agonist glycine binds to the GluN1 subunit, thereby triggering a cascade of ion channel opening, Ca^{2+} influx and the subsequent signal transduction [3]. Analogous to their vital physiological functions, NMDARs play a key role in several neuropathologies, such as Alzheimer's disease, Parkinson's disease and stroke, and thus they have been the focus of drug development efforts for decades [9,11–13]. Initially developed as a competitive allosteric GluN1/2B antagonist, ifenprodil served as the template for multiple subunit-selective contemporary antagonists such as CP-101,606 and CERC-301 for the treatment of various neuropathologies [14]. The attention garnered toward GluN1/2B ligands stems from the detrimental role of the GluN1/2B receptors, in addition to the side effects that result from the attenuation of physiological NMDAR functions by non-subtype selective NMDAR-targeted ion channel blockers [15]. Accordingly, it is envisioned that subtype-selective GluN1/2B antagonism has the potential to elicit efficacy while being well tolerated. Nonetheless, the clinical development of GluN1/2B antagonists has been hampered by the lack of efficacy and significant off-target activity, thus necessitating more sophisticated drug development strategies to ensure appropriate GluN1/2B selectivity [14,16]. [^{3}H]Ifenprodil is still hailed as the most commonly used radioligand for GluN1/2B competitive binding assays, despite the lack of selectivity over other CNS targets, especially regarding sigma receptors (K_i (σ1R) = 13 nM; K_i (σ2R) = 1.89 nM; K_i (GluN1/2B) = 11 nM) [17,18]. Research efforts to characterize the binding of [^{3}H]ifenprodil shed light on the effect of low temperatures of 4 °C on the preferential binding of ifenprodil toward NMDARs [19]. However, performing binding affinity assays at 4 °C is not ideal, and could influence the behavior of the ligands under investigation. In light of this information, the search for a highly selective tritiated GluN1/2B antagonist is vital. In addition to the competitive receptor binding assays, a suitable tritiated ligand can be used in autoradiography experiments and saturation assays to determine the receptor expression (B_{max}) levels.

Positron-emission tomography (PET) has become a vital tool in early drug development and has been used effectively to determine drug uptake, distribution, target engagement and receptor occupancy in vivo [20,21]. Previously published data suggested that 2,3,4,5-tetrahydro-1H-3-benzazepines constitute promising GluN1/2B ligands with regard to selectivity and potency [22–29]. A particularly auspicious ligand from this series is OF-NB1, which exhibits high GluN1/2B affinity and high selectivity over σ1Rs (Figure 1, K_i (GluN1/2B) = 10.4 ± 4.7 nM; K_i (σ1R) = 410 nM) [22]. This ligand was radiofluorinated with fluorine-18 for PET imaging studies and showed excellent accumulation in GluN1/2B-rich brain regions, namely the cortex, hippocampus, thalamus and striatum. Furthermore, it exhibited excellent specificity and selectivity when challenged with GluN1/2B and σ1R ligands. With regards to its physicochemical properties, OF-NB1 exhibited a log $D_{7.4}$ of 2.05 ± 0.08 (n = 4), which is optimal for brain penetration [29]. As such, we tritiated OF-NB1 with the ultimate goal of providing a highly selective probe to facilitate GluN1/2B-targeted drug development. To support this goal, blocking studies with the GluN1/2B antagonist CP-101,606 were conducted using the σ1R PET tracer, [^{18}F]fluspidine, in Wistar rats in vivo [30]. Subsequently, we performed docking studies of OF-NB1 and ifenprodil in order to substantiate the experimental results.

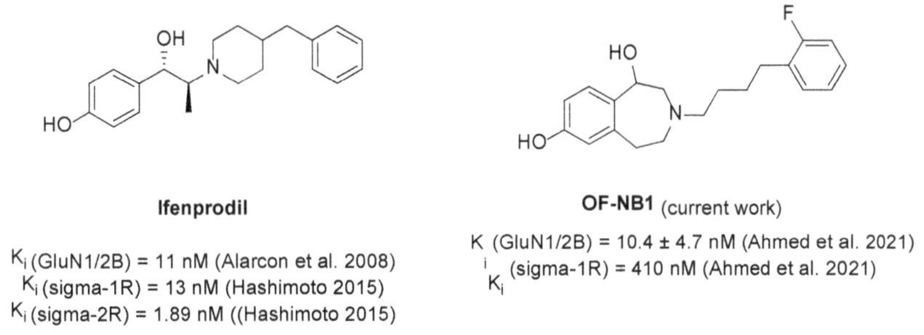

Ifenprodil

K_i (GluN1/2B) = 11 nM (Alarcon et al. 2008)
K_i (sigma-1R) = 13 nM (Hashimoto 2015)
K_i (sigma-2R) = 1.89 nM ((Hashimoto 2015)

OF-NB1 (current work)

K_i (GluN1/2B) = 10.4 ± 4.7 nM (Ahmed et al. 2021)
K_i (sigma-1R) = 410 nM (Ahmed et al. 2021)

Figure 1. Left: Structure of ifenprodil, first ligand described for binding to the ifenprodil binding site of the GluN1/2B. Right: Structure of OF-NB1, a selective GluN1/2B ligand that is investigated in the current work.

2. Results and Discussion

At the outset of the studies, a five-step synthetic route was devised to prepare the dibromine-bearing precursor for the radiosynthesis of [^{3}H]OF-NB1 (**7**) (Scheme 1). Sonogashira coupling of substituted iodobenzene **1** and alkyne **2** yielded benzalkyne alcohol **3** in 78% yield. Alcohol **3** was reduced with Raney nickel, subsequently treated with MsCl, and the resulting intermediate was used for the N-alkylation of commercially available bezazepine derivative **6** to give the N-alkylated product **7**, with a yield of 44% over three steps. The final demethylation of **7** using BBr$_3$ gave the desired precursor **8** with a yield of 26%. Tritium-labeling of precursor **8** yielded [^{3}H]OF-NB1 with 99% radiochemical purity and a molar activity of 1.79 GBq/μmol (48.3 Ci/mmol).

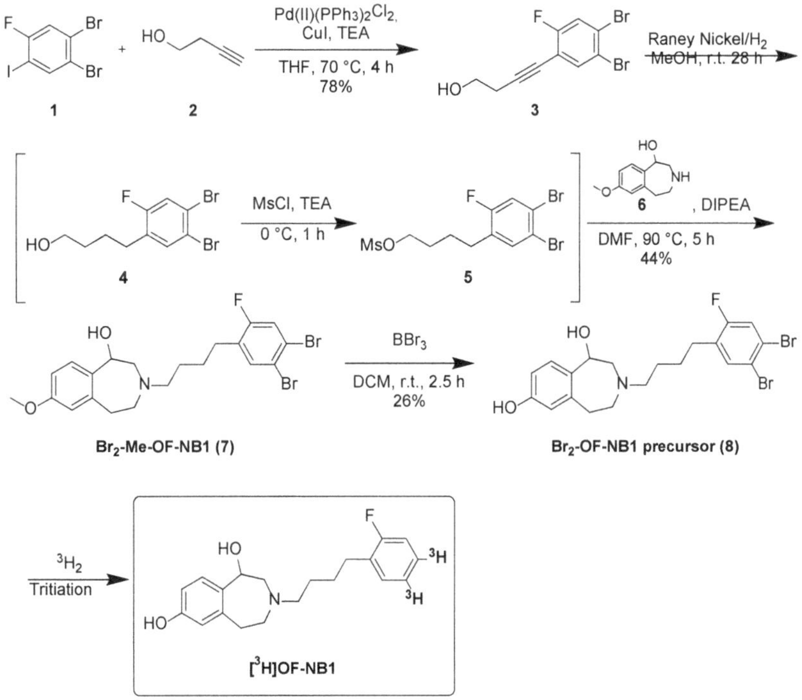

Scheme 1. Synthetic pathway toward dibromo precursor **8** and [^{3}H]OF-NB1.

In order to evaluate the viability and superiority of [³H]OF-NB1 relative to the commercially available [³H]ifenprodil, we determined the K_i (GluN1/2B) values of the GluN1/2B antagonist CP-101,606 and the σ1R ligand SA-4503 using both tritiated radioligands (Figure 2). Furthermore, we assessed the selectivity of the two drugs over σ1Rs using the σ1R radioligand, (+)-[³H]pentazocine. The results are summarized in Table 1.

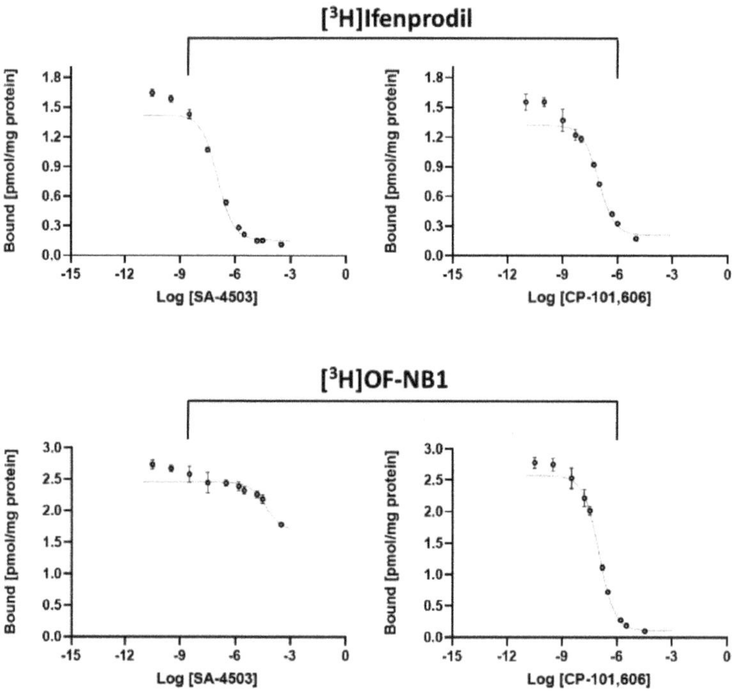

Figure 2. Representative saturation binding curves of the σ1R ligand SA-4503 (left) and the GluN1/2B antagonist CP-101,606 (right) using the two radioligands [³H]ifenprodil and [³H]OF-NB1.

The results revealed a two-digit nanomolar GluN1/2B affinity for CP-101,606 when using either of the two radioligands, [³H]OF-NB1 or [³H]ifenprodil. Both values were, however, considerably higher than the reported literature value of 16 nM [31]. Three notable differences comparing the assay at hand from the one described in literature, are as follows: (1) the absence of other agents that block non-NMDA receptors; (2) the usage of whole rat brain homogenates; and (3) the higher temperature of 25 °C.

Table 1. Results of in vitro GluN1/2B and σ1R binding affinity testing for CP-101,606 and SA-4503 in nM ± STD.

Radioligand	Test Compound	K_i (GluN1/2B)	K_i (σ1R)
[³H]OF-NB1	SA-4503	>100,000	
	CP-101,606	53 ± 4.3	
[³H]Ifenprodil	SA-4503	51 ± 13	
	CP-101,606	37 (16 *)	
[³H](+)-Pentazocine	SA-4503		3.8 (4.6 *)
	CP-101,606		94 ± 6

* Reported values in the literature [31,32].

The superiority of [³H]OF-NB1 over [³H]ifenprodil was evidently established when testing the σ1R ligand SA-4503. The K_i (GluN1/2B) value exceeded 100 μM (Table 1) when using [³H]OF-NB1 compared to 51 nM (Table 1) when using [³H]ifenprodil. The key reason for such a disparity is that SA-4503 is an extremely potent σ1R-selective ligand; however, ifenprodil binds to both GluN1/2B and σ1Rs indiscriminately. The K_i (σ1R) value of 3.8 nM for SA-4503 matched the published value of 4.6 nM [32]. On the other hand, the K_i (σ1R) value of 94 nM for CP-101,606 was considerably close to the published value of 60 nM using the σR ligand, [³H](3-(3-hydroxyphenyl)N-(1-propyl)-piperidine (3-PPP) [33]. This shows that CP-101,606 potentially exhibits a σ1R binding component, thereby providing a valid explanation for the higher K_i (GluN1/2B) value of CP-101,606 when using [³H]OF-NB1, given its low σ1R affinity as opposed to ifenprodil.

To verify our findings of CP-101,606 possessing significant σ1R binding, we conducted in vivo PET imaging studies with [¹⁸F]fluspidine, a σ1R radioligand which is known for its lack of uptake in the brain of σ1R knock-out mice [22,30]. These findings are depicted in Figures 3 and 4.

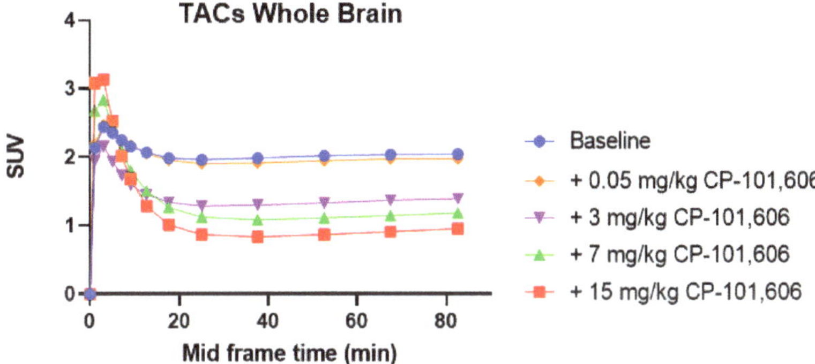

Figure 3. Time activity curves (TACs) of Wistar rat brain uptake of the σ1R tracer [¹⁸F]fluspidine under baseline and blockade conditions using the GluN1/2B antagonist CP-101,606 (0.05, 3, 7 and 15 mg/kg). Standard uptake values (SUVs) are averaged from 0–90 min.

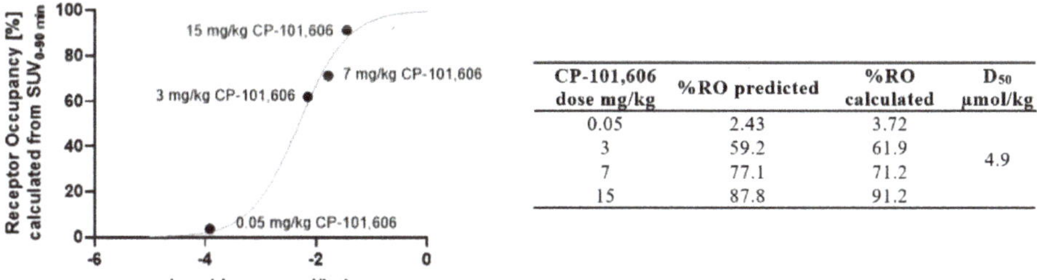

CP-101,606 dose mg/kg	%RO predicted	%RO calculated	D_{50} μmol/kg
0.05	2.43	3.72	
3	59.2	61.9	
7	77.1	71.2	4.9
15	87.8	91.2	

Figure 4. In vivo receptor occupancy (RO) of GluN1/2B antagonist CP-101,606 in Wistar rats in the presence of the σ1R tracer [¹⁸F]fluspidine. Plotting the calculated receptor occupancy against the different dose regimes allowed the calculation of D_{50} which is the administered dose of CP-101,606 that occupies 50% of the σ1Rs [23].

In vivo PET results pertaining to the off-target characteristics of CP-101,606 toward σ1Rs corroborated the results obtained from the in vitro binding study. A series of doses were investigated, ranging from 0.05–15 mg/kg. Importantly, the GluN1/2B ligand, CP-101,606, displayed a dose-dependent blockade similar to that of the σ1R ligand SA-4503 [22]. This allowed us to calculate the D_{50} (σ1R), which indicates the administered dose of CP-101,606 that occupies 50% of the σ1Rs (Figure 3). The D_{50} (σ1R) of CP-101,606 was calculated to be 4.9 μmol/kg, which is higher than the D_{50} (GluN1/2B) of 8.1 μmol/kg [26]. This strong σ1R binding behavior of CP-101,606 offsets its unwarranted reputation of being a selective GluN1/2B. There have been efforts to develop ligands with dual GluN1/2B and σ1R activity [34]. Nonetheless, for the purpose of developing selective GluN1/2B ligands, one has to consider fundamentally reformulating the development strategy.

With the aim to better understand the interactions of OF-NB1 on a molecular level, the binding mode was studied alongside the two target receptors, GluN1/2B and σ1Rs, in a retrospective in silico study. With regards to GluN1/2B, the binding of OF-NB1 was compared to ifenprodil, which was previously co-crystallized with GluN1/2B (PDB-ID: 3QEL). As shown in Figure 5A, the binding of ifenprodil can be credited to the existence of three H-bonds with Gln110 and Glu236. Additionally, the two phenyl rings of ifenprodil guided multiple hydrophobic interactions with amino acids from the two GluN subunits, such as Tyr109 (GluN1A) and Ile111 (GluN2B). While OF-NB1 was able to maintain common interactions as ifenprodil as shown in Figure 5B, it surprisingly managed to form two additional interactions. Specifically, it should be noted that OF-NB1 achieved a new H-bond with Glu106 via its secondary hydroxyl functionality. Furthermore, the distant phenyl ring of OF-NB1 makes van-der-Waals interactions with amino acids Phe176 and Leu135. These additional interactions are energetically favorable, thus contributing positively to the binding affinity of OF-NB1 toward GluN1/2B.

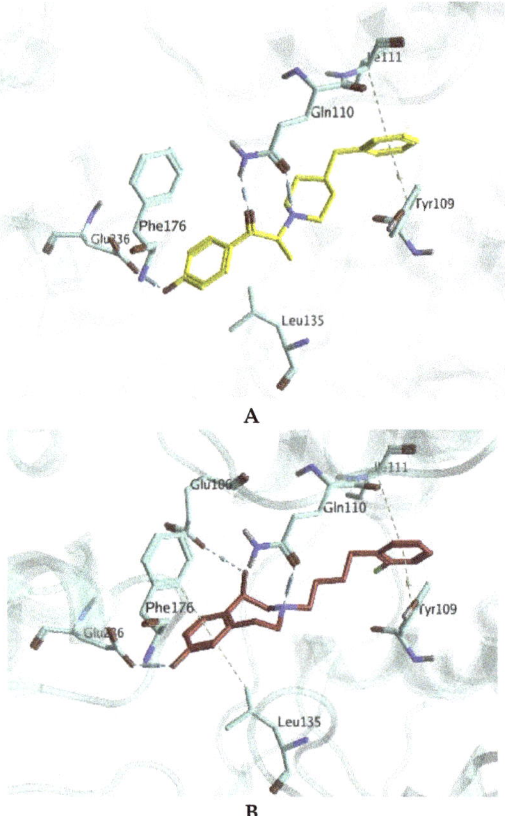

Figure 5. Predicted binding modes and possible interactions with GluN1/2B of ifenprodil (**A**) and OF-NB1 (**B**). Ifenprodil and OF-NB1 are depicted in yellow and maroon, respectively.

The co-crystal structure of σ1R with pentazocine was used (PDB-ID: 6DK1) for the docking of ifenprodil and OF-NB1 against σ1R. Both Schmidt et al. and the Glennon model for σ1R ligands highlighted the prominence of the presence of a positively charged nitrogen in all σ1R ligands, where it forms a salt-bridge with Glu172 [35]. Such an interaction was noted in the docking of ifenprodil, in addition to further interactions with Tyr103 and Asp126 (Figure 6A). Conversely, OF-NB1 was only able to maintain van-der-Waals interaction with Tyr103 and even failed to interact with the crucial Glu172, despite its positively charged nitrogen (Figure 6B). Such behavior could be explained by the odd orientation exhibited by OF-NB1 within the σ1R binding pocket. The failure of OF-NB1 to align with ifenprodil supports the observations from the in vitro saturation binding experiments.

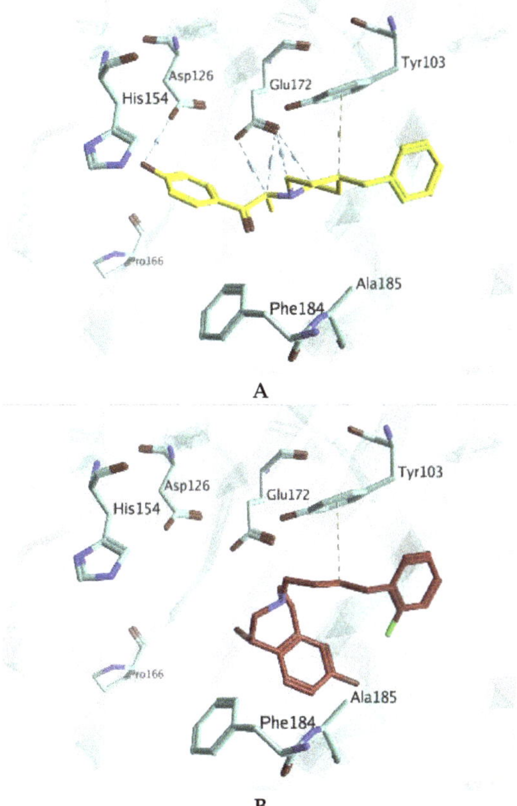

Figure 6. Predicted binding modes and possible interactions with σ1R of ifenprodil (**A**) and OF-NB1 (**B**). Ifenprodil and OF-NB1 are depicted in yellow and maroon, respectively.

3. Materials and Methods

3.1. General Methods

All non-aqueous reactions were performed under N_2 atmosphere using flame-dried glassware and standard syringe/septa methods unless stated otherwise. Reactions were magnetically stirred and further monitored by thin layer chromatography (TLC) performed on Merck TLC aluminum sheets (silica gel 60 F254). TLC spots were visualized using UV light (λ = 254 nm) or through staining with $KMnO_4$ solution. Chromatographic purification of products was performed using SiliaFlash P60 silica gel (Silicycle) for preparative column chromatography with a particle size of 40–63 μm (230–400 mesh). Reactions at −78 °C were cooled in a dry ice/acetone bath and reactions at 0 °C were cooled in an ice/water bath.

Chemicals were purchased from ABCR, Acros Organics, Amatek Chemical, Fluorochem, Merck, Perkin Elmer and Sigma Aldrich and used without further purification. Solvents for TLC, extraction and flash column chromatography were of technical grade. Extra dry solvents for all non-aqueous reactions were supplied by Acros Organics (puriss., dried over molecular sieves, water content <0.005%). Deuterated solvents (D, 99.9%) were purchased from Cambridge Isotope Laboratories, Inc. NMR spectra of compounds 3, 7 and 8 are presented in the Supplementary Material (Section S1). Tritium labeling was performed by RC Tritec AG (Teufen, Switzerland). Quality control chromatogram and mass spectrum are presented in the Supplementary Material (Section S2, Figures S1 and S2). Positron emission tomography (PET) imaging was performed in Wistar rats according to a previously published procedure [22]. PET images are presented in the Supplementary Materials (Section S3, Figure S3).

3.2. Mass Spectrometry

High-resolution mass spectra were obtained by the mass spectrometry service of the ETH Zürich Laboratorium für Organische Chemie on a Varian IonSpec FT-ICR (ESI), a Bruker Daltonics maXis ESI- QTOF spectrometer (ESI), a Bruker Daltonics SOLARIX spectrometer (MALDI), or a Bruker Daltonics UltraFlex II spectrometer (MALDI-TOF). For ESI (+MS) an enhanced quadratic calibration mode was used with the following reference mass peaks: 118.0863, 322.0481, 622.0290, 922.0098, 1221.9906, 1521.9715, 1821.9523, 2121.9332, 2421.9140, and 2721.8948.

3.3. NMR Spectroscopy

^{1}H, ^{19}F and ^{13}C nuclear magnetic resonance (NMR) were recorded at room temperature on a Bruker Avance FT-NMR (400 MHz) with $CDCl_3$ or methanol-d_4 as solvent. Chemical shifts values (δ) are reported in parts per million (ppm) relative to tetramethylsilane (0.00 ppm) as an internal standard and the appropriate $CDCl_3$ solvent signals (δ_H = 7.26 ppm and δ_C = 77.16). For ^{1}H NMR spectra, resonance multiplicities are abbreviated as s = singlet, d = doublet, t = triplet, m = multiplet. Coupling constants (J) are reported in hertz (Hz).

3.4. Synthetic Procedures

3.4.1. 4-(4,5-Dibromo-2-fluorophenyl)but-3yn-1-ol (3)

To a solution of 1,2-dibromo-4-fluoro-5-iodobenzene (1.0 eq., 1.58 g, 4.15 mmol) in THF (11.9 mL), but-3-yn-1-ol (1.2 eq., 0.38 mL, 4.98 mmol) and copper(I) iodide (0.2 eq., 0.16 g, 0.83 mmol) were added. To the stirring mixture, triethylamine (5 eq., 2.9 mL, 20.75 mmol) was added dropwise and left to stir for 15 min. Bis(triphenylphosphine)palladium(II)dichloride (0.1 eq., 0.33 g, 0.47 mmol) was added, and the mixture was refluxed for 4 h. After cooling to room temperature, the mixture was diluted with EtOAc (100 mL) and filtered through Celite using EtOAc (2 × 200 mL) as a solvent. The filtrate was concentrated under reduced pressure, and the resulting oily crude was purified by flash column chromatography (gradient elution with hexane:EtOAc, 90:10 to 70:30) to afford the title compound 3 (1.34 g, 4.15 mmol, 78%) as a white solid. ^{1}H NMR (400 MHz, $CDCl_3$) δ 7.65 (d, J = 6.9 Hz, 1H), 7.36 (d, J = 8.5 Hz, 1H), 3.83 (t, J = 6.3 Hz, 2H), 2.71 (t, J = 6.3 Hz, 2H), 1.84 (s, 1H). ^{13}C NMR (101 MHz, $CDCl_3$) δ 161.4 (d, J = 255.8 Hz), 137.3, 124.8 (d, J = 9.5 Hz), 121.0 (d, J = 26.1 Hz), 119.4, 113.2 (d, J = 17.4 Hz), 94.7, 73.9, 61.0, 24.1. ^{19}F NMR (376 MHz, $CDCl_3$) δ −110.8. HRMS (EI) calcd for $C_{10}H_7OBr_2F$ [M] 319.8848; found 319.8841.

3.4.2. 4-(4,5-Dibromo-2-fluorophenyl)butan-1-ol (4)

To a stirring solution of 4-(4,5-dibromo-2-fluorophenyl)but-3yn-1-ol (1.0 eq., 1.04 g, 3.23 mmol) in MeOH (30 mL), Raney nickel (436 μL, 3.76 mmol) was added under nitrogen atmosphere. The reaction mixture was hydrogenated at atmospheric pressure for 26 h. The mixture was filtered through a filter paper using EtOAc (3 × 20 mL) as a solvent.

The filtrate was concentrated under reduced pressure and the resulting oily crude (1.05 g, 3.22 mmol) was used in the next reaction without further purification.

3.4.3. 4-(4,5-Dibromo-2-fluorophenyl)butan-1-ol (**5**)

To a stirring solution of 4-(4,5-dibromo-2-fluorophenyl)butan-1-ol (1 eq., 1.05 g, 3.22 mmol) in DCM (33 mL), triethylamine (1.8 eq., 0.8 mL, 5.79 mmol) was added dropwise. The mixture was cooled to 0 °C and methanesulfonyl chloride (1.0 eq., 0.25 mL, 3.22 mmol) was added dropwise. The solution was allowed to come to room temperature and was stirred for 1.5 h. The mixture was quenched with water (250 mL) and extracted with DCM (3 × 250 mL). The combined organic layers were washed with brine (300 mL), dried over $MgSO_4$, filtered and concentrated under reduced pressure affording 4-(4,5-dibromo-2-fluorophenyl)butyl methanesulfonate (1.3 g, 3.22 mmol) as a yellow oily crude, which was used in the next reaction without further purification.

3.4.4. 3-(4-(4,5-Dibromo-2-fluorophenyl)butyl)-7-methoxy-2,3,4,5-tetrahydro-1h-benzo[d]azepin-1-ol (**7**)

To a stirring solution of 7-methoxy-2,3,4,5-tetrahydro-1H-benzo[d]azepin-1-ol 6 (1.0 eq., 0.519 g, 2.69 mmol) and 4-(4,5-dibromo-2-fluorophenyl)butyl methanesulfonate (1.2 eq. 1.3 g, 3.22 mmol) in DMF (17.3 mL), DIPEA (0.87 eq., 0.4 mL, 2.35 mmol) was added dropwise. The mixture was refluxed for 5 h. After cooling the solution to r.t., it was quenched with aq. NaOH (200 mL, 1 mM) and extracted with DCM (3 × 200 mL). The combined organic layers were washed with brine (300 mL), dried over Na_2SO_4, filtered and concentrated under reduced pressure to yield a brown oily crude. The crude was purified by flash column chromatography (gradient elution with hexane:EtOAc 80:20 to 50:50 each containing 0.1% ammonia) to afford the title compound **5** as a clear oil (589 mg, 1.18 mmol, 44%). ^{1}H NMR (400 MHz, $CDCl_3$) δ 7.44 (d, *J* = 7.3 Hz, 1H), 7.31 (d, *J* = 9.1 Hz, 1H), 7.11 (d, *J* = 8.0 Hz, 1H), 6.68–6.62 (m, 2H), 4.61 (d, *J* = 6.7 Hz, 1H), 3.78 (s, 3H), 3.27 (t, *J* = 12.8 Hz, 1H), 3.20–3.12 (m, 1H), 3.04–2.96 (m, 1H), 2.72–2.58 (m, 5H), 2.55 (d, *J* = 12.0 Hz, 1H), 2.44 (t, *J* = 11.9 Hz, 1H), 1.69–1.52 (m, 4H). ^{13}C NMR (101 MHz, $CDCl_3$) δ 159.9 (d, *J* = 249.1 Hz), 159.1, 141.2, 135.5, 134.8 (d, *J* = 6.4 Hz), 130.6 (d, *J* = 17.7 Hz), 129.8, 122.1 (d, *J* = 9.5 Hz), 120.8 (d, *J* = 27.4 Hz), 119.4, 116.7, 110.4, 72.4, 60.9, 59.5, 56.2, 55.4, 36.9, 28.5, 27.6, 26.6. ^{19}F NMR (376 MHz, $CDCl_3$) δ −118.1. HRMS (ESI) calcd for $C_{21}H_{25}Br_2FNO_2$ $[M + H]^+$ 500.0231; found 500.0230.

3.4.5. 3-(4-(4,5-Dibromo-2-fluorophenyl)butyl)-2,3,4,5-tetrahydro-1h-benzo[d]azepine-1,7-diol (Br_2-OF-NB1, **8**)

To a stirring solution of 3-(4-(4,5-dibromo-2-fluorophenyl)butyl)-7-methoxy-2,3,4,5-tetrahydro-1H-benzo[d]azepin-1-ol (1.0 eq., 589.5 mg, 1.18 mmol) in DCM (16.8 mL) tribromoborane (7.4 eq., 8.7 mL, 8.66 mmol) was added dropwise at −78 °C. The solution was stirred at −78 °C for 10 min and subsequently allowed to come to room temperature. Stirring was continued at room temperature for 3 h. The reaction was quenched with H_2O, and the pH was adjusted to 7 with NaOH (4 M). The mixture was diluted with H_2O (200 mL) and extracted with DCM (3 × 200 mL) and ethyl acetate (2 × 200 mL). The combined organic layers were back extracted with brine (500 mL), dried over Na_2SO_4, filtered and concentrated under reduced pressure. The crude was purified twice by flash column chromatography (isocratic elution with hexane:EtOAc 20:80 containing 0.1% ammonia) to afford the title compound **6** as a yellow solid (149.1 mg, 0.306 mmol, 26%). ^{1}H NMR (400 MHz, $CDCl_3$) δ 7.44 (d, *J* = 7.2 Hz, 1H), 7.31 (d, *J* = 9.1 Hz, 1H), 7.03 (d, *J* = 7.8 Hz, 1H), 6.60–6.53 (m, 2H), 4.60 (d, *J* = 6.7 Hz, 1H), 3.29–3.12 (m, 2H), 3.04–2.95 (m, 1H), 2.66–2.52 (m, 6H), 2.44 (t, *J* = 11.9 Hz, 1H), 1.68–1.49 (m, 4H). ^{13}C NMR (101 MHz, $CDCl_3$) δ 155.2, 141.5, 135.5, 134.8 (d, *J* = 6.1 Hz, 2C), 130.7, 130.1 (2C), 121.0, 120.7, 117.6, 112.5, 72.5, 60.9, 59.5, 56.2, 36.8, 28.5, 27.6, 26.7. ^{19}F NMR (376 MHz, $CDCl_3$) δ −118.1. HRMS (ESI) calcd for $C_{20}H_{22}Br_2FNNaO_2$ $[M + Na]^+$ 507.9894; found 507.9893.

3.5. In Vitro GluN1/2B Competitive Binding Assay

Competitive binding assays were performed as previously reported [23]. Briefly, IC_{50} binding affinity assays were conducted using Wistar rat brain homogenates and HEPES buffer (30 mM HEPES, 110 mM NaCl, 5 mM KCl, 2.5 mM $CaCl_2$ and 1.2 mM $MgCl_2$). A dilution series of the test ligands was prepared ranging from 30 µM to 300 pM. The cold ligand was displaced with 4.7 nM of either [³H]ifenprodil or [³H]OF-NB1. Total binding was measured without the cold ligand, and non-specific binding was measured with 100 µM CP101,606 instead of the cold ligand. Each measurement vial contained 0.5 mg/mL of brain homogenate proteins, 20 µL of cold ligand, 10 µL of radioligand and was diluted to 200 µL with HEPES buffer. The vials were incubated at 25 °C and 110 rpm for 1 h. The mixtures were quenched with buffer and filtered through Whatman® GF/C 25 mm filters soaked with 0.05% PEI solution. Filters were washed twice with cold buffer and placed in scintillation vials. Scintillation vials were filled with 2 mL Ultima Gold™ LSC cocktail and measured in a Beckmann LS6500 liquid scintillator counter.

3.6. In Vitro σ1R Competitive Binding Assay

The σ1R competitive binding assay was performed in line with previously reported procedures [23]. IC_{50} binding affinity assays were conducted using Wistar rat brain homogenates, and a HEPES buffer (30 mM HEPES, 110 mM NaCl, 5 mM KCl, 2.5 mM $CaCl_2$ and 1.2 mM $MgCl_2$). Dilution series were prepared with the test cold ligands in a concentration ranging from 30 µM to 300 pM. The cold ligand was displaced with 2.5 nM of (+)-[³H]pentazocine. Total binding was measured without the cold ligand, and non-specific binding was measured with 100 µM eliprodil instead of the cold ligand. Each measurement vial contained 0.75 mg/mL of brain homogenate proteins, 20 µL of cold ligand, 10 µL of radioligand and was diluted to 200 µL with HEPES buffer. The vials were incubated at 37 °C and 110 rpm for 2.5 h. The mixtures were quenched with buffer and filtered through Whatman® GF/C 25 mm filters soaked with 0.05% PEI solution. Filters were washed twice with buffer and placed in scintillation vials. Scintillation vials were filled with 2 mL Ultima Gold™ LSC cocktail and measured in Beckmann LS6500 liquid scintillator counter.

3.7. In Silico Simulation

3.7.1. Preparation of Co-Crystallized Protein Structure of 3QEL and 6DK1

The crystal structure of NMDA-GluN1b/GluN2B dimer in complex with ifenprodil (PDB ID: 3QEL) and σ1R bound to (+)-pentazocine (PDB ID: 6DK1) were selected for docking simulations. [35,36] The simulations were performed using Molecular Operating Environment (MOE 2019.0101, Chemical Computing Group, ULC, Montreal, QC, Canada, H3A 2R7, 2021) software. After loading the structures in MOE, they were prepared for docking simulations using the default parameters in the 'QuickPrep Panel', including the removal of water molecules 4.5 Å away from the ligand pocket, adding hydrogen atoms to the protein structure, adjusting protonation states, and ensuring the energy minimization of the protein structures with an Amber10:EHT force field.

3.7.2. Validation of the Docking Protocol

In order to validate the docking protocol for the two selected targets implemented in the study (described below), the co-crystallized ligands were re-docked in the receptor binding pocket. The ability to reproduce the reported interactions with a minimum root mean square distance (RMSD) value between the co-crystallized pose and docked pose validate the methodology. The RMSD values obtained from the re-docking of the co-crystal ligands of both 3QEL and 6DK1 were 0.257 Å and 0.250 Å, respectively.

3.7.3. Ligands Dataset Curation for Docking Simulations

The 'builder program' implemented in MOE was used to model both ifenprodil and OF-NB1, where all possible conformations at the physiological pH were obtained.

3.7.4. Docking Simulations

Docking of the obtained conformations of both ligands (ifenprodil and OF-NB1) was carried out using placement and refinement algorithms of the MOE program. Initial docking of the molecules in the active sites used the 'Triangle Matcher' placement method and the 'London dG' scoring function. Further postplacement refinement of docking poses was achieved by using the 'GBVI/WSA dG' scoring method. The poses with minimum energy were used for visualization of the binding interactions as well as occupancy of the binding site of both receptors. Docking overlays are presented in the Supplementary Material (Section S4, Figures S4 and S5).

4. Conclusions

In conclusion, we successfully synthesized and evaluated a novel GluN1/2B radioligand, [^{3}H]OF-NB1, for preclinical GluN1/2B ligand development. Its superiority over the current state of the art was demonstrated in a head-to-head comparison by in vitro binding assays, and we showed that systemic errors arise from the use of an unselective radioligand, such as [^{3}H]ifenprodil, in GluN1/2B binding affinity screening assays. Our aim is to raise awareness for the need to continuously improve the ligand development toolkit. Due to the high potential of GluN1/2B antagonists exhibiting off-target effects toward σRs and vice versa, we envision that the use of [^{3}H]OF-NB1 in GluN2B-targeted drug discovery will facilitate the identification of highly selective candidate drugs with the potential to hold up to expectations in clinical trials.

Supplementary Materials: The following supporting information can be downloaded at https://www.mdpi.com/article/10.3390/ph15080960/s1. NMR spectra of compounds 3, 7 and 8, Section S1. Quality control chromatogram and mass spectrum, Section S2, Figures S1 and S2. PET images, Section S3, Figure S3. Docking overlays, Section S4, Figures S4 and S5.

Author Contributions: H.A. designed the organic synthesis route and the competitive binding assay experiments, analyzed the competitive binding assay results, performed radiosynthesis and PET image analysis, and wrote the manuscript. L.G. performed organic synthesis and competitive binding experiments, analyzed the competitive binding assay results and participated in writing the manuscript. N.H.E. performed retrospective in silico binding studies, and participated in writing the manuscript. C.K. performed PET imaging experiments. W.S. contributed to the design of the in silico studies. S.H.L. contributed to the design of the study. A.H. was a co-investigator and coordinated the project. S.M.A. supervised the project and was the principal investigator. All authors have read and agreed to the published version of the manuscript.

Funding: This work was supported in part by the Swiss National Science Foundation grant numbers 310030E-160403/1 and 310030E-182872/1 to S.M.A.

Institutional Review Board Statement: Animal care and experiments were in accordance with Swiss Animal Welfare legislation. These experiments were authorized by the Veterinary Office of the Canton Zurich, Zurich, Switzerland (ZH028/18, approved on 6 Aug 2018).

Informed Consent Statement: Not applicable.

Data Availability Statement: The data generated and analyzed during our research are not available in any public database or repository but will be shared by the corresponding author upon reasonable request.

Conflicts of Interest: H.A., A.H. and S.M.A. are co-founders of Nemosia AG.

References

1. Paoletti, P.; Bellone, C.; Zhou, Q. NMDA receptor subunit diversity: Impact on receptor properties, synaptic plasticity and disease. *Nat. Rev. Neurosci.* **2013**, *14*, 383–400. [CrossRef] [PubMed]
2. Lau, C.G.; Zukin, R.S. NMDA receptor trafficking in synaptic plasticity and neuropsychiatric disorders. *Nat. Rev. Neurosci.* **2007**, *8*, 413–426. [CrossRef]
3. Paoletti, P.; Neyton, J. NMDA receptor subunits: Function and pharmacology. *Curr. Opin. Pharmacol.* **2007**, *7*, 39–47. [CrossRef]

4. Zhang, B.; Fang, W.; Ma, W.; Xue, F.; Ai, H.; Lu, W. Differential Roles of GluN2B in Two Types of Chemical-induced Long Term Potentiation-mediated Phosphorylation Regulation of GluA1 at Serine 845 in Hippocampal Slices. *Neuroscience* **2020**, *433*, 144–155. [CrossRef]

5. Yashiro, K.; Philpot, B.D. Regulation of NMDA receptor subunit expression and its implications for LTD, LTP, and metaplasticity. *Neuropharmacology* **2008**, *55*, 1081–1094. [CrossRef] [PubMed]

6. Goebel, D.J.; Poosch, M.S. NMDA receptor subunit gene expression in the rat brain: A quantitative analysis of endogenous mRNA levels of NR1Com, NR2A, NR2B, NR2C, NR2D and NR3A. *Mol. Brain Res.* **1999**, *69*, 164–170. [CrossRef]

7. Zhang, X.M.; Luo, J.H. GluN2A versus GluN2B: Twins, but quite different. *Neurosci. Bull.* **2013**, *29*, 761–772. [CrossRef]

8. Mony, L.; Kew, J.N.; Gunthorpe, M.J.; Paoletti, P. Allosteric modulators of NR2B-containing NMDA receptors: Molecular mechanisms and therapeutic potential. *Br. J. Pharmacol.* **2009**, *157*, 1301–1317. [CrossRef] [PubMed]

9. Hardingham, G.E.; Bading, H. Synaptic versus extrasynaptic NMDA receptor signalling: Implications for neurodegenerative disorders. *Nat. Rev. Neurosci.* **2010**, *11*, 682–696. [CrossRef]

10. Liu, Y.; Wong, T.P.; Aarts, M.; Rooyakkers, A.; Liu, L.; Lai, T.W.; Wu, D.C.; Lu, J.; Tymianski, M.; Craig, A.M.; et al. NMDA receptor subunits have differential roles in mediating excitotoxic neuronal death both in vitro and in vivo. *J. Neurosci.* **2007**, *27*, 2846–2857. [CrossRef]

11. Wang, R.; Reddy, P.H. Role of Glutamate and NMDA Receptors in Alzheimer's Disease. *J. Alzheimer's Dis.* **2017**, *57*, 1041–1048. [CrossRef] [PubMed]

12. Adell, A. Brain NMDA Receptors in Schizophrenia and Depression. *Biomolecules* **2020**, *10*, 947. [CrossRef] [PubMed]

13. Ahmed, H.; Haider, A.; Ametamey, S.M. *N*-Methyl-D-Aspartate (NMDA) receptor modulators: A patent review (2015-present). *Expert Opin. Ther. Pat.* **2020**, *30*, 743–767. [CrossRef]

14. Liu, W.; Jiang, X.; Zu, Y.; Yang, Y.; Liu, Y.; Sun, X.; Xu, Z.; Ding, H.; Zhao, Q. A comprehensive description of GluN2B-selective *N*-methyl-D-aspartate (NMDA) receptor antagonists. *Eur. J. Med. Chem.* **2020**, *200*, 112447. [CrossRef]

15. Garner, R.; Gopalakrishnan, S.; McCauley, J.A.; Bednar, R.A.; Gaul, S.L.; Mosser, S.D.; Kiss, L.; Lynch, J.J.; Patel, S.; Fandozzi, C.; et al. Preclinical pharmacology and pharmacokinetics of CERC-301, a GluN2B-selective *N*-methyl-D-aspartate receptor antagonist. *Pharmacol. Res. Perspect.* **2015**, *3*, e00198. [CrossRef] [PubMed]

16. Addy, C.; Assaid, C.; Hreniuk, D.; Stroh, M.; Xu, Y.; Herring, W.J.; Ellenbogen, A.; Jinnah, H.A.; Kirby, L.; Leibowitz, M.T.; et al. Single-dose administration of MK-0657, an NR2B-selective NMDA antagonist, does not result in clinically meaningful improvement in motor function in patients with moderate Parkinson's disease. *J. Clin. Pharmacol.* **2009**, *49*, 856–864. [CrossRef]

17. Alarcon, K.; Martz, A.; Mony, L.; Neyton, J.; Paoletti, P.; Goeldner, M.; Foucaud, B. Reactive derivatives for affinity labeling in the ifenprodil site of NMDA receptors. *Bioorg. Med. Chem. Lett.* **2008**, *18*, 2765–2770. [CrossRef] [PubMed]

18. Hashimoto, K. Activation of sigma-1 receptor chaperone in the treatment of neuropsychiatric diseases and its clinical implication. *J. Pharmacol. Sci.* **2015**, *127*, 6–9. [CrossRef] [PubMed]

19. Hashimoto, K.; Mantione, C.R.; Spada, M.R.; Neumeyer, J.L.; London, E.D. Further characterization of [^{3}H]ifenprodil binding in rat brain. *Eur. J. Pharmacol.* **1994**, *266*, 67–77. [CrossRef]

20. Ametamey, S.M.; Honer, M.; Schubiger, P.A. Molecular imaging with PET. *Chem. Rev.* **2008**, *108*, 1501–1516. [CrossRef] [PubMed]

21. Mu, L.; Krämer, S.D.; Ahmed, H.; Gruber, S.; Geistlich, S.; Schibli, R.; Ametamey, S.M. Neuroimaging with Radiopharmaceuticals Targeting the Glutamatergic System. *Chimia* **2020**, *74*, 960–967. [CrossRef] [PubMed]

22. Ahmed, H.; Wallimann, R.; Haider, A.; Hosseini, V.; Gruber, S.; Robledo, M.; Nguyen, T.A.N.; Herde, A.M.; Iten, I.; Keller, C.; et al. Preclinical Development of (18)F-OF-NB1 for Imaging GluN2B-Containing *N*-Methyl-d-Aspartate Receptors and Its Utility as a Biomarker for Amyotrophic Lateral Sclerosis. *J. Nucl. Med.* **2021**, *62*, 259–265. [CrossRef] [PubMed]

23. Ahmed, H.; Haider, A.; Varisco, J.; Stanković, M.; Wallimann, R.; Gruber, S.; Iten, I.; Häne, S.; Müller Herde, A.; Keller, C.; et al. Structure-Affinity Relationships of 2,3,4,5-Tetrahydro-1*H*-3-benzazepine and 6,7,8,9-Tetrahydro-5*H*-benzo [7]annulen-7-amine Analogues and the Discovery of a Radiofluorinated 2,3,4,5-Tetrahydro-1*H*-3-benzazepine Congener for Imaging GluN2B Subunit-Containing *N*-Methyl-d-aspartate Receptors. *J. Med. Chem.* **2019**, *62*, 9450–9470. [CrossRef] [PubMed]

24. Haider, A.; Iten, I.; Ahmed, H.; Herde, A.M.; Gruber, S.; Krämer, S.D.; Keller, C.; Schibli, R.; Wünsch, B.; Mu, L.; et al. Identification and Preclinical Evaluation of a Radiofluorinated Benzazepine Derivative for Imaging the GluN2B Subunit of the Ionotropic NMDA Receptor. *J. Nucl. Med.* **2019**, *60*, 259. [CrossRef]

25. Rischka, L.; Vraka, C.; Pichler, V.; Rasul, S.; Nics, L.; Gryglewski, G.; Handschuh, P.; Murgaš, M.; Godbersen, G.M.; Silberbauer, L.R.; et al. First-in-Humans Brain PET Imaging of the GluN2B-Containing *N*-methyl-d-aspartate Receptor with (*R*)-11C-Me-NB1. *J. Nucl. Med.* **2022**, *63*, 936. [CrossRef]

26. Haider, A.; Herde, A.M.; Krämer, S.D.; Varisco, J.; Keller, C.; Frauenknecht, K.; Auberson, Y.P.; Temme, L.; Robaa, D.; Sippl, W.; et al. Preclinical Evaluation of Benzazepine-Based PET Radioligands (*R*)- and (*S*)-(^{11})C-Me-NB1 Reveals Distinct Enantiomeric Binding Patterns and a Tightrope Walk Between GluN$_2$B- and σ$_1$-Receptor-Targeted PET Imaging. *J. Nucl. Med.* **2019**, *60*, 1167–1173. [CrossRef]

27. Smart, K.; Zheng, M.-Q.; Ahmed, H.; Fang, H.; Xu, Y.; Cai, L.; Holden, D.; Kapinos, M.; Haider, A.; Felchner, Z.; et al. Comparison of three novel radiotracers for GluN2B-containing NMDA receptors in non-human primates: (*R*)-[^{11}C]NR2B-Me, (*R*)-[^{18}F]of-Me-NB1, and (*S*)-[^{18}F]of-NB1. *J. Cereb. Blood Flow Metab.* **2022**, *42*, 1398–1409. [CrossRef]

28. Zheng, M.; Ahmed, H.; Smart, K.; Xu, Y.; Holden, D.; Kapinos, M.; Felchner, Z.; Haider, A.; Tamagnan, G.; Carson, R.E.; et al. Characterization in nonhuman primates of (*R*)-[^{18}F]OF-Me-NB1 and (*S*)-[^{18}F]OF-Me-NB1 for imaging the GluN2B subunits of the NMDA receptor. *Eur. J. Nucl. Med. Mol. Imaging* **2022**, *49*, 2153–2162. [CrossRef]
29. Ahmed, H.; Zheng, M.-Q.; Smart, K.; Fang, H.; Zhang, L.; Emery, P.R.; Gao, H.; Ropchan, J.; Haider, A.; Tamagnan, G.; et al. Evaluation of (rac)-, (R)- and (S)-18F-OF-NB1 for imaging GluN2B subunit-containing N-methyl-D-aspartate receptors in non-human primates. *J. Nucl. Med.* **2022**; *published ahead of print*. [CrossRef]
30. Weber, F.; Brust, P.; Laurini, E.; Pricl, S.; Wünsch, B. Fluorinated PET Tracers for Molecular Imaging of σ(1) Receptors in the Central Nervous System. *Adv. Exp. Med. Biol.* **2017**, *964*, 31–48. [CrossRef]
31. Grimwood, S.; Richards, P.; Murray, F.; Harrison, N.; Wingrove, P.B.; Hutson, P.H. Characterisation of *N*-methyl-D-aspartate receptor-specific [^{3}H]Ifenprodil binding to recombinant human NR1a/NR2B receptors compared with native receptors in rodent brain membranes. *J. Neurochem.* **2000**, *75*, 2455–2463. [CrossRef]
32. Lever, J.R.; Gustafson, J.L.; Xu, R.; Allmon, R.L.; Lever, S.Z. σ1 and σ2 receptor binding affinity and selectivity of SA4503 and fluoroethyl SA4503. *Synapse* **2006**, *59*, 350–358. [CrossRef] [PubMed]
33. Menniti, F.; Chenard, B.; Collins, M.; Ducat, M.; Shalaby, I.; White, F. CP-101,606, a potent neuroprotectant selective for forebrain neurons. *Eur. J. Pharmacol.* **1997**, *331*, 117–126. [CrossRef]
34. Zampieri, D.; Fortuna, S.; Calabretti, A.; Romano, M.; Menegazzi, R.; Schepmann, D.; Wünsch, B.; Collina, S.; Zanon, D.; Mamolo, M.G. Discovery of new potent dual sigma receptor/GluN2b ligands with antioxidant property as neuroprotective agents. *Eur. J. Med. Chem.* **2019**, *180*, 268–282. [CrossRef] [PubMed]
35. Schmidt, H.R.; Betz, R.M.; Dror, R.O.; Kruse, A.C. Structural basis for σ$_1$-receptor ligand recognition. *Nat. Struct. Mol. Biol.* **2018**, *25*, 981–987. [CrossRef] [PubMed]
36. Karakas, E.; Simorowski, N.; Furukawa, H. Subunit arrangement and phenylethanolamine binding in GluN1/GluN2B NMDA receptors. *Nature* **2011**, *475*, 249–253. [CrossRef] [PubMed]

Article

Fluorine-18 Labeled Urea-Based Ligands Targeting Prostate-Specific Membrane Antigen (PSMA) with Increased Tumor and Decreased Renal Uptake

Falguni Basuli [1,*], Tim E. Phelps [2], Xiang Zhang [1], Carolyn C. Woodroofe [1], Jyoti Roy [2], Peter L. Choyke [2], Rolf E. Swenson [1] and Elaine M. Jagoda [2]

[1] Chemistry and Synthesis Center, National Heart, Lung, and Blood Institute, National Institutes of Health, Bethesda, MD 20892, USA; xiang.zhang2@nih.gov (X.Z.); carolyn.woodroofe@nih.gov (C.C.W.); rolf.swenson@nih.gov (R.E.S.)
[2] Molecular Imaging Branch, National Cancer Institute, Bethesda, MD 20892, USA; tim.phelps@nih.gov (T.E.P.); jyotiroy14@gmail.com (J.R.); pchoyke@mail.nih.gov (P.L.C.); ejagoda@mail.nih.gov (E.M.J.)
* Correspondence: bhattacharyyaf@mail.nih.gov; Tel.: +1-301-827-1107

Abstract: High expression of prostate-specific membrane antigen (PSMA) in prostate cancers prompted the development of the PSMA-targeted PET-imaging agent [18F]DCFPyL, which was recently approved by the FDA. Fluorine-18-labeled Lys–Urea–Glu-based oxime derivatives of [18F]DCFPyL were prepared for the comparison of their in vitro and in vivo properties to potentially improve kidney clearance and tumor targeting. The oxime radiotracers were produced by condensation of an aminooxy functionalized PSMA-inhibitor Lys–Urea–Glu scaffold with fluorine-18-labeled aldehydes. The radiochemical yields were between 15–42% (decay uncorrected) in 50–60 min. In vitro saturation and competition binding assays with human prostate cancer cells transfected with PSMA, PC3(+), indicated similar high nM binding affinities to PSMA for all radiotracers. In vivo biodistribution studies with positive control PC3(+) tumor xenografts showed that the kidneys had the highest uptake followed by tumors at 60 min. The PC3(+) tumor uptake was blocked with non-radioactive DCFPyL, and PC3(−) tumor xenograft (negative control) tumor uptake was negligible indicating that PSMA targeting was preserved. The most lipophilic tracer, [18F]**2a**, displayed comparable tumor-targeting to [18F]DCFPyL and a desirable alteration in pharmacokinetics and metabolism, resulting in significantly lower kidney uptake with a shift towards hepatobiliary clearance and increased liver uptake.

Keywords: fluorine-18; PET; oxime; PSMA; lipophilicity; biodistribution

Citation: Basuli, F.; Phelps, T.E.; Zhang, X.; Woodroofe, C.C.; Roy, J.; Choyke, P.L.; Swenson, R.E.; Jagoda, E.M. Fluorine-18 Labeled Urea-Based Ligands Targeting Prostate-Specific Membrane Antigen (PSMA) with Increased Tumor and Decreased Renal Uptake. *Pharmaceuticals* **2022**, *15*, 597. https://doi.org/10.3390/ph15050597

Academic Editor: Xuyi Yue

Received: 5 April 2022
Accepted: 11 May 2022
Published: 13 May 2022

Publisher's Note: MDPI stays neutral with regard to jurisdictional claims in published maps and institutional affiliations.

Copyright: © 2022 by the authors. Licensee MDPI, Basel, Switzerland. This article is an open access article distributed under the terms and conditions of the Creative Commons Attribution (CC BY) license (https://creativecommons.org/licenses/by/4.0/).

1. Introduction

Prostate cancer (PC) is the most common malignancy in men in the United States and Europe [1–3]. In recent decades, prostate cancer survival rates have improved; however, it is still a significant cause of death. Local PC is usually diagnosed with screening for prostate serum antigen (PSA serum testing), clinical examination, and imaging such as magnetic resonance imaging (MRI) followed by a biopsy of the prostate. Advanced PC, however, is commonly staged with computed tomography (CT), bone scans and positron emission tomography (PET), frequently using prostate-specific membrane antigen (PSMA)-targeted radioligands. Due to the higher sensitivity of PET over the other techniques, it is becoming more widely accepted as a diagnostic approach to identify sites of extra-prostatic disease. The metabolic radiotracer, 2-deoxy-2-[18F]fluoro-D-glucose, [18F]FDG, although commonly used in other cancers, has proven less useful in PC [4,5]. Carbon-11 or fluorine-18-labeled choline PET/CT showed promising results for the detection of bone metastases. However, these agents have limitations in terms of sensitivity and specificity [6]. This unmet clinical need led to the development of another class of radiotracers targeting the transmembrane protein PSMA, which is expressed in approximately 95% of PC cases including both primary and metastatic disease [7–9]. PSMA is a cell surface glycoprotein with carboxypeptidase and folate hydrolase enzymatic activities that has emerged as an important biomarker for PC and prompted the development of small-molecule inhibitors [10–12]. These smallmolecule inhibitors have proven to be suitable platforms for PET imaging with faster clearance rates and lower backgrounds.

The gallium-68 labeled PET tracer, Glu-NH-CO-Lys-(Ahx)-[68Ga]Ga-N,N′-Bis(2-hydroxy-5-(ethylene-betacarboxy)benzyl)ethylenediamine N,N′-diacetic acid ([68Ga]Ga-PSMA-11 (also named [68Ga]Ga-PSMA-HBED-CC)), is the most widely studied PSMA radiotracer [13–17]. It was first reported by Eder et al. in 2012 [18]. The initial clinical PET imaging study with this tracer demonstrated a significant advantage compared to conventional imaging used for the detection of recurrent PC [13]. [68Ga]Ga-PSMA-11 was recently approved by the Food and Drug Administration (FDA) for PET imaging of PSMA-positive lesions in men with prostate cancer [19]. However, the longer half-life of fluorine-18 (110 min) compared to gallium-68 (68 min) enables sufficient time for central production and local distribution of the tracers which is more pragmatic for most medical facilities. The extended imaging time with fluorine-18-labeled PSMA radiotracers may further increase the overall detection rate in patients with PC [20,21]. Moreover, fluorine-18 offers comparatively lower positron energy (fluorine-18, 633 keV vs. gallium-68, 1899 keV) with a resultant shorter positron range in the tissue, which may also improve image resolution [22,23]. Thus, the growing demand for PSMA-targeted PET imaging is likely to be better met by fluorine-18labeled radiotracers. Recently, Gust et al. proposed a molecular absorption spectrometry (MAS) method that uses fluorination as tool to improve bioanalytical labeling and suggested it as a potential alternative to ^{18}F-PET [24].

A variety of fluorine-18labeled PSMA-targeted PET radiotracers have been developed for PC imaging [25–29]. The most extensively studied tracers of these classes are urea-based small molecule inhibitors, e.g., N-[N-[(S)-1,3-dicarboxypropyl]carbamoyl]-4-[^{18}F]fluorobenzyl-L-cysteine ([^{18}F] DCFBC), 2-(3-(1-carboxy-5-[(6-[^{18}F]fluoropyridine-3-carbonyl)-amino]-pentyl)-ureido)-pentanedioic acid ([^{18}F] DCFPyL), Glu-NH-CO-Lys-(Ahx)-[^{18}F]AlF-N,N′-Bis(2-hydroxy-5-(ethylene-betacarboxy)benzyl)ethylenediamine N,N′-diacetic acid ([^{18}F]-PSMA-11), and (2S)-2-[[(1S)-1-carboxy-5-[[(2S)-2-[[4-[[[(2S)-4-carboxy-2-[[(2S)-4-carboxy-2-[(6-[^{18}F]fluoranylpyridine-3-carbonyl)amino]butanoyl]amino]butanoyl]amino]methyl]benzoyl]amino]-3-naphthalen-2-ylpropanoyl]amino]pentyl]carbamoylamino]pentanedioic acid ([^{18}F] PSMA-1007) [30–33]. The clinical studies with the first-generation PSMA ligand [^{18}F]DCFBC demonstrated slow clearance with high background activity [34]. The second-generation ligands, [^{18}F] DCFPyL and [^{18}F] PSMA-1007 showed high tumor: background ratios and favorable pharmacokinetics compared to other small molecules [31,35–38]. [^{18}F]-DCFPyL was approved by the FDA in 2021 for the detection of possible early metastatic PC involvement [39]. A wide range of prosthetic groups and linkers have been introduced to improve pharmacokinetics and detection rates with PSMA PET [18,40–43]. These studies demonstrated favorable binding properties for more lipophilic compounds and inspired us to develop oxime derivatives with increased lipophilicity (Scheme 1). Herein, we report the synthesis of the precursor, radiolabeling, and biological evaluation of these oxime derivatives in comparison with previously reported tracers [^{18}F]DCFPyL and [^{18}F]**1a**. The biological evaluations include in vitro binding studies to assess the affinity (K_d) of these compounds for PSMA and in vivo biodistribution studies with PSMA-positive tumor mouse models to determine tumor targeting and differences in pharmacokinetics and metabolism.

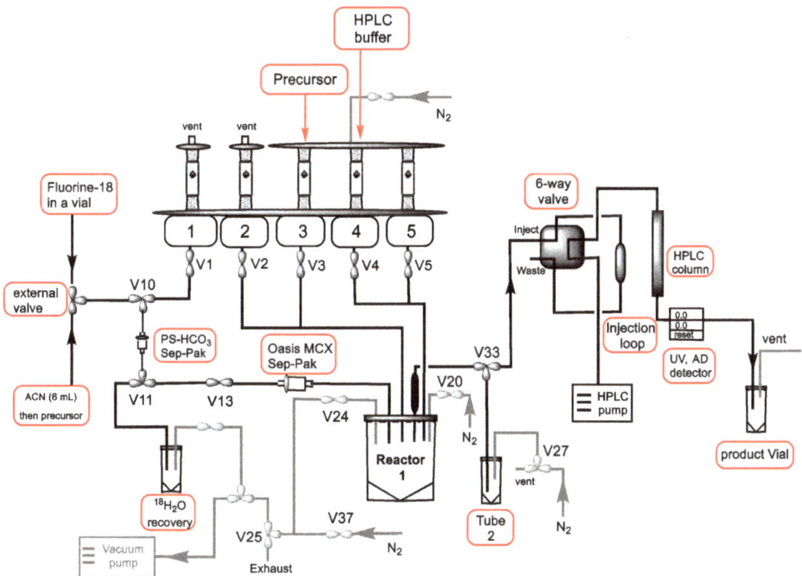

Scheme 1. Structures of the compounds investigated in the current study.

2. Results

2.1. Radiochemistry

Oxime formation of the aminooxy functionalized lysine-urea-glutamate scaffold (**1** and **2**, Scheme 1) with fluorine-18labeled aldehydes produced radiotracers [^{18}F]**1b**, [^{18}F]**2a** and [^{18}F]**2b**. The radiosyntheses of the tracers were performed either manually ([^{18}F]**1a** and [^{18}F]**2a**) or by automated synthesis method ([^{18}F]**1b** and [^{18}F]**2b**). For the automated synthesis method, an external three-way valve was added to V10 (Figure 1) on the GE Tracerlab module (GE FX-N Pro) to accomplish the Sep-Pak fluorination of 6-[^{18}F]fluoronicotinaldehyde. The overall radiochemical yield (2 steps) of the synthesis was 15–42% (n >10, decay uncorrected) in a 50–60 min procedure. The radiochemical purity was >98% with a molar activity of 300–360 GBq/μmol. The identities of the products were confirmed by comparing their HPLC retention times with co-injected, authentic non-radioactive standards. A representative HPLC profile for compound [^{18}F]**1b** is shown in Figure 2. The calculated logP values indicated the lipophilicity order of the tracers are [^{18}F]**2a** (1.45) > [^{18}F]**2b** (0.58) > [^{18}F]**1a** (0.04) > [^{18}F]**1b** (−0.83) > [^{18}F]DCFPyL (−0.94).

Figure 1. Schematic diagram of the automated synthesis on GE FX-N Pro Module.

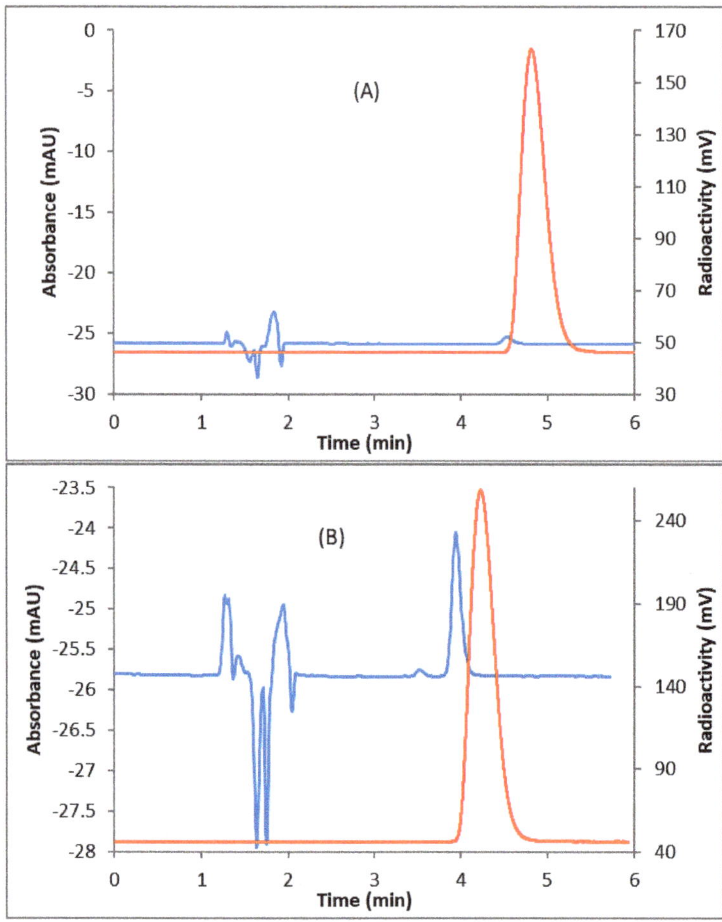

Figure 2. A representative HPLC profile for (**A**) compound [^{18}F]**1b**; (**B**) co-injected with the non-radioactive standard. Solid line, in-line radiodetector; blue line, UV detector at 254 nm.

2.2. In Vitro Cell Binding Studies

All tracers exhibited high specific binding (B$_{sp}$; 85–98%) with sub-nM affinity for PSMA using PC3(+) tumor membrane preparations (Figure 3A; Table 1). The K$_d$ of [^{18}F]DCFPyL (0.402 ± 0.121 nM, n = 5) was not significantly different from the K$_d$ of [^{18}F]**1a–b**, **2a–b** (Table 1) indicating that the addition of alkyl linkers to a fluorine-18labeled arene or heteroarene oxime moiety did not alter the binding affinity. A significant decrease of ~4-fold (p = 0.006) was observed in the K$_i$ of [^{18}F]**1a** (0.1 nM) compared to the K$_i$ of [^{18}F]DCFPyL (0.398 ± 0.055 nM, n = 3) suggesting that [^{18}F]**1a** may have higher affinity than [^{18}F]DCFPyL (Figure 3B; Table 1). PC3(+) tumor membrane preparations exhibited high PSMA expression levels [B$_{max}$ = 13.95 ± 1.60 fmol/µg of protein, n = 5] with [^{18}F]DCFPyL which compared favorably with the B$_{max}$ values determined from similar saturation assays with the other four tracers, [^{18}F]**1a–b**, **2a–b**.

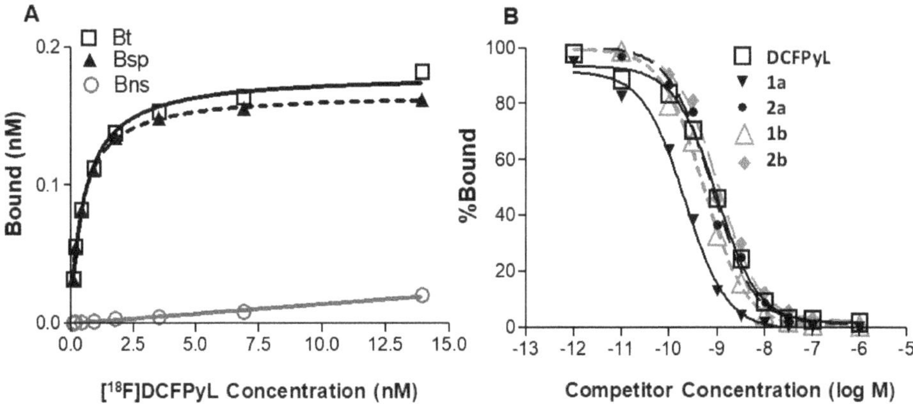

Figure 3. (**A**) Representative plot of [^{18}F]DCFPyL in vitro saturation binding assay using tumor membrane preparations from PC3(+); B_t = Bound total; B_{ns} = Bound non-specific; B_{sp} = Bound specific (B_t-B_{ns} = B_{sp}); K_d = 0.457 ± 0.011 nM (SE); n = 8 points). (**B**) Representative plot of [^{18}F]DCFPyL (with a concentration of 0.8 nM) in vitro competition assay with DCFPyL, **1a–b** and **2a–b**.

Table 1. Comparison of in vitro parameters of radioligands.

Radioligand	[^{18}F]DCFPyL	[^{18}F]1a	[^{18}F]1b	[^{18}F]2a	[^{18}F]2b
[1] K_d (nM)	0.402 (0.121)	0.238 (0.075)	0.458 (0.065)	0.362 (0.077)	0.489 (0.106)
Ligand	**DCFPyL**	**1a**	**1b**	**2a**	**2b**
[2] K_i (nM)	0.398 (0.055)	0.101 * (0.007)	0.312 (0.073)	0.621 (0.159)	0.901 (0.267)

[1] K_d values: mean (SE) (n = 3–5); Derived from saturation assays. [2] K_i values: mean (SE) (n = 3); Derived from competition assays with [^{18}F]DCFPyL as the radioligand. * $p < 0.05$ (n = 3 per group, student *t*-test) represents a significant decrease in K_i of **1a** compared to K_i of DCFPyL.

2.3. In Vivo Biodistribution

The biodistribution of [^{18}F]DCFPyL was determined in nude mice bearing human prostate cancer tumors transfected with PSMA (PC3(+) xenografts) at 30, 60, 90 and 120 min post-injection (Figure 4A,B). [^{18}F]DCFPyL distributed rapidly and cleared from the blood and non-target tissues except for the tumor over the 120 min time course (Figure 4A). The kidney exhibited the highest uptake (133%ID/g to 50%ID/g) at all time points and decreased by 65% from 15 to 120 min. All other tissue uptakes except tumor were > 30 fold lower than kidneys at all times indicating that [^{18}F]DCFPyL is dominated by renal clearance, as expected from published results [31]. The next highest uptakes after the kidneys occurred in the PC3(+) tumor in which [^{18}F]DCFPyL was highly retained from 15 (18.6%ID/g) to 120 min (20.8%ID/g; Figure 4A). The tumor tissue to muscle ratio (T:M) steadily increased over the time course with an 11-fold increase from 15 (24 T:M) to 120 min (260 T:M). These tumor T:M increases are reflective of an increased rate of clearance from the muscle while in tumors the majority of the radioactivity was retained (Figure 4B). The retention of [^{18}F]DCFPyL in the tumor with an accompanying increase in tumor T:M over time would indicate high-affinity binding to PSMA. However, this was not the case with the salivary glands in which 92% of the radioactivity had been cleared at 120 min and T:M decreased 34% from 15 (2.7 T:M) to 120 min (2.0 T:M). This lack of retention of [^{18}F]DCFPyL and the low T:M ratios suggest that salivary gland PSMA expression levels in mouse, known to be lower than in humans, are insufficient to render a meaningful biodistribution model. With this in mind, salivary glands were not included in further biodistributions with the ^{18}F-labeled analogues [35].

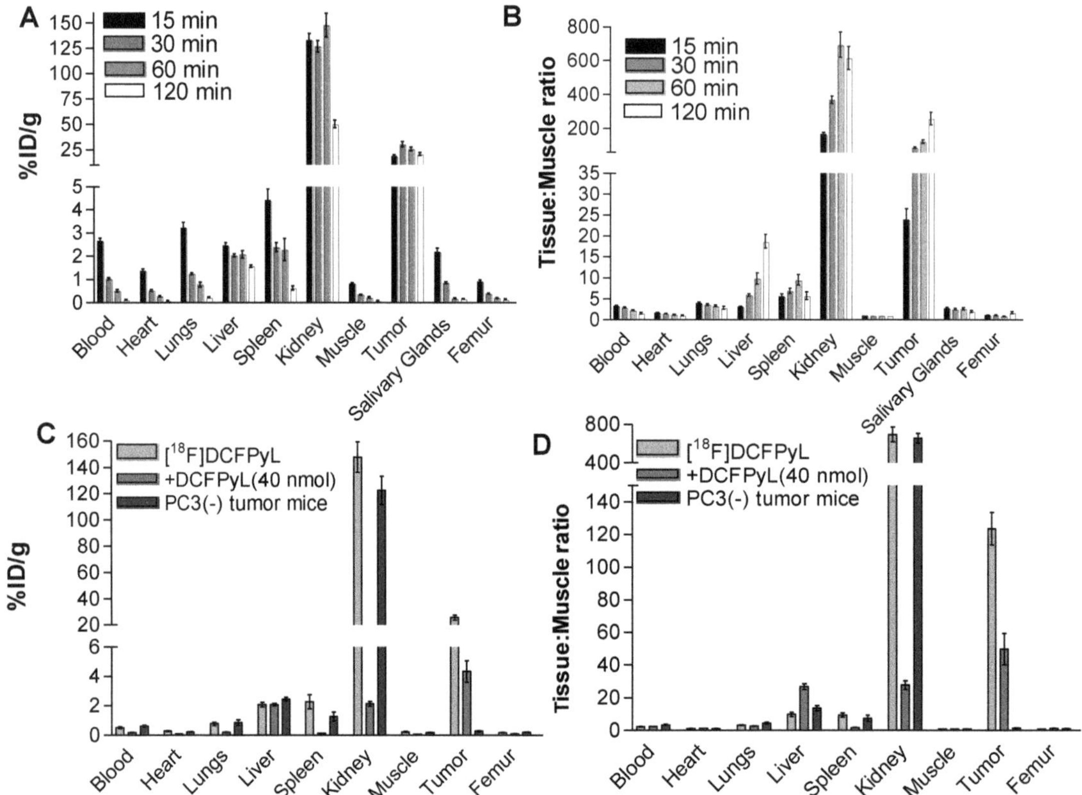

Figure 4. (**A**) Biodistribution [%ID/g (normalized to 20 g mouse)] of [^{18}F]DCFPyL in PC3(+) xenografts from 15 to 120 min. Each bar represents the mean %ID/g $\pm$ SE (*n* = 8–10 per time point); (**B**) Tissue (%ID/g) to Muscle (%ID/g) ratios of [^{18}F]DCFPyL in PC3(+) xenografts from 15 to 120 min. (**C,D**) Biodistribution [%ID/g (normalized to 20 g mouse); Tissue (%ID/g) to Muscle (%ID/g) ratios (T:M)]of [^{18}F]DCFPyL at 60 min in PC3(−) xenografts and PC3 (+) xenografts injected with [^{18}F]DCFPyL only or coinjected with DCFPyL (1000×; 40 nmoles). Each bar in the graph represents the mean (%ID/g or T:M) $\pm$ SE (*n* = 8–10 per group).

Additional [^{18}F]DCFPyL biodistribution studies at 60 min were carried out with PC3(−) xenografts and PC3(+) xenograft groups that received [^{18}F]DCFPyL alone or a coinjection with a blocking dose of non-radioactive DCFPyL [1000×; 40 µg] (Figure 4C,D). The biodistributions of the PC3(−) xenografts were comparable to the PC3(+) xenografts for the blood and all tissues except for the PC3(−) tumors (0.3%ID/g) which represented <2% of the uptake observed in PC3(+) tumors (25.6%ID/g). In the blocking studies with the PC3(+) xenografts, the group receiving DCFPyL exhibited >50% reduction in blood and most tissue uptakes (%ID/g) compared to the [^{18}F]DCFPyL- only group. T:Ms were calculated to take into account these alterations in the input function and metabolism caused by the DCFPyL blocking dose. The PC3(+) tumor T:M (50:1) of the blocked group was significantly decreased (60%) compared to the [^{18}F]DCFPyL-only group (141:1 T:M; *p* < 0.0001). This blocking taken together with the lack of uptake in PC3(−) tumors would indicate that tumor uptake represents specific PSMA binding. Other significant decreases in T:M ratios occurred in the kidney (96%) and spleen (81%) compared to the [^{18}F]DCFPyL-only group. These decreases most likely are not entirely attributable to PSMA specific binding but could be a result of the altered metabolism in the kidney or cross-reactivity with glutamate carboxypeptidase III (GPCIII) in the spleen, respectively [44]. The only increase in T:M occurred in the liver (27:1 T:M; 2.7 fold) compared to the [^{18}F]DCFPyL only group which most likely is due to a shift from renal towards hepatobiliary metabolism.

Initially, the biodistributions of the tracers ([^{18}F]**1a–b**, [^{18}F]**2a–b**) were evaluated at 60 min in PC3(−) xenografts and PC3(+) xenograft groups with or without a blocking dose of DCFPyL for direct comparison to the reference compound [^{18}F]DCFPyL (Figure 5A–C). For all the tracers the kidneys and tumors exhibited the highest uptakes as was observed with [^{18}F]DCFPyL, although differences in the blood and other tissues were observed indicating some alterations in pharmacokinetics and metabolism (Figure 5A). Radioactivity in the blood of [^{18}F]**1b** (0.309%ID/g; $p = 0.005$) and [^{18}F]**2b** (0.377 %ID/g; $p = 0.034$) was decreased significantly (40% and 27%, respectively) compared to [^{18}F]DCFPyL (0.519%ID/g) indicating faster blood clearance. Conversely, [^{18}F]**2a** in the blood (0.8297%ID/g; $p = 0.034$) was significantly increased by 1.6-fold vs. [^{18}F]DCFPyL. With these significant changes in the blood input function between [^{18}F]DCFPyL and these analogues, T:Ms were determined to assess the differences in PSMA targeting and metabolism (Figure 5B; Table 2). For [^{18}F]**1b,** PC3(+) tumor T:M (204:1 T:M) was significantly increased by 1.6 fold compared to [^{18}F]DCFPyL (124:1 T:M), whereas [^{18}F]**2a** (74:1 T:M) and [^{18}F]**2b** (90:1 T:M) significantly decreased by 40% and 27%, respectively. The PC3(−) tumor T:Ms for all the analogues were <1.0 (Table 2) and comparable to the PC3(−) tumor T:M of [^{18}F]DCFPyL (Figure 4D), indicating that the uptake in the PC3(+) tumor is reflective of PSMA expression levels. In the DCFPyL blocking studies, PC3(+) tumor T:Ms of all four analogues were decreased compared to the non-blocking groups, with significant decreases occurring in [^{18}F]**2b** PC3(+) tumor T:M (85%; $p < 0.0001$) and [^{18}F]**2a** PC3(+) tumor T:M (48%; $p < 0.0008$, Figure 5C). These results indicate that PSMA targeting has been preserved for all the analogues compared to [^{18}F]DCFPyL with [^{18}F]**1a** exhibiting the highest tumor T:M ratios. Although [^{18}F]**1b** had improved PC3(+) tumor targeting the kidney T:M was increased 1.4 fold compared to [^{18}F]DCFPyL. [^{18}F]**2a** was the only analogue in which kidney T:M was significantly decreased (55%; $p < 0.0001$) compared to [^{18}F]DCFPyL (Figure 5B). Liver T:Ms were increased by >2-fold for all four analogues compared to [^{18}F]DCFPyL with the greatest increase (6.8-fold) observed with [^{18}F]**2a**.

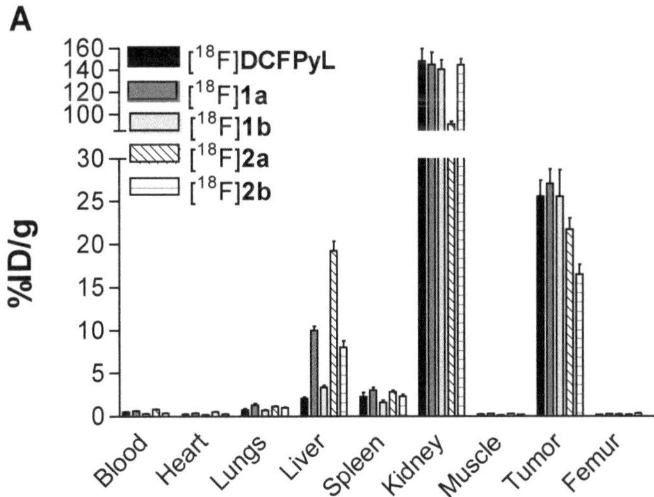

Figure 5. *Cont.*

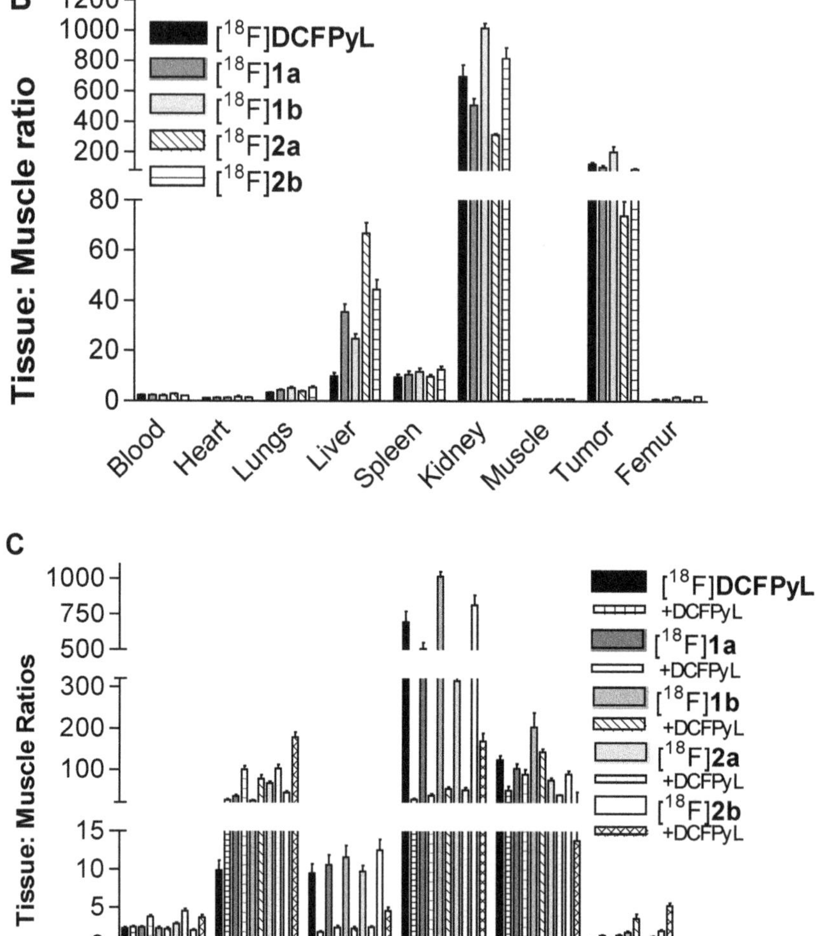

Figure 5. (**A,B**) Biodistribution [%ID/g (normalized to 20 g mouse); Tissue (%ID/g) to Muscle (%ID/g) ratios (T:M)] of [18F]DCFPyL and analogues in PC3(+) xenografts at 60 min. Each bar in the graph represents the mean (%ID/g or T:M) ± SE (*n* = 8–10 per group); (**C**) Tissue (%ID/g) to Muscle (%ID/g) ratios of [18F]DCFPyL and analogues determined from PC3(+) xenograft biodistributions at 60 min injected with [18F]DCFPyL only or coinjected with DCFPyL (1000×; ~10 to 80 nmol). Each bar in the graph represents the mean T:M ± SE (*n* = 8–10 per group, except *n* = 5 for [**18F]1b** + DCFPyL and [**18F]2a** + DCFPyL groups).

Table 2. Comparison of Tissue: Muscle ratios (T:M) of radioligands from PC3($-$) and PC3($+$) xenografts at 60 min.

T:M [1]	[^{18}F]DCFPyL	[^{18}F]1a	[^{18}F]1b	[^{18}F]2a	[^{18}F]2b
PC3($+$) Tumor	123. 6 ± 10.0	103.0 ± 10.9	203.7 ± 34.5	73.9 ± 5.8	89.9 ± 7.6
PC3($-$) Tumor	1.50 ± 0.17	1.07 ± 0.08	1.43 ± 0.29	1.42 ± 0.13	1.65 ± 0.20
Blood	2.29 ± 0.15	2.39 ± 0.14	2.26 ± 0.22	2.85 ± 0.15	2.04 ± 0.11
Kidney	695.1 ± 76.0	507.8 ± 44.0	1013 ± 34	314.8 ± 8.3	814.2 ± 71.9
Liver	9.88 ± 7.4	35.7 ± 3.1	24.9 ± 2.0	66.9 ± 4.2	44.6 ± 3.8
Femur	0.90 ± 0.09	0.92 ± 0.09	1.80 ± 0.17	0.73 ± 0.04	2.00 ± 0.18
%Parent compound in blood (TLC analysis) [2]	*49.9 ± 10.7*	*50.2 ± 1.4*	*74.4 ± 7.6*	*84.8 ± 0.6*	*78.8 ± 2.6*

[1] T:M values: mean ± SE (n = 8–10). [2] %Parent values: mean ± SD (n = 2–5).

Further pharmacokinetic studies were performed with [^{18}F]2a which had appropriate tumor targeting and lower kidney uptake relative to [^{18}F]DCFPyL (Figure 6A–D). Overall [^{18}F]2a cleared rapidly from the blood to the kidneys which had the highest uptakes at all time points (Figure 6A). The tumor uptakes (%ID/g) were the next highest at the later times of 60 and 120 min. The liver uptake of [^{18}F]2a was higher at the earlier times of 15 and 30 min whereas for [^{18}F]DCFPyL the tumor uptakes were the next highest at all times (Figure 6B). [^{18}F]2a was highly retained in the tumors (%ID/g) and tumor retention significantly increased (1.4-fold) from 15 to 120 min. Similarly, tumor T:Ms increased ~14-fold from 15 to 120 min indicating tumor retention and muscle clearance (Figure 6D). [^{18}F]2a tumor T:M was comparable to [^{18}F]DCFPyL at 15 and 120 min but decreased ~2-fold at 30 and 60 min compared to the [^{18}F]DCFPyL tumor T:M. These modest decreases in [^{18}F]2a tumor T:Ms most likely are attributable to changes in the blood input function and alterations in the kidney and liver metabolism compared to [^{18}F]DCFPyL. [^{18}F]2a blood radioactivity content (%ID/g) was higher at all time points compared to [^{18}F]DCFPyL indicating slower rates of clearance of [^{18}F]2a from the blood and other non-target tissues (Figure 6A,C). Further differences were observed in clearance of [^{18}F]2a to the kidney (%ID/g; T:M) which significantly decreased (40% to 54%) at all time points except 120 min compared to [^{18}F]DCFPyL while liver uptake (%ID/g; T:M) increased significantly (7 to 15-fold) over the time course (Figure 6B,D). Femur uptakes (%ID/g; T:M) of [^{18}F]2a and [^{18}F]DCFPyL were comparably low (<1%ID/g: <1.8 T:M) at all times except at 120 min in which [^{18}F]DCFPyL femur T:M was 1.8 fold greater than [^{18}F]2a, suggesting insignificant in vivo defluorination.

Since differences were observed in metabolism between [^{18}F]DCFPyL and the other tracers, the fraction of radioactivity that represented intact tracer (%Parent) in blood was determined at 60 min by TLC (Table 2). Compared to [^{18}F]DCFPyL and the other analogues, [^{18}F]2a exhibited the greatest in vivo blood stability at 60 min with 85% (parent) of the total blood radioactivity remaining intact. Additional TLC analysis was performed with [^{18}F]2a and [^{18}F]DCFPyL to determine the fraction of parent remaining over the time course from 15 to 120 min in blood and kidneys. The % of parent [^{18}F]2a in blood was relatively constant from 15 to 120 min (85 to 83% parent) which was greater than [^{18}F]DCFPyL(50 to 40% parent) over the same time period. In contrast, the majority of radioactivity in the kidney was metabolites of both [^{18}F]2a, and [^{18}F]DCFPyL, however, the % of parent [^{18}F]2a (29 to 15%), was greater than the % of parent [^{18}F]DCFPyL(16 to 7%) at all time points.

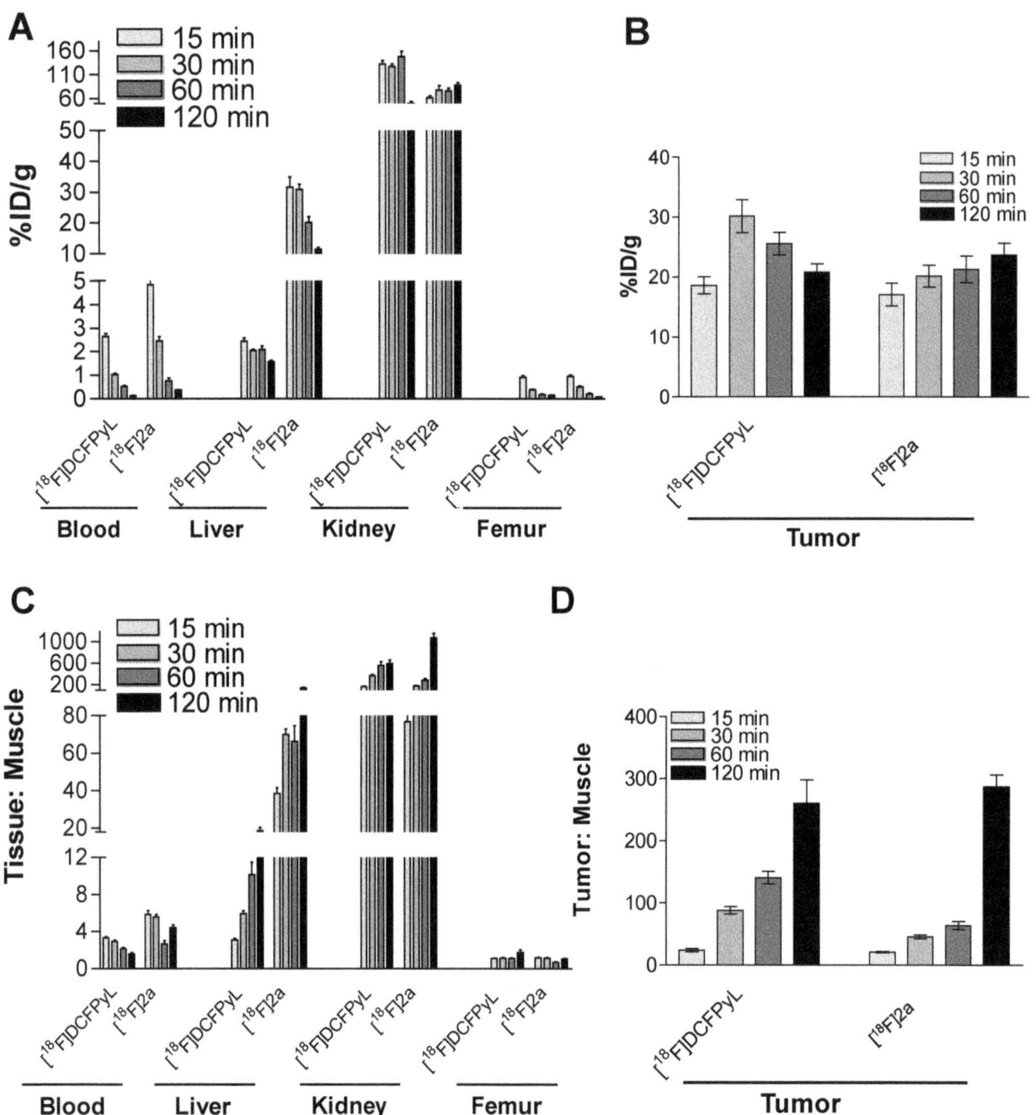

Figure 6. (**A,B**) Pharmacokinetic comparisons of PC3(+) xenograft biodistributions in blood, non-target and target (tumor) tissues [%ID/g (normalized to 20 g mouse)] of [^{18}F]DCFPyL and [^{18}F]**2a** from 15 to 120 min. (**C,D**) Pharmacokinetic comparisons of Tissue (%ID/g) to Muscle (%ID/g) ratios (T:M) of [^{18}F]DCFPyL to the analogue [^{18}F]**2a** determined from PC3(+) xenograft biodistributions from 15 to 120 min. Each bar represents the mean (%ID/g or T:M) ± SE [n = 8–10 ([^{18}F]DCFPyL); n = 5–6 per group ([^{18}F]**2a**)].

In Table 3, [^{18}F]**2a** and [^{18}F]DCFPyL uptake (%ID/g) in the blood and kidney before and after correction for metabolites (parent) were compared over the 15 to 120 min time course. At all times parent [^{18}F]**2a** in the blood was 3 to 7-fold greater than parent [^{18}F]DCFPyL, indicating [^{18}F]**2a** had increased stability in the blood and slower clearance. The kidney uptake (%ID/g) of the parent [^{18}F]**2a** and [^{18}F]DCFPyL was comparable over the time course except for 120 min in which [^{18}F]**2a** was increased by 4-fold (Table 3). Parent [^{18}F]**2a** kidney uptake was relatively unchanged over the time course indicating that this fraction of retained parent [^{18}F]**2a** may be representative of specific binding to PSMA in the proximal tubules of the kidneys (Table 3) [44]. Similarly [^{18}F]DCFPyL parent kidney uptake was retained from 15 to 60 min although a decrease was observed at 120 min.

Table 3. Comparison of [^{18}F]DCFPyL and [^{18}F]**2a** total radioactive uptake (%ID/g) and parent (corrected for metabolites) uptake (%ID/g) in the blood and kidneys after 15, 30, 60 and 120 min. Each value represents the mean (%ID/g) $\pm$ SE [n = 8–10 ([^{18}F]DCFPyL); n = 5–6 per group ([^{18}F]**2a**)].

Uptake (%ID/g)		Time Post-Injection (min)			
		15	30	60	120
Blood	[^{18}F]DCFPyL				
	Total radioactivity	2.66 $\pm$ 0.12	1.03 $\pm$ 0.05	0.52 $\pm$ 0.06	0.13 $\pm$ 0.01
	Parent	*1.06 $\pm$ 0.05*	*0.40 $\pm$ 0.02*	*0.26 $\pm$ 0.02*	*0.045 $\pm$ 0.003*
	[^{18}F]**2a**				
	Total radioactivity	4.83 $\pm$ 0.57	2.47 $\pm$ 0.18	0.77 $\pm$ 0.10	0.37 $\pm$ 0.03
	Parent	*4.10 $\pm$ 0.49*	*2.08 $\pm$ 0.15*	*0.65 $\pm$ 0.87*	*0.31 $\pm$ 0.03*
Kidney	[^{18}F]DCFPyL				
	Total radioactivity	132.9 $\pm$ 6.6	126.9 $\pm$ 5.6	147.7 $\pm$ 11.6	50.2 $\pm$ 3.6
	Parent	*17.05 $\pm$ 0.84*	*20.18 $\pm$ 0.72*	*17.58 $\pm$ 1.48*	*3.41 $\pm$ 0.46*
	[^{18}F]**2a**				
	Total radioactivity	62.2 $\pm$ 3.4	78.4 $\pm$ 9.1	75.8 $\pm$ 6.3	87.8 $\pm$ 5.1
	Parent	*11.84 $\pm$ 0.64*	*22.66 $\pm$ 1.73*	*15.63 $\pm$ 1.19*	*13.04 $\pm$ 0.98*

2.4. PET Imaging Studies

Small animal PET imaging studies were performed with PC3(+) tumor xenograft mice at 60 min post-injection of [^{18}F]DCFPyL or [^{18}F]**2a** [3.7 to 7.4 MBq (100–200 µCi)] (Figure 7). Tumors and kidneys were easily visualized from PET images of xenografts injected with both [^{18}F]DCFPyL and [^{18}F]**2a** whereas the liver was only apparent in the [^{18}F]**2a** image. From ROI analysis of the images the PC3(+) tumor uptakes of [^{18}F]DCFPyL or [^{18}F]**2a** were determined to be 19%ID/g and 14%ID/g, respectively, which were comparable to [^{18}F]DCFPyL and [^{18}F]**2a** tumor uptakes from the biodistribution studies. From similar ROI analysis the [^{18}F]**2a** kidney uptake (98%ID/g) was reduced by ~50% compared to the [^{18}F]DCFPyL kidney uptake (187%ID/g) whereas [^{18}F]**2a** liver uptake (29%ID/g) was increased ~10-fold compared to the [^{18}F]DCFPyL liver uptake (3%ID/g). These [^{18}F]**2a** and [^{18}F]DCFPyL quantitative imaging results were in agreement with the biodistribution results for the respective tissues.

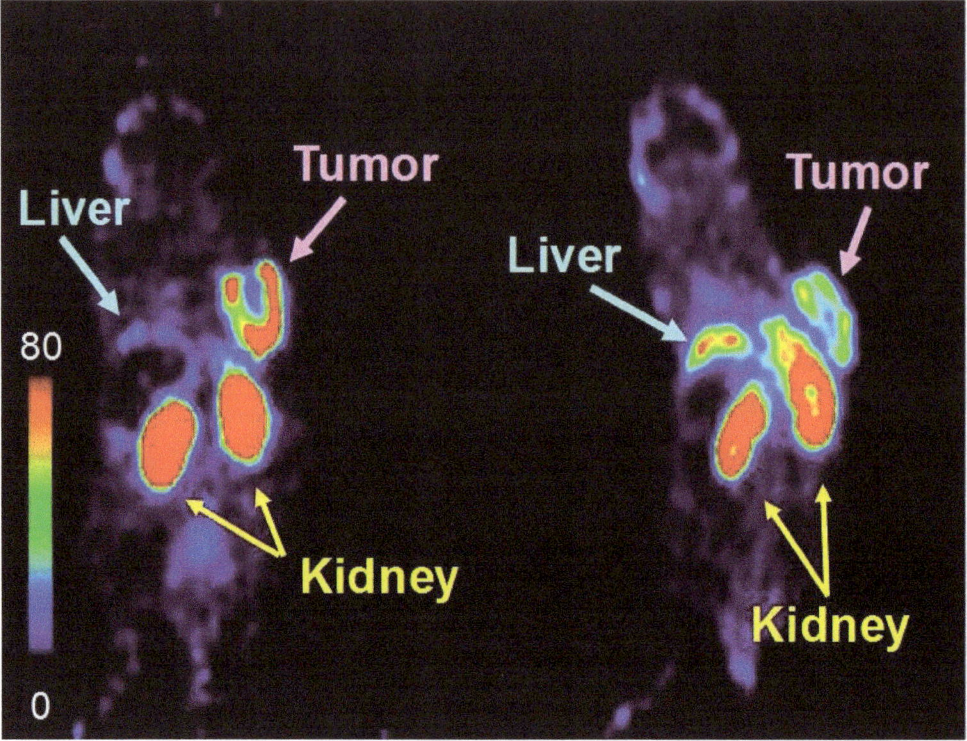

Figure 7. Representative coronal PET images of, PC3(+) tumor xenograft mice at 60 min post-injection of [¹⁸F]DCFPyL or [¹⁸F]2a [3.7–7.4 MBq (100–200 µCi)]. Tumors appear on the right shoulder (pink arrows).

3. Discussion

Recently, Bouvet et al. reported the influence of different prosthetic groups on PSMA-targeted radiotracers (DCFPyL analogues) with improved tumor uptake and clearance profile [41]. The highest tumor uptake in their study was achieved with the most lipophilic compound (**1a**, Scheme 1), prepared via the oxime formation of 4-[¹⁸F]fluorobenzaldehyde with aminooxy precursor **1**. Moreover, they suggested that the low tumor uptake of the [¹⁸F]FDG linked oxime tracer could be due to the combination of high hydrophilicity and steric crowding [36]. This result inspired us to further investigate the effect of adding an alkyl chain between the PSMA-inhibitor lysine-urea-glutamate scaffold and labeled prosthetic groups to increase lipophilicity and decrease steric crowding of the labeled PSMA probe. Therefore, compounds [¹⁸F]**1b** and [¹⁸F]**2a–b** were designed with alkyl chains of various lengths, and an arene or heteroarene substituent.

The biological evaluation of all tracers found that PSMA targeting was preserved both in vitro and in vivo and for the most part was comparable to [¹⁸F]DCFPyL. In vitro, the labeled tracers and non-radioactive standards had retained specific and high nM binding to PSMA, with [¹⁸F]**1a** tending to have higher affinity than [¹⁸F]DCFPyL. All four analogues exhibited in vivo PSMA tumor-targeting comparable to [¹⁸F]DCFPyL with tumor uptakes and T:M ratios, ranging from 27 to 17%ID/g and 203 to 74 T:M, respectively, which were at least 8-fold greater than non-target tissues except for the kidney and liver.

Tumor uptakes of [^{18}F]**1a** tended to be higher than [^{18}F]DCFPyL comparing favorably with previous findings [41]. Therefore, the modification of the canonical amide bond of DCFPyL to include an alkyl chain and oxime-linked [^{18}F]fluorobenzyl or -pyridinyl substituent minimally affected in vivo tumor targeting to PSMA, indicating that in human patients, all four [^{18}F]DCFPyL analogues would be expected to identify PSMA expressing lesions as has been clinically observed with [^{18}F]DCFPyL [45].

In human patients [^{18}F]DCFPyL has demonstrated favorable dosimetry within acceptable limits for diagnostic PET tracers, however high accumulation and retention in the kidneys and salivary glands could limit use in other clinical applications such as radionuclide therapies [35,46]. In a retrospective clinical trial Barber et.al reported 25% renal injury in patients treated with [^{177}Lu]Lu-PSMA-617 and currently sufficient kidney function is an important criterion for patient eligibility for this recently FDA-approved therapy [47–49]. In addition, retrospectively, xerostomia was found in 24% of patients by Heck et.al. which most likely is an underreported adverse effect resulting from high PSMA expression levels in the salivary glands [50]. The uptake of PSMA targeted imaging agents in human salivary glands is specific and consistent with known high PSMA expression levels which is not the case for mice. The mouse PSMA, a homolog of human PSMA, has a 12-fold lower expression level in salivary glands with higher levels of non-specific binding and therefore, may not be as reliable to detect changes in specific PSMA uptake [51,52]. In contrast, the elevated kidney uptake observed in mice is comparable to humans representing both specific PSMA binding in the renal cortex and non-specific radioactivity in the urinary tract due to excretion [44,53]. In studies investigating other fluorine-18labeled PSMA inhibitors the physiochemical properties of the labeled prosthetic groups were found to affect the biological clearance profiles, therefore, modification of the ^{18}F-labeled prosthetic group of [^{18}F]DCFPyL may offer a strategy to lower kidney uptake [41]. In these pre-clinical studies only [^{18}F]**2a** displayed a desirable alteration in pharmacokinetics and metabolism resulting in greater in vivo stability and significantly lower kidney uptake with higher liver uptake compared to [^{18}F]DCFPyL. [^{18}F]**2a** liver uptake significantly decreased over the 2 h time course whereas kidney uptake remained unchanged. This shift to hepatobiliary clearance by [^{18}F]**2a** may, in part, be explained by an increase in lipophilicity (logP = 1.45) compared to the other analogues. It is interesting to note that the other analogue which had the next highest logP (0.58), [^{18}F]**2b**, had higher liver uptake as observed with [^{18}F]**2a** but comparable kidney uptake to [^{18}F]DCFPyL. Similarly, liver uptake was increased with [^{18}F]**1a** compared to [^{18}F]DCFPyL, as reported by Bouvet et al. [41]. The rank order of the liver uptake (%ID/g) of all the tracers indicated a switch to hepatobiliary clearance that corresponded to the tracers logP rank order suggesting that the lipophilicity plays a role in determining the clearance profile of the tracer. These results suggest that [^{18}F]**2a** may offer an alternative PSMA-targeting agent with decreased renal clearance in clinical applications.

4. Materials and Methods

The aminooxy precursor, (((S)-5-(((aminooxy)carbonyl)amino)-1-carboxypentyl)carbamoyl)-L-glutamic acid (**1**), non-radioactive standards, (((S)-5-(2-(aminooxy)acetamido)-1-carboxypentyl)carbamoyl)-L-glutamic acid 4-fluorobenzaldehyde oxime (**1a**) and (((S)-5-(2-(aminooxy)acetamido)-1-carboxypentyl)carbamoyl)-L-glutamic acid 6-fluoronicotinaldehyde oxime (**1b**), were prepared according to literature methods [41]. Fluorine-18 radiolabeled DCFPyL, 4-[^{18}F]fluorobenzaldehyde, and 6-[^{18}F]fluoronicotinaldehyde were prepared following a recently published method [54,55]. PBS 1X buffer (Gibco) was obtained from Life Technologies (Carlsbad, CA, USA). All other chemicals and solvents were received from Sigma-Aldrich (St. Louis, MO, USA) and used without further purification. Fluorine-18 in target water was obtained from the National Institutes of Health cyclotron facility (Bethesda, MD, USA). Chromafix 30-PS-HCO$_3$ anion-exchange cartridges were purchased from Macherey-Nagel (Düren, Germany). Columns and all other Sep-Pak cartridges used in this synthesis were obtained from Agilent Technologies (Santa Clara, CA, USA) and Waters (Milford, MA, USA), respectively. Oasis MCX Plus cartridges were conditioned with 5 mL anhydrous acetonitrile. The thin-layer chromatography (TLC) plates for phosphorimaging were obtained from Miles Scientific (Newark, DE, USA). Phosphorimaging plates were read using a Fuji FLA5100 and the data was analyzed using Image Gauge V4.0. Flash chromatography was performed on a Teledyne Isco Combiflash Rf+ instrument using hexane:ethyl acetate gradients. NMR spectra were obtained on a 400 MHz Varian NMR and processed using MestReNova software. LC/MS data for small molecules were acquired on an Agilent Technologies 1290 Infinity HPLC system using a 6130 quadrupole LC/MS detector and a Poroshell 120 SB-C18 2.7 um column (4.6 × 50 mm). HRMS data were acquired on a Waters XEVO G2-XS QTOF running MassLynx version 4.1. Semi-prep HPLC purification and analytical HPLC for radiochemical work were performed on an Agilent 1200 Series instrument equipped with multi-wavelength detectors.

4.1. Precursor and Non-Radioactive Standard

4.1.1. (((. S)-5-(6-(Aminooxy)hexanamido)-1-carboxypentyl)carbamoyl)-L-glutamic acid (2)

di-*tert*-Butyl-(((S)-6-amino-1-(*tert*-butoxy)-1-oxohexan-2-yl)carbamoyl)-L-glutamate (1.00 g, 2.05 mmol) was combined with 6-(N-*tert*-butyloxycarbonyl)aminooxyhexanoic acid (500 mg, 2.02 mmol) and triethylamine (0.42 mL, 3 mmol) in 30 mL of dichloromethane [56]. HBTU (1.15 g, 3.03 mmol) was added, and the mixture was allowed to stir overnight. The reaction was diluted with dichloromethane, then washed sequentially with 50 mL each of saturated $NaHCO_3$, 1 N HCl, and water. The organic layer was dried over anhydrous Na_2SO_4 and evaporated under reduced pressure. The resulting residue was purified by flash chromatography with a gradient from 50 to 100% ethyl acetate in hexanes to yield 1.00 g (68%) of a colorless residue. ^{1}H NMR ($CDCl_3$): δ 7.64 (s, 1H), 6.37 (s, 1H), 5.33 (br s, *J* = 27.3 Hz, 2H), 4.31 (ddd, *J* = 19.4, 8.4, 4.6 Hz, 2H), 4.12 (q, *J* = 7.2 Hz, 2H), 3.88 (t, *J* = 6.4 Hz, 2H), 3.38 (m, 2H), 3.12 (m, 2H), 2.81 (s, 6H), 2.38–2.30 (m, 2H), 2.25 (t, *J* = 8.0, 2H); 2.04 (s, 2H), 1.89–1.51 (m, 10H), 1.47 (s, 9H); 1.46 (s, 9H); 1.45 (s, 9H); 1.44 (s, 9H); 1.26 (t, *J* = 7.1 Hz, 2H). MS: Calculated for $C_{35}H_{65}N_4O_{11}$ (M+H): 717.5; found 717.5.

The intermediate, tri-*tert*-butyl-(((S)-5-(6-((t-butoxycarbonyl)aminooxy)hexanamido)-1-carboxypentyl)carbamoyl)-L-glutamic acid was dissolved in 2 mL of dichloromethane. TIPSH (0.2 mL), followed by TFA (2 mL) were added. The reaction was stirred for 3 h at RT, then concentrated under reduced pressure. The resulting residue was dissolved in water and purified by reverse phase HPLC using a semi-preparative HPLC column and gradient mobile phase from 0–30% acetonitrile in water. Both phases contained 0.05% TFA. The desired product (**2**) was isolated as a white solid (TFA salt, 495 mg, 65%) after lyophilizing the relevant fractions. ^{1}H NMR (400 MHz, Methanol-d_4) δ 4.31 (dd, *J* = 8.6, 5.1 Hz, 1H), 4.25 (dd, *J* = 8.5, 4.9 Hz, 1H), 4.02 (t, *J* = 6.4 Hz, 2H), 3.18 (t, *J* = 6.8 Hz, 2H), 2.50–2.33 (m, 2H), 2.24–2.08 (m, 3H), 1.96–1.78 (m, 2H), 1.76–1.59 (m, 5H), 1.59–1.48 (m, 2H), 1.48–1.38 (m, 4H). HRMS: Calculated for $C_{18}H_{33}N_4O_9$ (M+H): 449.2248, found 449.2239.

4.1.2. (((. S)-5-(6-(Aminooxy)hexanamido)-1-carboxypentyl)carbamoyl)-L-glutamic acid 4-fluorobenzaldehyde oxime (2a)

Compound 2 (20 mg, 36.7 μmol) was dissolved in 2 mL of MeOH, triethylamine (50 μL) and 4-benzaldehyde (11.1 mg, 89.2 μmol) were added. The reaction was stirred for 2 h at RT, then concentrated under reduced pressure. The desired major product was isolated by preparative HPLC (10–70% MeCN in H_2O) followed by lyophilization of the relevant fractions (white solid, 15.4 mg, 62% yield). ^{1}H NMR (400 MHz, Methanol-d4) δ 8.08 (s, 1H), 7.67–7.58 (m, 2H), 7.18–7.06 (m, 2H), 4.31 (dd, *J* = 8.6, 5.0 Hz, 1H), 4.26 (dd, *J* = 8.4, 4.9 Hz, 1H), 4.14 (t, *J* = 6.5 Hz, 2H), 3.16 (t, *J* = 6.8 Hz, 2H), 2.50–2.32 (m, 2H), 2.24–2.08 (m, 3H), 1.96–1.79 (m, 2H), 1.79–1.68 (m, 2H), 1.68–1.60 (m, 3H), 1.55–1.51 (m, 2H), 1.51–1.38 (m, 4H). HRMS: Calculated for $C_{25}H_{36}FN_4O_9$ (M+H): 555.2466, found 555.2463.

4.1.3. (((. S)-5-(6-(Aminooxy)hexanamido)-1-carboxypentyl)carbamoyl)-L-glutamic acid 6-fluoronicotinaldehyde oxime (2b)

Compound **2** (20 mg, 36.7 μmol) was dissolved in 2 mL of MeOH, and triethylamine (50 μL) and 6-fluoronicotinaldehyde (11.2 mg, 89.2 μmol) were added. The reaction was stirred for 2 h at RT, then concentrated under reduced pressure. The desired major product was isolated by preparative HPLC (10–70% MeCN in H$_2$O) followed by lyophilization of the relevant fractions (white solid, 11.4 mg, 46% yield).

^{1}H NMR (400 MHz, Methanol-d_4) δ 8.36 (d, *J* = 2.5 Hz, 1H), 8.26–8.19 (m, 1H), 8.17 (d, *J* = 0.5 Hz, 1H), 7.13–7.07 (m, 1H), 4.31 (dd, *J* = 8.6, 5.0 Hz, 1H), 4.26 (dd, *J* = 8.3, 4.8 Hz, 1H), 4.18 (t, *J* = 6.5 Hz, 2H), 3.20–3.13 (m, 2H), 2.41 (ddd, *J* = 8.5, 6.8, 3.5 Hz, 2H), 2.24–2.08 (m, 3H), 1.96–1.78 (m, 2H), 1.77–1.62 (m, 5H), 1.58–1.49 (m, 2H), 1.47–1.39 (m, 4H). HRMS: Calculated for C$_{24}$H$_{35}$FN$_5$O$_9$ (M+H): 556.2419, found 556.2418.

4.2. Radiochemical Syntheses

All radiochemical syntheses were performed according to the following two general procedures described below.

General Method

Procedure 1: Manual syntheses for compounds [^{18}F]**1a** and [^{18}F]**2a**

Fluorine-18 in target water (3700–7400 MBq) was diluted with 2 mL water and passed through an anion-exchange cartridge (Chromafix 30-PS-HCO$_3$). The cartridge was washed with anhydrous acetonitrile (6 mL) and dried for 1 min. The [^{18}F]fluoride from the cartridge was slowly eluted (0.5 mL/min) with its 4-formyl-N,N,N-trimethylbenzenaminium triflate precursor (5–7 mg) in 0.5 mL 1:4 acetonitrile: *t*-butanol. The Sep-Pak was further eluted with 0.5 mL acetonitrile and the eluent was collected in the same vial. The reaction mixture was heated at 120 °C for 2 min to produce 4-[^{18}F]fluorobenzaldehyde. The radiolabeled intermediate was purified by passing the reaction mixture through a pre-conditioned Oasis MCX Plus cartridge and collected in a vial containing aminooxy precursor **1** or **2** (5 mg) in 0.2 mL water. The cartridge was flushed with 1 mL acetonitrile and the eluent was collected in the same vial. The solution was stirred for 10 min at 70 °C and the solvent was evaporated under N$_2$ and reduced pressure. The HPLC buffer (3 mL) was added via a syringe. The mixture was injected into the HPLC for purification. The collected product was buffered to pH ~7 with 45 mM sodium phosphate. The identity and purity of the product were confirmed by analytical HPLC.

Procedure 2: Automated syntheses for compounds [^{18}F]**1b** and [^{18}F]**2b** on a GE Tracerlab FX-N Pro module

Fluorine-18 in target water (3700–7400 MBq) was diluted with 2 mL water and passed through an anion-exchange cartridge (Chromafix 30-PS-HCO$_3$) followed by anhydrous acetonitrile (6 mL) and the cartridge was dried for 3 min under vacuum. The [^{18}F]fluoride from the Sep-Pak was eluted with 5-formyl-N,N,N-trimethylpyridin-2-aminium triflate precursor (5–7 mg) in 0.5 mL 1:4, acetonitrile: *t*-butanol (in a syringe) via an external three-way valve. The mixture was passed through a pre-conditioned Oasis MCX Plus cartridge (incorporated between V13 and Reactor 1). The cartridge was flushed with 1 mL acetonitrile through the external three-way valve and the eluent was collected in the same vial (Reactor 1). To this solution in Reactor 1 was added the aminooxy precursor, **1** or **2**, (5 mg) in water (0.5 mL) from Vial 3. The solution was stirred for 10 min at 70 °C and the solvent was then evaporated under N$_2$ and vacuum. The HPLC buffer (3 mL) was added from Vial 4. The mixture was transferred to Tube 2 and injected into the HPLC for purification. The collected product was buffered to pH ~7 with 45 mM sodium phosphate. The identity and purity of the product were confirmed by analytical HPLC.

HPLC conditions for purification: Agilent Eclipse plus C18 column (9.4 × 250 mm, 10 μm), mobile phase: B = ethanol, A = 50 mM phosphoric acid, flow rate of 3.5 mL/min.

HPLC conditions for analysis: Agilent Eclipse plus C18 (4.6 × 150 mm, 3.5 μm), B = acetonitrile, A = 0.1 M aqueous ammonium formate pH adjusted to 3.5 with trifluoroacetic acid, flow rate of 1 mL/min.

[^{18}F]**1a**: The radiochemical yield was 21–32% (uncorrected, *n* > 5) in 50 min with a molar activity of 300–330 GBq/μmol. HPLC conditions for purification: 30% B in A, t_R = ~18 min. HPLC conditions for analysis: 20% B in A, t_R = ~5 min.

[^{18}F]**1b**: The radiochemical yield was 37–40% (uncorrected, $n = 9$) in 45 min with a molar activity of 300–360 GBq/μmol. HPLC conditions for purification: 20% B in A, t_R = ~20 min. HPLC conditions for analysis: 15% B in A, t_R = ~4 min.

[^{18}F]**2a**: The radiochemical yield was 15–27% (uncorrected, $n > 5$) in 50 min with a molar activity of 300 -330 GBq/μmol. HPLC conditions for purification: 50% B in A, t_R = ~22 min. HPLC conditions for analysis: 20% B in A, t_R = ~9 min.

[^{18}F]**2b**: The radiochemical yield was 36–42% (uncorrected, $n = 9$) in 45 min with a molar activity of 300–360 GBq/μmol. HPLC conditions for purification: 35% B in A, t_R = ~15 min. HPLC conditions for analysis: 20% B in A, t_R = ~7 min.

4.3. Lipophilicity

The lipophilicities of the molecules were determined by calculating the value of the partition coefficient (logP) using ChemDraw 2019. The logP values of the molecules, [^{18}F]DCFPyL, [^{18}F]**1a**, [^{18}F]**1b**, [^{18}F]**2a**, [^{18}F]**2b** were −0.94, 0.04, −0.83, 1.45, 0.58, respectively.

4.4. Cell Lines and Human Tumor Xenograft Mouse Models

PC3(−) (wildtype human prostate cancer cell line, PSMA negative) and PC3(+) (transfected with human PSMA) were provided by Dr. Hisataka Kobayashi [55,57]. Cell lines were grown at 37 °C in 5% CO_2 in RPMI-1640 supplemented with 10% FBS, 2 mM L-glutamine and Pen/Strep/Amphotericin B. PC3(+) and PC3(−) cell suspensions from in vitro cell culture were subcutaneously implanted (right shoulder) into athymic mice (Athymic NCr- nu/nu, Charles River Laboratory, 4 weeks old) for use as positive and negative controls, respectively, for in vitro or in vivo studies. When tumors reached the appropriate size (>100 mg) the xenograft mice were used for in vivo biodistributions and imaging studies or the tumors were excised and further processed to obtain membrane preparations for in vitro assays as described previously [58].

4.5. In Vitro Binding Studies

In vitro saturation studies were performed to determine binding affinities (K_d) and PSMA expression levels (B_{max}) using tumor membrane preparations from PC3(−) and PC3(+) PSMA xenografts (human prostate cancer cell line transfected with human PSMA; PC3(+)). A constant aliquot of the tumor membrane preparation was added to increasing concentrations of the tracers (0.5–70 nM) in duplicate (total bound activity (B_t)); non-specific binding (B_{nsp}) was determined by adding non-radioactive DCFPyL (10^{-6} M) to another set of duplicates. For competition assays a constant concentration of [^{18}F]DCFPyL (0.5 to 1.0 nM) and increasing concentrations (0–1000 nM) of competitors (non-radioactive standards; DCFPyL, **1a–b**, or **2a–b**) were added to membrane aliquots. After incubation (2 h at RT) separation of bound [^{18}F]DCFPyL from free was accomplished by filtration using GF/C filter papers followed by 2 washes with saline. Filter papers were collected, and the radioactive content was quantified by gamma counting (PerkinElmer 2480 Wizard3). From the saturation studies, the K_d and B_{max} were determined from 6- 8 concentrations of the radiolabeled tracers and analyzed using non-linear regression curve fitting (one-site specific binding hyperbola); from the competition studies, inhibitory constants (K_i)'s were determined from 8–10 competitor concentrations of non-radioactive standard/DCFPyL [PRISM (version 7.0 Windows), GraphPad software, San Diego, CA]. Aliquots of each membrane preparation were taken for the determination of the protein concentration (Bradford method).

4.6. Biodistributions

Tumor-bearing mice (tumor weights: 0.1–0.8 g) were injected while awake via the tail vein with each of the tracers [0.74–3.7 MBq (20 to 100 μCi), 10 to 80 pmol] and euthanized (via CO_2 inhalation) at selected times. For the blocking studies, mice were coinjected with one of the tracers [0.74–3.7 MBq (20 to 100 μCi), 10 to 80 pmol] + DCFPyL (1000×: ~10 to 80 nmol) and euthanized at 60 min post-injection. Blood samples and tissues were excised from each animal, weighed, and radioactivity content was determined (Perkin Elmer 2480 Wizard3). Radioactivity content in the blood and each tissue was expressed as % injected dose per gram of tissue [%ID/g; (Formula (1))] and then normalized for body weight to a 20 g mouse (Formula (2)) from which Tissue:Muscle ratios [T:M; (Formula (3))] were determined as follows:

$$\%\mathrm{ID/g} = \frac{[\text{counts per minute (cpm)}_{\text{tissue}} \, / \, \text{tissue weight (g)}] \times 100}{\mathrm{cpm}_{\text{total injected dose}}} \tag{1}$$

$$\%\mathrm{ID/g} \text{ (normalized to a 20 g mouse)} = (\%\mathrm{ID/g}) \times (\text{body weight}/20 \text{ g}) \tag{2}$$

$$\mathrm{T:M} = \frac{\%\mathrm{ID/g}_{\text{tissue}}}{\%\mathrm{ID/g}_{\text{muscle}}} \tag{3}$$

Statistical analysis of the differences between the 2 groups was carried out using the Student's *t*-test with $p < 0.05$ as significant (GraphPad In Stat 3 for Windows).

In some cases, additional blood samples and/or tissue samples after gamma counting were taken for determining the fraction of intact radiolabeled tracer (parent) using thin-layer chromatography (TLC). For these TLC determinations: tissues were placed in equal volumes of acetonitrile and homogenized, or serum was obtained from the blood samples and mixed with an equal volume of acetonitrile. Following centrifugation of the samples, supernatants were collected and the radioactive content of the supernatants and pellets were determined. The supernatants were then applied to thin-layer chromatography (TLC) plates. The TLC plates were developed [solvent system: ethyl acetate (80%), methanol (10%) and acetic acid (10%)], and exposed on a phosphorimaging plate which was scanned the next day.

4.7. PET Imaging Studies

Tumor-bearing mice were anesthetized with isoflurane/O_2 (1.5–3% v/v) and imaged at various times after intravenous injection of each tracer [2.6 to 3.7 MBq (70 to 100 μCi)]. Whole body static PET images were obtained at 2 bed positions (FOV = 2.0 cm, total imaging time: 10 min) using a BioPET scanner (Bioscan Inc., Washington, DC, USA). The images were reconstructed by a 3-dimensional ordered-subsets expectation maximum (3D-OSEM).

5. Conclusions

Fluorine-18-labeled urea-based PSMA inhibitors were prepared either manually or automatically in high radiochemical yield using the prosthetic group 4-[18F]fluorobenzalde-hyde or 6-[18F]fluoronicotinaldehyde. [18F]**2a** displayed a desirable alteration in pharma-cokinetics and metabolism resulting in significantly lowering the kidney uptake while maintaining high-affinity binding to PSMA compared to [18F]DCFPyL. Therefore, [18F]**2a** may be of use in clinical applications to reduce the radioactive dose to kidneys while maintaining high tumor uptake.

Author Contributions: Conceptualization, F.B. and E.M.J.; Data curation, F.B., T.E.P., X.Z., C.C.W., J.R. and E.M.J.; Formal analysis, F.B., X.Z. and E.M.J.; Methodology, F.B., T.E.P., X.Z., C.C.W., J.R. and E.M.J.; Supervision, R.E.S. and P.L.C.; Writing—original draft, F.B. and E.M.J.; Writing—review and editing, F.B., T.E.P., X.Z., C.C.W., J.R., E.M.J., P.L.C. and R.E.S. All authors have read and agreed to the published version of the manuscript.

Funding: This project has been funded in whole or in part with federal funds from the National Cancer Institute, National Institutes of Health, under Contract No. HHSN261200800001E.

Institutional Review Board Statement: All animal studies were performed in compliance with the protocols (MIP006) approved by NIH Animal Care and Use Committee.

Informed Consent Statement: Not applicable.

Data Availability Statement: Data is contained within the article.

Conflicts of Interest: The authors declare no conflict of interest.

References

1. Rawla, P. Epidemiology of Prostate Cancer. *World J. Oncol.* **2019**, *10*, 63–89. [CrossRef] [PubMed]
2. Mirzaei-Alavijeh, M.; Ahmadi-Jouybari, T.; Vaezi, M.; Jalilian, F. Prevalence, Cognitive and Socio-Demographic Determinants of Prostate Cancer Screening. *Asian Pac. J. Cancer Prev.* **2018**, *19*, 1041–1046. [CrossRef] [PubMed]
3. Vellky, J.E.; Ricke, W.A. Development and prevalence of castration-resistant prostate cancer subtypes. *Neoplasia* **2020**, *22*, 566–575. [CrossRef] [PubMed]
4. Jadvar, H. Molecular imaging of prostate cancer with 18F-fluorodeoxyglucose PET. *Nat. Rev. Urol.* **2009**, *6*, 317–323. [CrossRef]
5. Minamimoto, R.; Senda, M.; Jinnouchi, S.; Terauchi, T.; Yoshida, T.; Murano, T.; Fukuda, H.; Iinuma, T.; Uno, K.; Nishizawa, S.; et al. The current status of an FDG-PET cancer screening program in Japan, based on a 4-year (2006–2009) nationwide survey. *Ann. Nucl. Med.* **2013**, *27*, 46–57. [CrossRef]
6. Mapelli, P.; Incerti, E.; Ceci, F.; Castellucci, P.; Fanti, S.; Picchio, M. 11C- or 18F-Choline PET/CT for Imaging Evaluation of Biochemical Recurrence of Prostate Cancer. *J. Nucl. Med.* **2016**, *57*, 43s–48s. [CrossRef]
7. Ceci, F.; Fanti, S. PSMA-PET/CT imaging in prostate cancer: Why and when. *Clin. Transl. Imaging* **2019**, *7*, 377–379. [CrossRef]
8. Hupe, M.C.; Philippi, C.; Roth, D.; Kümpers, C.; Ribbat-Idel, J.; Becker, F.; Joerg, V.; Duensing, S.; Lubczyk, V.H.; Kirfel, J.; et al. Expression of Prostate-Specific Membrane Antigen (PSMA) on Biopsies Is an Independent Risk Stratifier of Prostate Cancer Patients at Time of Initial Diagnosis. *Front. Oncol.* **2018**, *8*, 623. [CrossRef]
9. Queisser, A.; Hagedorn, S.A.; Braun, M.; Vogel, W.; Duensing, S.; Perner, S. Comparison of different prostatic markers in lymph node and distant metastases of prostate cancer. *Mod. Pathol.* **2015**, *28*, 138–145. [CrossRef]
10. Wang, F.; Li, Z.; Feng, X.; Yang, D.; Lin, M. Advances in PSMA-targeted therapy for prostate cancer. *Prostate Cancer Prostatic Dis.* **2022**, *25*, 11–26. [CrossRef]
11. Awenat, S.; Piccardo, A.; Carvoeiras, P.; Signore, G.; Giovanella, L.; Prior, J.O.; Treglia, G. Diagnostic Role of 18F-PSMA-1007 PET/CT in Prostate Cancer Staging: A Systematic Review. *Diagnostics* **2021**, *11*, 552. [CrossRef] [PubMed]
12. Farolfi, A.; Calderoni, L.; Mattana, F.; Mei, R.; Telo, S.; Fanti, S.; Castellucci, P. Current and Emerging Clinical Applications of PSMA PET Diagnostic Imaging for Prostate Cancer. *J. Nucl. Med.* **2021**, *62*, 596–604. [CrossRef] [PubMed]
13. Afshar-Oromieh, A.; Malcher, A.; Eder, M.; Eisenhut, M.; Linhart, H.G.; Hadaschik, B.A.; Holland-Letz, T.; Giesel, F.L.; Kratochwil, C.; Haufe, S.; et al. PET imaging with a [68Ga]gallium-labelled PSMA ligand for the diagnosis of prostate cancer: Biodistribution in humans and first evaluation of tumour lesions. *Eur. J. Nucl. Med. Mol. Imaging* **2013**, *40*, 486–495. [CrossRef] [PubMed]
14. Perera, M.; Papa, N.; Christidis, D.; Wetherell, D.; Hofman, M.S.; Murphy, D.G.; Bolton, D.; Lawrentschuk, N. Sensitivity, Specificity, and Predictors of Positive (68)Ga-Prostate-specific Membrane Antigen Positron Emission Tomography in Advanced Prostate Cancer: A Systematic Review and Meta-analysis. *Eur. Urol.* **2016**, *70*, 926–937. [CrossRef]
15. Einspieler, I.; Rauscher, I.; Düwel, C.; Krönke, M.; Rischpler, C.; Habl, G.; Dewes, S.; Ott, A.; Wester, H.-J.; Schwaiger, M.; et al. Detection Efficacy of Hybrid[68] Ga-PSMA Ligand PET/CT in Prostate Cancer Patients with Biochemical Recurrence After Primary Radiation Therapy Defined by Phoenix Criteria. *J. Nucl. Med.* **2017**, *58*, 1081–1087. [CrossRef]
16. Schwarzenboeck, S.M.; Rauscher, I.; Bluemel, C.; Fendler, W.P.; Rowe, S.P.; Pomper, M.G.; Afshar-Oromieh, A.; Herrmann, K.; Eiber, M. PSMA Ligands for PET Imaging of Prostate Cancer. *J. Nucl. Med.* **2017**, *58*, 1545–1552. [CrossRef]
17. Hofman, M.S.; Hicks, R.J.; Maurer, T.; Eiber, M. Prostate-specific Membrane Antigen PET: Clinical Utility in Prostate Cancer, Normal Patterns, Pearls, and Pitfalls. *Radiographics* **2018**, *38*, 200–217. [CrossRef]
18. Eder, M.; Schäfer, M.; Bauder-Wüst, U.; Hull, W.-E.; Wängler, C.; Mier, W.; Haberkorn, U.; Eisenhut, M. 68Ga-Complex Lipophilicity and the Targeting Property of a Urea-Based PSMA Inhibitor for PET Imaging. *Bioconjugate Chem.* **2012**, *23*, 688–697. [CrossRef]
19. Hennrich, U.; Eder, M. [68Ga]Ga-PSMA-11: The First FDA-Approved 68Ga-Radiopharmaceutical for PET Imaging of Prostate Cancer. *Pharmaceuticals* **2021**, *14*, 713. [CrossRef]
20. Hohberg, M.; Kobe, C.; Täger, P.; Hammes, J.; Schmidt, M.; Dietlein, F.; Wild, M.; Heidenreich, A.; Drzezga, A.; Dietlein, M. Combined Early and Late [(68)Ga]PSMA-HBED-CC PET Scans Improve Lesion Detectability in Biochemical Recurrence of Prostate Cancer with Low PSA Levels. *Mol. Imaging Biol.* **2019**, *21*, 558–566. [CrossRef]
21. Schmuck, S.; Nordlohne, S.; von Klot, C.A.; Henkenberens, C.; Sohns, J.M.; Christiansen, H.; Wester, H.J.; Ross, T.L.; Bengel, F.M.; Derlin, T. Comparison of standard and delayed imaging to improve the detection rate of [(68)Ga]PSMA I&T PET/CT in patients with biochemical recurrence or prostate-specific antigen persistence after primary therapy for prostate cancer. *Eur. J. Nucl. Med. Mol. Imaging* **2017**, *44*, 960–968. [CrossRef] [PubMed]
22. Dietlein, F.; Kobe, C.; Neubauer, S.; Schmidt, M.; Stockter, S.; Fischer, T.; Schomäcker, K.; Heidenreich, A.; Zlatopolskiy, B.D.; Neumaier, B.; et al. PSA-Stratified Performance of (18)F- and (68)Ga-PSMA PET in Patients with Biochemical Recurrence of Prostate Cancer. *J. Nucl. Med.* **2017**, *58*, 947–952. [CrossRef] [PubMed]
23. Ferreira, G.; Iravani, A.; Hofman, M.S.; Hicks, R.J. Intra-individual comparison of (68)Ga-PSMA-11 and (18)F-DCFPyL normal-organ biodistribution. *Cancer Imaging* **2019**, *19*, 23. [CrossRef] [PubMed]

24. Baecker, D.; Obermoser, V.; Kirchner, E.A.; Hupfauf, A.; Kircher, B.; Gust, R. Fluorination as tool to improve bioanalytical sensitivity and COX-2-selective antitumor activity of cobalt alkyne complexes. *Dalton Trans.* **2019**, *48*, 15856–15868. [CrossRef] [PubMed]
25. Maurer, T.; Eiber, M.; Schwaiger, M.; Gschwend, J.E. Current use of PSMA-PET in prostate cancer management. *Nat. Rev. Urol.* **2016**, *13*, 226–235. [CrossRef]
26. Werner, R.A.; Derlin, T.; Lapa, C.; Sheikbahaei, S.; Higuchi, T.; Giesel, F.L.; Behr, S.; Drzezga, A.; Kimura, H.; Buck, A.K.; et al. (18)F-Labeled, PSMA-Targeted Radiotracers: Leveraging the Advantages of Radiofluorination for Prostate Cancer Molecular Imaging. *Theranostics* **2020**, *10*, 1–16. [CrossRef]
27. Zimmerman, M.E.; Meyer, A.R.; Rowe, S.P.; Gorin, M.A. Imaging of prostate cancer with positron emission tomography. *Clin. Adv. Hematol. Oncol.* **2019**, *17*, 455–463.
28. Hillier, S.M.; Maresca, K.P.; Femia, F.J.; Marquis, J.C.; Foss, C.A.; Nguyen, N.; Zimmerman, C.N.; Barrett, J.A.; Eckelman, W.C.; Pomper, M.G.; et al. Preclinical evaluation of novel glutamate-urea-lysine analogues that target prostate-specific membrane antigen as molecular imaging pharmaceuticals for prostate cancer. *Cancer Res.* **2009**, *69*, 6932–6940. [CrossRef]
29. Chen, Y.; Foss, C.A.; Byun, Y.; Nimmagadda, S.; Pullambhatla, M.; Fox, J.J.; Castanares, M.; Lupold, S.E.; Babich, J.W.; Mease, R.C.; et al. Radiohalogenated prostate-specific membrane antigen (PSMA)-based ureas as imaging agents for prostate cancer. *J. Med. Chem.* **2008**, *51*, 7933–7943. [CrossRef]
30. Mease, R.C.; Dusich, C.L.; Foss, C.A.; Ravert, H.T.; Dannals, R.F.; Seidel, J.; Prideaux, A.; Fox, J.J.; Sgouros, G.; Kozikowski, A.P.; et al. N-[N-[(S)-1,3-Dicarboxypropyl]carbamoyl]-4-[18F]fluorobenzyl-L-cysteine, [18F]DCFBC: A new imaging probe for prostate cancer. *Clin. Cancer Res.* **2008**, *14*, 3036–3043. [CrossRef]
31. Chen, Y.; Pullambhatla, M.; Foss, C.A.; Byun, Y.; Nimmagadda, S.; Senthamizhchelvan, S.; Sgouros, G.; Mease, R.C.; Pomper, M.G. 2-(3-{1-Carboxy-5-[(6-[18F]fluoro-pyridine-3-carbonyl)-amino]-pentyl}-ureido)-pentanedioic acid, [18F]DCFPyL, a PSMA-based PET imaging agent for prostate cancer. *Clin. Cancer Res.* **2011**, *17*, 7645–7653. [CrossRef] [PubMed]
32. Malik, N.; Baur, B.; Winter, G.; Reske, S.N.; Beer, A.J.; Solbach, C. Radiofluorination of PSMA-HBED via Al(18)F(2+) Chelation and Biological Evaluations In Vitro. *Mol. Imaging Biol.* **2015**, *17*, 777–785. [CrossRef] [PubMed]
33. Cardinale, J.; Schäfer, M.; Benešová, M.; Bauder-Wüst, U.; Leotta, K.; Eder, M.; Neels, O.C.; Haberkorn, U.; Giesel, F.L.; Kopka, K. Preclinical Evaluation of (18)F-PSMA-1007, a New Prostate-Specific Membrane Antigen Ligand for Prostate Cancer Imaging. *J. Nucl. Med.* **2017**, *58*, 425–431. [CrossRef] [PubMed]
34. Wester, H.J.; Schottelius, M. PSMA-Targeted Radiopharmaceuticals for Imaging and Therapy. *Semin. Nucl. Med.* **2019**, *49*, 302–312. [CrossRef]
35. Szabo, Z.; Mena, E.; Rowe, S.P.; Plyku, D.; Nidal, R.; Eisenberger, M.A.; Antonarakis, E.S.; Fan, H.; Dannals, R.F.; Chen, Y.; et al. Initial Evaluation of [(18)F]DCFPyL for Prostate-Specific Membrane Antigen (PSMA)-Targeted PET Imaging of Prostate Cancer. *Mol. Imaging Biol.* **2015**, *17*, 565–574. [CrossRef]
36. Dietlein, M.; Kobe, C.; Kuhnert, G.; Stockter, S.; Fischer, T.; Schomäcker, K.; Schmidt, M.; Dietlein, F.; Zlatopolskiy, B.D.; Krapf, P.; et al. Comparison of [(18)F]DCFPyL and [(68)Ga]Ga-PSMA-HBED-CC for PSMA-PET Imaging in Patients with Relapsed Prostate Cancer. *Mol. Imaging Biol.* **2015**, *17*, 575–584. [CrossRef]
37. Giesel, F.L.; Knorr, K.; Spohn, F.; Will, L.; Maurer, T.; Flechsig, P.; Neels, O.; Schiller, K.; Amaral, H.; Weber, W.A.; et al. Detection Efficacy of (18)F-PSMA-1007 PET/CT in 251 Patients with Biochemical Recurrence of Prostate Cancer After Radical Prostatectomy. *J. Nucl. Med.* **2019**, *60*, 362–368. [CrossRef]
38. Giesel, F.L.; Will, L.; Lawal, I.; Lengana, T.; Kratochwil, C.; Vorster, M.; Neels, O.; Reyneke, F.; Haberkon, U.; Kopka, K.; et al. Intraindividual Comparison of (18)F-PSMA-1007 and (18)F-DCFPyL PET/CT in the Prospective Evaluation of Patients with Newly Diagnosed Prostate Carcinoma: A Pilot Study. *J. Nucl. Med.* **2018**, *59*, 1076–1080. [CrossRef]
39. Keam, S.J. Piflufolastat F 18: Diagnostic First Approval. *Mol. Diagn. Ther.* **2021**, *25*, 647–656. [CrossRef]
40. Zlatopolskiy, B.D.; Endepols, H.; Krapf, P.; Guliyev, M.; Urusova, E.A.; Richarz, R.; Hohberg, M.; Dietlein, M.; Drzezga, A.; Neumaier, B. Discovery of (18)F-JK-PSMA-7, a PET Probe for the Detection of Small PSMA-Positive Lesions. *J. Nucl. Med.* **2019**, *60*, 817–823. [CrossRef]
41. Bouvet, V.; Wuest, M.; Bailey, J.J.; Bergman, C.; Janzen, N.; Valliant, J.F.; Wuest, F. Targeting Prostate-Specific Membrane Antigen (PSMA) with F-18-Labeled Compounds: The Influence of Prosthetic Groups on Tumor Uptake and Clearance Profile. *Mol. Imaging Biol.* **2017**, *19*, 923–932. [CrossRef] [PubMed]
42. Dannoon, S.; Ganguly, T.; Cahaya, H.; Geruntho, J.J.; Galliher, M.S.; Beyer, S.K.; Choy, C.J.; Hopkins, M.R.; Regan, M.; Blecha, J.E.; et al. Structure-Activity Relationship of (18)F-Labeled Phosphoramidate Peptidomimetic Prostate-Specific Membrane Antigen (PSMA)-Targeted Inhibitor Analogues for PET Imaging of Prostate Cancer. *J. Med. Chem.* **2016**, *59*, 5684–5694. [CrossRef] [PubMed]
43. Kularatne, S.A.; Zhou, Z.; Yang, J.; Post, C.B.; Low, P.S. Design, Synthesis, and Preclinical Evaluation of Prostate-Specific Membrane Antigen Targeted 99mTc-Radioimaging Agents. *Mol. Pharm.* **2009**, *6*, 790–800. [CrossRef] [PubMed]
44. Bařinka, C.; Rojas, C.; Slusher, B.; Pomper, M. Glutamate carboxypeptidase II in diagnosis and treatment of neurologic disorders and prostate cancer. *Curr. Med. Chem.* **2012**, *19*, 856–870. [CrossRef] [PubMed]
45. Rowe, S.P.; Macura, K.J.; Mena, E.; Blackford, A.L.; Nadal, R.; Antonarakis, E.S.; Eisenberger, M.; Carducci, M.; Fan, H.; Dannals, R.F.; et al. PSMA-Based [(18)F]DCFPyL PET/CT Is Superior to Conventional Imaging for Lesion Detection in Patients with Metastatic Prostate Cancer. *Mol. Imaging Biol.* **2016**, *18*, 411–419. [CrossRef]

46. Chakravarty, R.; Siamof, C.M.; Dash, A.; Cai, W. Targeted α-therapy of prostate cancer using radiolabeled PSMA inhibitors: A game changer in nuclear medicine. *Am. J. Nucl. Med. Mol. Imaging* **2018**, *8*, 247–267.

47. Barber, T.W.; Singh, A.; Kulkarni, H.R.; Niepsch, K.; Billah, B.; Baum, R.P. Clinical Outcomes of (177)Lu-PSMA Radioligand Therapy in Earlier and Later Phases of Metastatic Castration-Resistant Prostate Cancer Grouped by Previous Taxane Chemotherapy. *J. Nucl. Med.* **2019**, *60*, 955–962. [CrossRef]

48. Seitzer, K.E.; Seifert, R.; Kessel, K.; Roll, W.; Schlack, K.; Boegemann, M.; Rahbar, K. Lutetium-177 Labelled PSMA Targeted Therapy in Advanced Prostate Cancer: Current Status and Future Perspectives. *Cancers* **2021**, *13*, 3715. [CrossRef]

49. Mullard, A. FDA approves first PSMA-targeted radiopharmaceutical. *Nat. Rev. Drug Discov.* **2022**, *21*, 327. [CrossRef]

50. Heck, M.M.; Tauber, R.; Schwaiger, S.; Retz, M.; D'Alessandria, C.; Maurer, T.; Gafita, A.; Wester, H.-J.; Gschwend, J.E.; Weber, W.A.; et al. Treatment Outcome, Toxicity, and Predictive Factors for Radioligand Therapy with 177Lu-PSMA-I&T in Metastatic Castration-resistant Prostate Cancer. *Eur. Urol.* **2019**, *75*, 920–926.

51. Roy, J.; Warner, B.M.; Basuli, F.; Zhang, X.; Wong, K.; Pranzatelli, T.; Ton, A.T.; Chiorini, J.A.; Choyke, P.L.; Lin, F.I.; et al. Comparison of Prostate-Specific Membrane Antigen Expression Levels in Human Salivary Glands to Non-Human Primates and Rodents. *Cancer Biother. Radiopharm.* **2020**, *35*, 284–291. [CrossRef] [PubMed]

52. O'Keefe, D.S.; Bacich, D.J.; Molloy, P.L.; Heston, W.D.W. *Prostate Specific Membrane Antigen*; Humana Press: Totowa, Nj, USA, 2008.

53. Mhawech-Fauceglia, P.; Zhang, S.; Terracciano, L.; Sauter, G.; Chadhuri, A.; Herrmann, F.R.; Penetrante, R. Prostate-specific membrane antigen (PSMA) protein expression in normal and neoplastic tissues and its sensitivity and specificity in prostate adenocarcinoma: An immunohistochemical study using mutiple tumour tissue microarray technique. *Histopathology* **2007**, *50*, 472–483. [CrossRef] [PubMed]

54. Basuli, F.; Zhang, X.; Woodroofe, C.C.; Jagoda, E.M.; Choyke, P.L.; Swenson, R.E. Fast indirect fluorine-18 labeling of protein/peptide using the useful 6-fluoronicotinic acid-2,3,5,6-tetrafluorophenyl prosthetic group: A method comparable to direct fluorination. *J. Label. Comp. Radiopharm.* **2017**, *60*, 168–175. [CrossRef] [PubMed]

55. Basuli, F.; Zhang, X.; Jagoda, E.M.; Choyke, P.L.; Swenson, R.E. Rapid synthesis of maleimide functionalized fluorine-18 labeled prosthetic group using "radio-fluorination on the Sep-Pak" method. *J. Label. Comp. Radiopharm.* **2018**, *61*, 599–605. [CrossRef] [PubMed]

56. Jones, D.S.; Branks, M.J.; Campbell, M.A.; Cockerill, K.A.; Hammaker, J.R.; Kessler, C.A.; Smith, E.M.; Tao, A.; Ton-Nu, H.T.; Xu, T. Multivalent poly(ethylene glycol)-containing conjugates for in vivo antibody suppression. *Bioconjug. Chem.* **2003**, *14*, 1067–1076. [CrossRef] [PubMed]

57. Nakajima, T.; Mitsunaga, M.; Bander, N.H.; Heston, W.D.; Choyke, P.L.; Kobayashi, H. Targeted, activatable, in vivo fluorescence imaging of prostate-specific membrane antigen (PSMA) positive tumors using the quenched humanized J591 antibody-indocyanine green (ICG) conjugate. *Bioconjug. Chem.* **2011**, *22*, 1700–1705. [CrossRef] [PubMed]

58. Roy, J.; White, M.E.; Basuli, F.; Opina, A.C.L.; Wong, K.; Riba, M.; Ton, A.T.; Zhang, X.; Jansson, K.H.; Edmondson, E.; et al. Monitoring PSMA Responses to ADT in Prostate Cancer Patient-Derived Xenograft Mouse Models Using [(18)F]DCFPyL PET Imaging. *Mol. Imaging Biol.* **2021**, *23*, 745–755. [CrossRef]

 *pharmaceuticals*

Article

Pharmacokinetic Imaging Using ^{99m}Tc-Mebrofenin to Untangle the Pattern of Hepatocyte Transporter Disruptions Induced by Endotoxemia in Rats

Solène Marie [1,2,3], Irene Hernández-Lozano [4], Marc Le Vée [5], Louise Breuil [1], Wadad Saba [1], Maud Goislard [1], Sébastien Goutal [1], Charles Truillet [1], Oliver Langer [4], Olivier Fardel [6] and Nicolas Tournier [1,*]

[1] Université Paris-Saclay, CEA, CNRS, Inserm, Laboratoire d'Imagerie Biomédicale Multimodale, BIOMAPS, Service Hospitalier Frédéric Joliot, 4 Place du Général Leclerc, 91401 Orsay, France; solene.marie@aphp.fr (S.M.); louise.breuil@universite-paris-saclay.fr (L.B.); wadad.saba@cea.fr (W.S.); maud.goislard@universite-paris-saclay.fr (M.G.); sebastien.goutal@universite-paris-saclay.fr (S.G.); charles.truillet@cea.fr (C.T.)

[2] Faculté de Pharmacie, Université Paris-Saclay, 92296 Châtenay-Malabry, France

[3] AP-HP, Université Paris-Saclay, Hôpital Bicêtre, Pharmacie Clinique, 94270 Le Kremlin Bicêtre, France

[4] Department of Clinical Pharmacology, Medical University of Vienna, 1090 Vienna, Austria; irene.hernandezlozano@meduniwien.ac.at (I.H.-L.); oliver.langer@meduniwien.ac.at (O.L.)

[5] Univ. Rennes, Inserm, EHESP, Irset (Institut de Recherche en Santé, Environnement et Travail)-UMR_S 1085, 35043 Rennes, France; marc.levee@univ-rennes1.fr

[6] Univ. Rennes, CHU Rennes, Inserm, EHESP, Irset (Institut de Recherche en Santé, Environnement et Travail)-UMR_S 1085, 35043 Rennes, France; olivier.fardel@univ-rennes1.fr

* Correspondence: n.tournier@universite-paris-saclay.fr; Tel.: +33-(0)1-69-86-77-02

Citation: Marie, S.; Hernández-Lozano, I.; Le Vée, M.; Breuil, L.; Saba, W.; Goislard, M.; Goutal, S.; Truillet, C.; Langer, O.; Fardel, O.; et al. Pharmacokinetic Imaging Using ^{99m}Tc-Mebrofenin to Untangle the Pattern of Hepatocyte Transporter Disruptions Induced by Endotoxemia in Rats. *Pharmaceuticals* **2022**, *15*, 392. https://doi.org/10.3390/ph15040392

Academic Editors: Xuyi Yue and Irina Velikyan

Received: 13 February 2022
Accepted: 19 March 2022
Published: 24 March 2022

Publisher's Note: MDPI stays neutral with regard to jurisdictional claims in published maps and institutional affiliations.

Copyright: © 2022 by the authors. Licensee MDPI, Basel, Switzerland. This article is an open access article distributed under the terms and conditions of the Creative Commons Attribution (CC BY) license (https://creativecommons.org/licenses/by/4.0/).

Abstract: Endotoxemia-induced inflammation may impact the activity of hepatocyte transporters, which control the hepatobiliary elimination of drugs and bile acids. ^{99m}Tc-mebrofenin is a non-metabolized substrate of transporters expressed at the different poles of hepatocytes. ^{99m}Tc-mebrofenin imaging was performed in rats after the injection of lipopolysaccharide (LPS). Changes in transporter expression were assessed using quantitative polymerase chain reaction of resected liver samples. Moreover, the particular impact of pharmacokinetic drug–drug interactions in the context of endotoxemia was investigated using rifampicin (40 mg/kg), a potent inhibitor of hepatocyte transporters. LPS increased ^{99m}Tc-mebrofenin exposure in the liver (1.7 ± 0.4-fold). Kinetic modeling revealed that endotoxemia did not impact the blood-to-liver uptake of ^{99m}Tc-mebrofenin, which is mediated by organic anion-transporting polypeptide (Oatp) transporters. However, liver-to-bile and liver-to-blood efflux rates were dramatically decreased, leading to liver accumulation. The transcriptomic profile of hepatocyte transporters consistently showed a downregulation of multidrug resistance-associated proteins 2 and 3 (Mrp2 and Mrp3), which mediate the canalicular and sinusoidal efflux of ^{99m}Tc-mebrofenin in hepatocytes, respectively. Rifampicin effectively blocked both the Oatp-mediated influx and the Mrp2/3-related efflux of ^{99m}Tc-mebrofenin. The additive impact of endotoxemia and rifampicin led to a 3.0 ± 1.3-fold increase in blood exposure compared with healthy non-treated animals. ^{99m}Tc-mebrofenin imaging is useful to investigate disease-associated change in hepatocyte transporter function.

Keywords: ABC-transporter; drug-induced liver injury; hepatotoxicity; organic anion-transporting polypeptide; pharmacokinetics; liver function; SLC-transporter

1. Introduction

Inflammation is a common feature of many pathophysiological states, including infection, systemic diseases, and cancer [1]. The physiological changes accompanying inflammation may alter liver function, with consequences for the hepatobiliary elimination of endogenous and exogeneous compounds and lead to liver injury [2,3] or impact the pharmacokinetics (PK) of medications [4].

From a molecular perspective, liver function can be estimated by the activity of the membrane transporters expressed in hepatocytes, which work together to mediate the hepatobiliary elimination of bile salts, drugs, and metabolites [5]. Several solute carrier (SLC) influx transporters, including the organic anion-transporting polypeptides (OATP)1B1/3 (*SLCO1B1/1B3* corresponding to Oatp4/*Slco1b2* in rodents [6]), are expressed in the sinusoidal (blood-facing) membrane of hepatocytes, where they mediate the uptake of their substrates from blood into the liver (Figure 1) [7]. Adenosine triphosphate-binding cassette (ABC) efflux transporters expressed in the canalicular (bile-facing) membrane of hepatocytes, such as the human multidrug resistance-associated protein (MRP)2/*ABCC2* (Mrp2/*Abcc2* in rodents) or the P-glycoprotein (P-gp, encoded by the multidrug resistance 1 gene (*MDR1*/*ABCB1*) in humans and both *Mdr1a*/*Abcb1a* and *Mdr1b*/*Abcb1b* in rodents), control the biliary excretion of solutes and bile acids (Figure 1). Other ABC transporters such as human MRP3/*ABCC3* (Mrp3/*Abcc3* gene in rodents) mediate the sinusoidal efflux of drugs and metabolites from hepatocytes into the blood (Figure 1) [7].

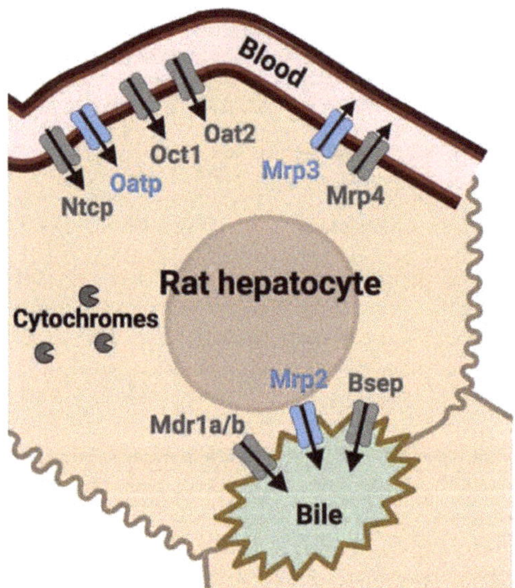

Figure 1. Membrane transporters expressed in rat hepatocytes. Transporters known to be involved in the hepatobiliary transport of ^{99m}Tc-mebrofenin are highlighted in blue. Bsep: bile salt export pump; Mdr: multidrug resistance; Mrp: multidrug resistance-associated protein; Ntcp: Na+-taurocholate co-transporting polypeptide; Oat: organic anion transporter; Oatp: organic anion-transporting polypeptide; Oct: organic cation transporter.

Disruption of hepatocyte transporter function by inflammation is increasingly considered as a cause of disease-related changes in liver function, hepatobiliary elimination of drugs, PK, and toxicity [3,8]. Dysregulation of hepatocyte transporters and cytochromes has been reported during endotoxin-mediated inflammation induced by lipopolysaccharide (LPS) in vitro [9] and ex vivo, in several animal species [10–13]. In animal models, the impaired activity of ABC transporters during inflammation has been linked to cholestasis and/or abnormal liver accumulation of drugs and metabolites, which may provide a mechanistical explanation for drug-induced liver injury (DILI) [14]. A similar disruption in hepatocyte transporter function is likely to occur in patients and may lead to severe hepatotoxicity outcomes. It remains, however, technically difficult to untangle and estimate the intrinsic importance of each individual transporter in vivo, apart from cytochrome activity, in mediating inflammation-related changes in PK in animal models and patients [4].

Hepatocyte transporters are targets for pharmacokinetic drug–drug interactions (DDI) [15]. The antituberculosis drug rifampicin (rifampin) is known to inhibit the function of several hepatocyte transporters in vitro, including human OATPs, MRP2, and MRP3, and rat Oatp2/*Slco1a4*, Oatp4, and Mrp3 [16–19]. Single dose rifampicin (600 mg) was shown to enhance plasma exposure to OATP substrates [20], and is therefore recommended by regulatory agencies as a model inhibitor drug to conduct pharmacokinetic DDI studies in animals and humans [15]. However, the in vivo impact of rifampicin on each individual transporter function and the consequences of liver exposure to solutes cannot be assessed from the plasma PK. This particularly holds for transporters expressed at the canalicular pole of hepatocytes distant from the plasma compartment [21]. Moreover, most non-clinical/clinical PK and DDI studies are conducted in healthy subjects. This may not recapitulate the pathological context, and underestimates the impact of DDI in patients with inflammatory diseases [22].

^{99m}Tc-mebrofenin is a radiopharmaceutical used for hepatobiliary scintigraphy, routinely used in hepatobiliary disorders to perform imaging of the liver, bile, and gallbladder to assess hepatic function and/or biliary excretion [23]. ^{99m}Tc-mebrofenin is well characterized as a non-metabolized substrate of human OATP1B1, OATP1B3, MRP2, and MRP3, which govern its hepatobiliary clearance [23]. Preclinical studies conducted in Oatp- and Mrp2-deficient animals confirmed the transport of ^{99m}Tc-mebrofenin by the rodent orthologs of these transporters [24,25]. Our team and other teams have shown that dynamic ^{99m}Tc-mebrofenin imaging, aided by kinetic modeling, can be used for molecular imaging of hepatocyte transporter function at the different poles of hepatocytes in animals and humans [23,26–28] (Figure 1). Moreover, ^{99m}Tc-mebrofenin allows for selectively and simultaneously assessing the impact of transporter-mediated DDIs at the sinusoidal or the canalicular level in vivo [29].

In this study, the impact of LPS-induced inflammation on liver transporter expression and function was investigated in rats, using quantitative transcriptomics and ^{99m}Tc-mebrofenin imaging. Moreover, we hypothesized that inflammation may exacerbate the impact of transporter-mediated DDIs precipitated by rifampicin in terms of PK and/or liver exposure.

2. Results

2.1. Quantitative Transcriptomics

A significant decrease in the mRNA expression of several transporters was observed after LPS-treatment, including Mrp2 and Mrp3 (Figure 2 and Table S1). Among the tested ABC transporters, the most important decrease was observed for the genes encoding P-gp, with a significant −68% decrease for *Mdr1a* and a −75% decrease for *Mdr1b*. The mRNA expression of most hepatic SLC transporters was also significantly decreased after LPS-treatment, except for organic anion transporter (Oat)2/*Slc22a7*. The decrease in Oatp1/*Slco1a1* mRNA expression was less pronounced than for Oatp2/*Slco1a4*. The decrease in Oatp4/*Slco1b2* expression was also not statistically significant. Moreover, qPCR showed a significant decrease in mRNA expression coding for cytochromes (Cyp), with a −74% decrease of *Cyp3a1* mRNA expression (Figure 2).

2.2. ^{99m}Tc-Mebrofenin Imaging

Figure 3 shows the mean time activity curves (TACs) obtained in each studied group. The mean area under the curve (AUC) for each TAC is reported in Figure 4 and Table S2. Compared with the healthy group, LPS significantly increased exposure to ^{99m}Tc-mebrofenin in the liver (1.7 ± 0.4-fold, $p < 0.01$). A significant 1.5 ± 0.3-fold decrease of radioactivity in the intestine was also observed, suggesting reduced biliary excretion ($p < 0.01$).

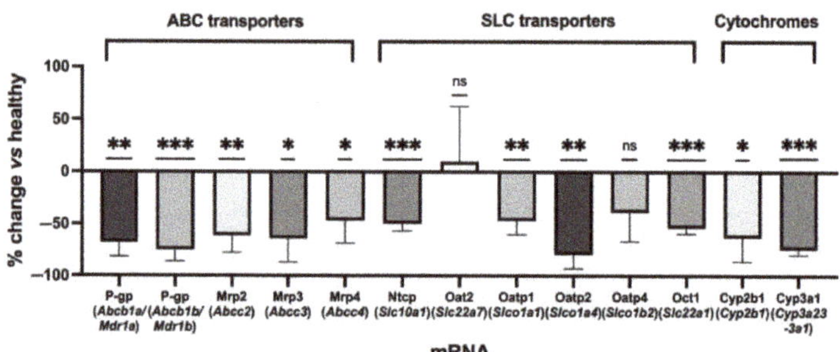

Figure 2. Changes in transporter and cytochrome mRNA expression in the liver of LPS-treated rats (24 h after exposure to LPS) compared to healthy rats estimated by qPCR ($n = 4$ both in healthy and LPS-treated groups). Data are mean $\pm$ SD. * $p \leq 0.05$, ** $p \leq 0.01$, *** $p \leq 0.001$, ns indicates not significant, one-sample *t*-test after a Shapiro–Wilk Normality test. Cyp: cytochrome; LPS: lipopolysaccharide; Mdr: multidrug resistance; Mrp: multidrug-resistance associated protein; Ntcp: Na+-taurocholate cotransporting polypeptide; Oat: organic anion transporter; Oatp: organic anion-transporting polypeptide; Oct: organic cation transporter.

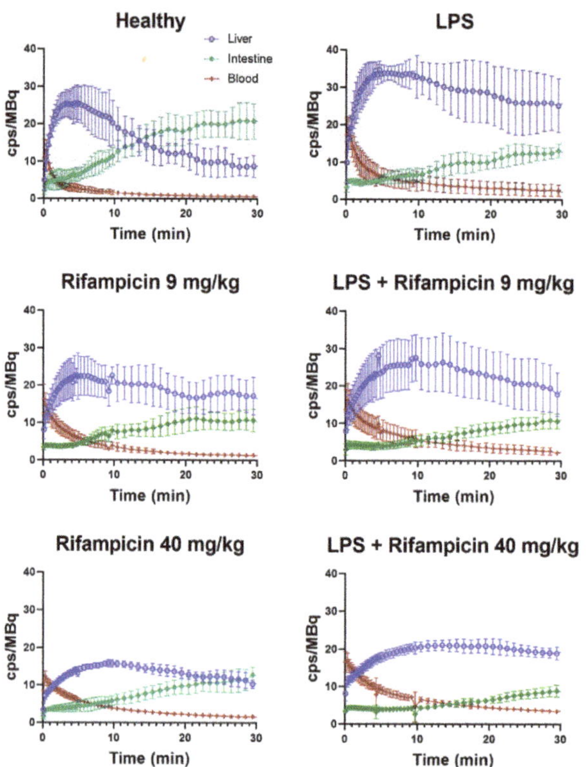

Figure 3. Mean ($\pm$SD) time–activity curves of ^{99m}Tc-mebrofenin in the liver ($\bigcirc$), intestine ($\blacklozenge$), and blood (+) of healthy and LPS-treated rats (24 h after exposure to LPS) under baseline conditions and after treatment with 9 or 40 mg/kg rifampicin ($n = 5$ for control animals and $n = 6$ for LPS-treated animals). Radioactivity is expressed as counts per second (cps) normalized to the injected dose (MBq). LPS: lipopolysaccharide. Data obtained in the healthy and Rifampicin 40 mg/kg groups have already been presented in a previous study [29].

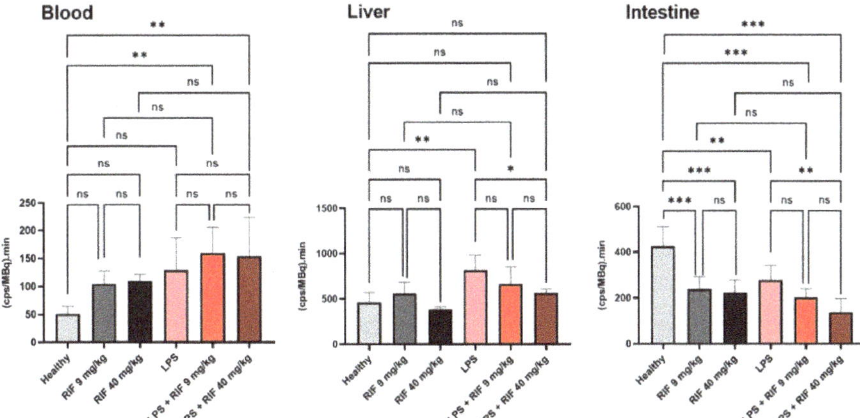

Figure 4. Area under the time–activity curves (AUC) of ^{99m}Tc-mebrofenin in the blood pool (imaged-derived), liver, and intestine obtained in healthy (n = 5 per group) and in lipopolysaccharide (LPS)-treated rats (n = 6 per group). Impact of pre-treatment with 9 or 40 mg/kg of rifampicin (RIF), a potent Oatp/Mrp2/Mrp3 inhibitor is reported. Radioactivity is expressed in counts per second (cps) in each region of interest, normalized to the injected radioactivity amount in each animal (cps/MBq). Data are mean ± SD. * $p < 0.05$, ** $p < 0.01$, *** $p < 0.001$, ns indicates not significant, ordinary one-way ANOVA followed by a Tukey's post hoc test for multiple comparison. LPS: lipopolysaccharide; RIF: rifampicin. Data obtained in the healthy and RIF 40 mg/kg groups have already been presented in a previous study [29].

High dose rifampicin (40 mg/kg) significantly reduced the amount of ^{99m}Tc-mebrofenin in the intestines of healthy animals (1.9 ± 0.4-fold, $p < 0.001$) and LPS-treated animals (1.2 ± 0.3-fold, $p < 0.001$). High dose rifampicin did not impact liver exposure neither in healthy nor in LPS-treated animals. Rifampicin (9 or 40 mg/kg) increased blood exposure in healthy animals (unpaired t-test, $p < 0.01$), although no significant difference was found using a one-way ANOVA ($p > 0.05$, Figure 4). Rifampicin did not significantly increase blood exposure in LPS-treated animals, regardless of statistical analysis ($p > 0.05$, Figure 4). Impact of endotoxemic inflammation and rifampicin-perpetrated DDI on overall blood clearance were additive: blood exposure of ^{99m}Tc-mebrofenin in LPS/rifampicin-treated animals was 3.0 ± 1.3-fold higher than for healthy non-treated animals ($p < 0.01$).

A pharmacological dose of rifampicin (9 mg/kg) significantly reduced the amount of ^{99m}Tc-mebrofenin in the intestines in healthy animals ($p < 0.001$), reaching similar levels as for the high dose rifampicin-treated animals. The pharmacological dose of rifampicin did not further decrease intestinal radioactivity in LPS-treated animals ($p > 0.05$). The pharmacological dose of rifampicin did not impact liver exposure in healthy or LPS-treated animals ($p > 0.05$). The pharmacological dose of rifampicin did not significantly exacerbate the impact of LPS on blood exposure ($p > 0.05$, Figure 4).

Visually, the implemented PK model provided good fits for both the observed liver and intestinal radioactivity (Figure S1), and the parameter precision (assessed by percentage coefficient of variation, %CV) was, in general, acceptable (with a %CV of less than 40% for most of the subjects, Table S3). A higher %CV was observed for four subjects in the estimation of k_2 and k_3, especially in the situation of complete inhibition. This probably reflects the difficulty to accurately estimate the extremely low values of the transfer rate constants (Table S3). These subjects were not excluded from the statistical analysis. The rate constant k_1 defining the radiotracer transfer from blood into liver was not significantly different between healthy and LPS-treated rats (Figure 5). Rifampicin dose-dependently decreased k_1 to a similar extent for both healthy and LPS-treated animals (Figure 5). The sinusoidal efflux rate constant k_2 was significantly lower in the LPS-treated animals compared with the healthy controls. Pharmacological dose of rifampicin significantly decreased k_2 in healthy animals but caused no further k_2 decrease in LPS-treated rats. A maximal decrease in k_2 was observed after treatment with a high dose of rifampicin, in both healthy and LPS-treated animals. LPS exposure strikingly decreased the liver-to-bile transfer rate constant (k_3) to a minimal level, which was not further decreased by rifampicin (Figure 5). The non-parametric test showed no significant differences in the values of k_5, which estimates the rate transfer between the bile ducts and intestine, except between healthy and LPS-treated animals when treated with a pharmacological dose of rifampicin (Figure 5). However, the parameter precision was not satisfying for some subjects (Table S3) and a high intra-group variability was observed (Figure 5).

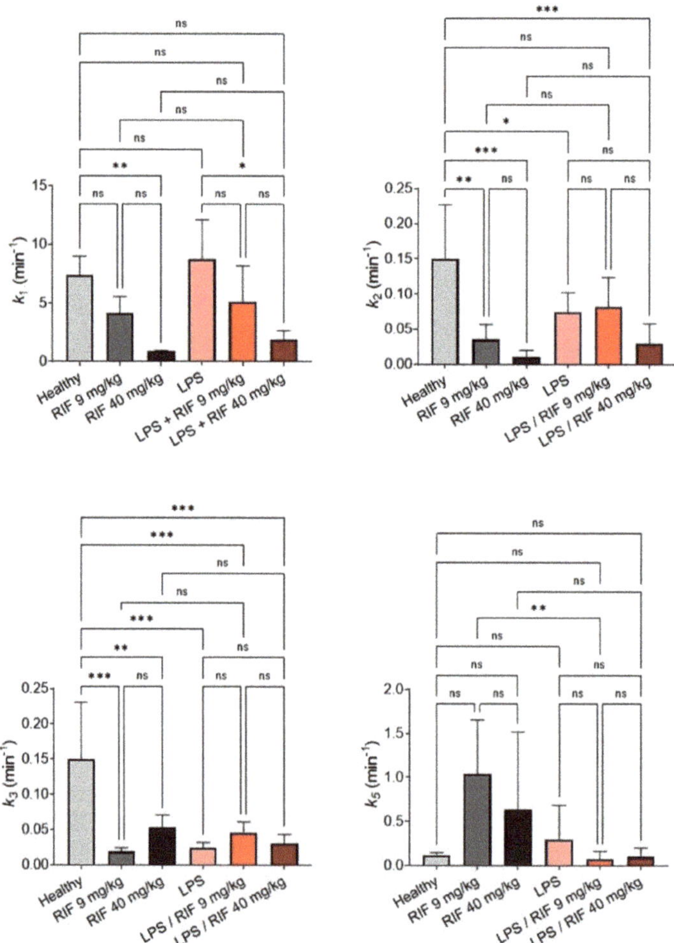

Figure 5. Pharmacokinetic model outcome parameters describing the hepatobiliary transport of ^{99m}Tc-mebrofenin in healthy and LPS-treated rats (24 h after exposure to LPS) under baseline conditions and after treatment with 9 or 40 mg/kg rifampicin ($n = 5$ for control animals and $n = 6$ for LPS-treated animals). Data are mean ± SD. * $p \leq 0.05$, ** $p \leq 0.01$, *** $p \leq 0.001$, ns not significant, ordinary one-way ANOVA followed by a Tukey's post hoc test for k_2 and k_3, and Kruskal–Wallis test for comparison of k_1 and k_3 data. LPS: lipopolysaccharide; RIF: rifampicin. The data of healthy and RIF 40 mg/kg groups have already been presented in a previous study [29].

3. Discussion

The present study highlights the dramatic impact of LPS-induced inflammation on hepatocyte transporter expression and function in rats. The consequences for carrier-mediated hepatobiliary clearance and for the magnitude of transporter-mediated DDIs were assessed in vivo using ^{99m}Tc-mebrofenin imaging and kinetic modeling.

Quantitative transcriptomics confirmed the important disruption of the expression of several hepatocyte transporters 24 h after LPS exposure. This is consistent with previous rat studies, in which the mRNA and protein levels of most hepatic transporters were dramatically decreased 6–12 h after LPS administration, although an increase in Mrp3 expression has been reported [10–12,30]. The concomitant decrease in the expression of cytochrome P450 justified the use of a metabolically-stable imaging probe for quantitative and selective determination of the functional impact of this downregulation on transporter expression [31]. The mechanisms for the dysregulation of transporters and cytochromes by LPS are thought to be mediated by xenobiotic receptors, such as pregnane X receptor (PXR), aryl hydrocarbon receptor (AhR), glucocorticoid receptor (GR), and constitutive androstane receptor (CAR) [32].

From a functional perspective, our results point to significant changes in the kinetics of ^{99m}Tc-mebrofenin in LPS-treated rats, with a pronounced increase in liver exposure and a decreased bile content (Figure 4). The mechanistic importance of each individual transporter system was estimated using compartmental modeling. LPS significantly decreased both the canalicular (k_3) and basolateral (k_2) efflux rate of ^{99m}Tc-mebrofenin, which may reflect the activity of MRP2 and MRP3, respectively (Figure 5). This is consistent with the observed decrease in the *Mrp2* and *Mrp3* expression (Figure 2). These changes led to a greater exposure to ^{99m}Tc-mebrofenin in the liver of LPS-treated rats compared with healthy rats. Interestingly, LPS did not impact the Oatp-mediated uptake of ^{99m}Tc-mebrofenin (estimated by k_1), consistent with the limited and non-significant impact of LPS treatment on Oatp4 mRNA expression (Figure 2).

The impact of two different doses of the OATP/MRP inhibitor rifampicin on hepatocyte transporter function was also assessed using ^{99m}Tc-mebrofenin. In healthy rats, high dose rifampicin (40 mg/kg) almost completely inhibited the activity of Oatp, Mrp2, and Mrp3 inferred from the reduction in k_1, k_3, and k_2, respectively, compared to the healthy untreated group (Figure 5). In rats, pharmacological dose rifampicin (9 mg/kg) did not significantly decrease Oatp function (although a tendency towards a decrease was observed), while basolateral Mrp3 and canalicular Mrp2 were almost entirely inhibited at this dose. This suggests differences in vulnerability to the inhibition of Oatp and Mrp2/3 at the different poles of hepatocytes by a same dose of rifampicin in vivo. This suggests that liver excretion of MRP2/3 substrates may be hindered, while uptake transport remains active during concomitant rifampicin therapy. This may lead to liver accumulation of concomitant drugs and bile acids, and account for the rifampicin-associated DILI observed in clinical practice [33,34].

The rifampicin-challenge was also tested in LPS-treated rats to investigate the magnitude of transporter-mediated DDIs during endotoxemia. In the presence of high-dose rifampicin (40 mg/kg), transfer rate constants were similar in healthy and LPS-treated animals (Figure 5). This may be explained by the almost complete inhibition of investigated transporters, achieved with the administration of high dose rifampicin, which was not further modulated by their LPS-mediated repression (Figure 5). However, the blood exposure to ^{99m}Tc-mebrofenin in the situation of complete transporter inhibition (40 mg/kg rifampicin) was significantly higher in LPS-treated compared with healthy controls. This suggests a potentiation of the rifampicin perpetrated DDI in the context of endotoxemia, in an additive manner (Figure 4). This may be explained by the multi-organ dysfunction induced by endotoxemia, which may also impact the non-hepatic clearance of ^{99m}Tc-mebrofenin, a phenomenon that is revealed when the carrier-mediated hepatobiliary route is blocked. Interestingly, the absence of an additional impact of high-dose rifampicin on Mrp2 and Mrp3 function in LPS-treated rats suggests that almost complete suppression of the transporter activity is obtained during endotoxemia. This supports a transporter-based mechanism for cholestasis observed during endotoxemia in animals and patients [10,35,36]. Pharmacological dose rifampicin did not exacerbate the impact of LPS on hepatocyte Mrp2 and Mrp3 function, which were already almost completely suppressed by LPS. Partial inhibition of the blood-to-liver transfer of ^{99m}Tc-mebrofenin by a pharmacological dose of rifampicin was not exacerbated in LPS-treated animals, which further supports the negligible impact of endotoxemia on Oatp function (Figure 5).

In the absence of adequate methods, studies to address the impact of systemic inflammation or sepsis on PK are still rare and are mainly focused on metabolic enzyme activity [3,37]. ^{99m}Tc-mebrofenin imaging in combination with kinetic modeling offers an appealing method to be safely translated into a clinical/hospital set-up to non-invasively investigate the impact of inflammation on the function of important hepatocyte transporters. ^{9m}Tc-mebrofenin imaging may thus help guide precision medicine with better dose-selection for the numerous drugs, whose elimination depends on these particular transporters. This work illustrates the great translational potential of molecular imaging techniques to untangle the impact of drug transporters in controlling blood and liver exposure in healthy volunteers and patients [31].

4. Materials and Methods

4.1. Chemicals and Radiochemicals

LPS (from *Escherichia coli* serotype 0111:B4), was purchased from Sigma-Aldrich (Saint Quentin Fallavier, France) and rifampicin was used as the commercial drug Rifadine (Sanofi, Paris, France). Commercial kits of mebrofenin (Cholediam) were gifted from Mediam (Marcq en Baroeul, France). Each kit was labeled with a sodium ^{99m}Tc-pertechnetate (^{99m}Tc-TcO$_4$Na) eluate (~750 MBq/mL) obtained from a sterile ^{99}Mo/^{99m}Tc generator (Tekcis, GE Healthcare, Vélizy-Villacoublay, France), followed by quality control according to the manufacturer's recommendations.

4.2. Animals

A total of 41 male rats (Wistar, weight = 261 ± 108 g) aged of 5–7 weeks were used for this study. All animal experiments were in accordance with the recommendations of the European Community for animal experiments (2010/63/UE) and the French National Committees (law 2013-118) for the care and use of laboratory animals. The experimental protocol was approved by a local ethics committee for animal use (CETEA) and by the French ministry of agriculture (APAFIS#5375-20l60513 17426342, and 13 December 2018). Animals were housed in a controlled environment with access to food and water ad libitum. Half of the animals received an intraperitoneal injection of an extemporaneously prepared solution of LPS in physiological saline (2 mg/mL) at a dose of 4 mg/kg body weight. The mean weight loss observed 24 h after treatment by LPS was $-18.8 \pm 11.6\%$ of the initial weight.

4.3. Transcriptomics in LPS-Treated Rats

The expression of the transcript of selected membrane transporters and metabolic enzymes was determined using quantitative polymerase chain reaction (qPCR) analysis, as previously described [38]. Twenty-four hours after LPS injection, eight animals were sacrificed by injecting pentobarbital, and the livers were excised and frozen in liquid nitrogen. The total RNAs were extracted from frozen liver fragments using the Nucleospin RNA kit (Macherey-Nagel, Düren, Germany) and the TissueLyser LT (Qiagen, Courtaboeuf, France) at 50 Hz for 2 min after RNA quantification using the NanoDrop ND-1000 Spectrophotomer (Thermo Fisher Scientific, Illkirch-Graffenstaden, France). RNAs were reversed transcribed using the Applied Biosystems cDNA Reverse Transcription kit (Thermo Fisher Scientific, Illkirch-Graffenstaden, France). Real-time quantitative PCR was performed using the fluorescent SYBR Green dye (Thermo Fisher Scientific, Illkirch-Graffenstaden, France) and the CFX384 TouchTM Real-time PCR Detection System (Bio-Rad, Marnes la Coquette,

France). The primer sequences used for qPCR are detailed in Table S4. Two independent liver fragments were processed and analyzed for each animal, and each measurement was performed in duplicate. The specificity of each gene amplification was verified at the end of quantitative PCR reactions, through analysis of the dissociation curves of the PCR products. Amplification curves were analyzed with CFX Manager Software (Bio-Rad, Marnes la Coquette, France), using the comparative cycle threshold method. Relative quantification of the steady-state target mRNA levels was calculated after normalization of the total amount of cDNA tested to the 18S rRNA endogenous reference, using the $2^{(-\Delta\Delta Ct)}$ method. Data were expressed as arbitrary units (a.u.) relative to the 18S rRNA contents, as previously reported [38]. The percentage change in mRNA expression in each individual LPS-treated rat (n = 4) was calculated compared with the mean gene expression obtained in four healthy rats.

4.4. ^{99m}Tc-Mebrofenin Imaging

Imaging was performed in LPS-treated animals (24 h after LPS injection) and healthy animals (no LPS). In both groups, ^{99m}Tc-mebrofenin scintigraphy was performed either under the baseline condition (no transporter inhibition) or after transporter inhibition using rifampicin. Two different doses of rifampicin were tested to investigate the different levels of transporter inhibition: 9 mg/kg (corresponding to a human pharmacological dose ~600 mg/70 kg) or 40 mg/kg (assuming maximal transporter inhibition at this dose) [29]. The number of investigated animals under each tested condition is reported in Tables S2 and S3.

Rats were anesthetized with isoflurane (3.5% and 1.5–2% in oxygen for induction and maintenance, respectively). ^{99m}Tc-mebrofenin scintigraphy was performed using a clinical SPECT-CT camera (Symbia, Siemens, Knoxville, TN, USA) with a Low Energy High Resolution (LEHR) collimator. In each session, three rats were placed in a row on the scanner bed and catheters were inserted in the caudal vein. Scintigraphy imaging is not an absolute quantitative technique [31]. The position of the detectors relative to the scanner bed was standardized to limit the variability associated with in counting efficiency. Rifampicin was injected intravenously (i.v.), immediately (<5 s) before ^{99m}Tc-mebrofenin injection. Dynamic planar scintigraphy acquisitions started with ^{99m}Tc-mebrofenin injection (37.4 $\pm$ 5.0 MBq, i.v.) for 40 min, followed by an X-ray CT-scan. Dynamic images were reconstructed in 54 frames with time durations of 20×0.25 min, 10×0.5 min, 20×1 min, and 5×2 min.

4.5. Imaging Data Analysis

Images were analyzed with PMOD software (version 3.9, PMOD Technologies LLC, Zurich, Switzerland), as described in a previous rat study [29]. Regions of interest (ROI) were drawn on planar images over the liver and intestine (assumed to represent excreted bile). Standardized ROIs consisted in the largest part possible of each organ, excluding the overlapping region. The whole-heart was delineated to derive an image-derived blood input function, as previously described [29]. Corresponding TACs were generated by plotting the mean radioactivity counts (counts per second (cps)) in each region of interest normalized to the injected radioactivity amount in each animal (cps/MBq) versus time.

A previously developed four-compartment pharmacokinetic model, which had already been applied to describe the transporter-mediated hepatobiliary disposition of ^{99m}Tc-mebrofenin in rats [29,39], was used to estimate the rate constants that describe the transfer of the radiotracer between the blood and hepatocytes (k_1 and k_2, min^{-1}), from hepatocytes to the intrahepatic bile ducts (k_3, min^{-1}), and from the intrahepatic bile ducts to bile excreted into the intestine (k_5, min^{-1}). The blood concentration was estimated from the total amount of radioactivity in the heart divided by a standard heart volume (V_{heart} = 2.55 mL), which was measured on an X-ray CT scan in a previous study in rats [29]. The model assumes that all radioactivity present in the intestine corresponds to excreted bile. In addition, the model includes a dual blood input function, which accounts for the radiotracer delivery to the liver via the hepatic artery and the portal vein. The hepatic artery concentration was obtained from the image-derived input function, while the concentration in the portal vein was mathematically estimated from the arterial blood curve and the liver and intestine scintigraphy data during the modeling process, as described previously [39]. The final flow-weighted dual-input blood curve was generated using a hepatic arterial flow fraction of 0.17 [29].

4.6. Statistical Analysis

Statistical analysis was performed using Prism 8 (GraphPad Software method, La Jolla, CA, USA) and *R* package (http://www.R-project.org/, accessed on 3 March 2022). Transcriptomic profiles were compared using a one-sample *t*-test after confirmation of normal distribution by a Shapiro–Wilk Normality test. Differences in pharmacokinetic parameters between study groups were assessed by ordinary one-way ANOVA, followed by a Tukey's post hoc test for multiple comparison. Homoscedasticity was checked using the Levene's test. Homoscedasticity was not confirmed for k_1 and k_5, for which a non-parametric Kruskal–Wallis test was performed. The level of statistical significance was set to a p-value < 0.05. Data are reported as mean $\pm$ standard deviation (S.D.).

5. Conclusions

^{99m}Tc-mebrofenin imaging in rats unveiled a pattern of dramatic disruptions of hepatocyte transporters during LPS-induced endotoxemia in rats, particularly for Mrp2 and Mrp3 which control the biliary excretion and sinusoidal efflux of their substrates, respectively. This led to decreased hepatobiliary clearance and increased liver accumulation. High dose rifampicin almost completely blocked the carrier-mediated hepatobiliary transport of ^{99m}Tc-mebrofenin, respectively. However, the impact of endotoxemia and transporter-mediated DDI induced by a high dose rifampicin were additive, suggesting the importance of multi-organ dysfunction in controlling blood exposure.

Supplementary Materials: The following supporting information can be downloaded at: https://www.mdpi.com/article/10.3390/ph15040392/s1. Table S1: Expression levels of transporter mRNA of selected genes in the liver of control and LPS-treated rats (24 h after exposure to LPS). Table S2: Area under the time-activity curves (AUC) of ^{99m}Tc-mebrofenin in the blood pool (imaged-derived), liver and intestine obtained in all the studied rat groups. Table S3: Transfer rate constants obtained with the four-compartment model describing the hepatobiliary disposition of ^{99m}Tc-mebrofenin in all the studied rat groups. Table S4: Primer sequences used for qPCR assays. Figure S1: Time-activity curves of ^{99m}Tc-mebrofenin in the liver and intestine for observed data and model fits in a representative subject of each studied rat group.

Author Contributions: Conceptualization, O.F., N.T. and S.M.; methodology, I.H.-L., M.G., L.B., S.G., M.L.V., O.F. and S.M.; software, I.H.-L., S.G. and S.M.; validation, O.F., N.T. and O.L.; formal analysis, I.H.-L., S.G., M.L.V., O.F., O.L., N.T. and S.M.; investigation, M.G., L.B., W.S., C.T., M.L.V. and S.M.; resources, O.F. and N.T.; writing—original draft preparation, I.H.-L. and S.M.; writing—review and editing, O.F., O.L. and N.T.; visualization, S.M. and N.T.; supervision, S.M.; project administration, S.M. and N.T.; funding acquisition, O.F. and N.T. All authors have read and agreed to the published version of the manuscript.

Funding: Louise Breuil and Nicolas Tournier received funding from the French National Research Agency (grant number ANR-19-CE17-0027). This work was performed on a platform member of the France Life Imaging network (grant number ANR-11-INBS-0006).

Institutional Review Board Statement: All of the animal experiments were in accordance with the recommendations of the European Community for animal experiments (2010/63/UE) and the French National Committees (law 2013-118) for the care and use of laboratory animals. The experimental protocol was approved by a local ethics committee for animal use (CETEA) and by the French ministry of agriculture (APAFIS#5375-20160513 17426342, and 13 December 2018).

Informed Consent Statement: Not applicable.

Data Availability Statement: Data is contained within the article and Supplementary Material.

Acknowledgments: The authors would like to thank Vincent Brulon, Yoan Fontyn, and Kevin Phansavath for technical assistance.

Conflicts of Interest: The authors declare no conflict of interest.

References

1. Furman, D.; Campisi, J.; Verdin, E.; Carrera-Bastos, P.; Targ, S.; Franceschi, C.; Ferrucci, L.; Gilroy, D.; Fasano, A.; Miller, G.; et al. Chronic Inflammation in the Etiology of Disease across the Life Span. *Nat. Med.* **2019**, *25*, 1822–1832. [CrossRef] [PubMed]
2. Christensen, H.; Hermann, M. Immunological Response as a Source to Variability in Drug Metabolism and Transport. *Front. Pharmacol.* **2012**, *3*, 8. [CrossRef] [PubMed]
3. Stanke-Labesque, F.; Gautier-Veyret, E.; Chhun, S.; Guilhaumou, R. Inflammation Is a Major Regulator of Drug Metabolizing Enzymes and Transporters: Consequences for the Personalization of Drug Treatment. *Pharmacol. Ther.* **2020**, *215*, 107627. [CrossRef] [PubMed]
4. Dunvald, A.; Järvinen, E.; Mortensen, C.; Stage, T. Clinical and Molecular Perspectives on Inflammation-Mediated Regulation of Drug Metabolism and Transport. *Clin. Pharmacol. Ther.* **2021**. Available online: https://pubmed.ncbi.nlm.nih.gov/34605009/ (accessed on 3 March 2022). [CrossRef]
5. Jetter, A.; Kullak-Ublick, G.A. Drugs and Hepatic Transporters: A Review. *Pharmacol. Res.* **2020**, *154*, 104234. [CrossRef]
6. Roth, M.; Obaidat, A.; Hagenbuch, B. OATPs, OATs and OCTs: The Organic Anion and Cation Transporters of the SLCO and SLC22A Gene Superfamilies. *Br. J. Pharmacol.* **2012**, *165*, 1260. [CrossRef] [PubMed]
7. Patel, M.; Taskar, K.; Zamek-Gliszczynski, M. Importance of Hepatic Transporters in Clinical Disposition of Drugs and Their Metabolites. *J. Clin. Pharmacol.* **2016**, *56*, S23–S39. [CrossRef] [PubMed]
8. Evers, R.; Piquette-Miller, M.; Polli, J.W.; Russel, F.G.M.; Sprowl, J.A.; Tohyama, K.; Ware, J.A.; de Wildt, S.N.; Xie, W.; Brouwer, K.L.R. Disease-Associated Changes in Drug Transporters May Impact the Pharmacokinetics and/or Toxicity of Drugs: A White Paper from the International Transporter Consortium. *Clin. Pharmacol. Ther.* **2018**, *104*, 900–915. [CrossRef]
9. Diao, L.; Li, N.; Brayman, T.; Hotz, K.; Lai, Y. Regulation of MRP2/ABCC2 and BSEP/ABCB11 Expression in Sandwich Cultured Human and Rat Hepatocytes Exposed to Inflammatory Cytokines TNF-α, IL-6, and IL-1β. *J. Biol. Chem.* **2010**, *285*, 31185–31192. [CrossRef]
10. Cherrington, N.; Slitt, A.; Li, N.; Klaassen, C. Lipopolysaccharide-Mediated Regulation of Hepatic Transporter MRNA Levels in Rats. *Drug Metab. Dispos.* **2004**, *32*, 734–741. [CrossRef]
11. Kubitz, R.; Wettstein, M.; Warskulat, U.; Häussinger, D. Regulation of the Multidrug Resistance Protein 2 in the Rat Liver by Lipopolysaccharide and Dexamethasone. *Gastroenterology* **1999**, *116*, 401–410. [CrossRef] [PubMed]
12. Donner, M.G.; Warskulat, U.; Saha, N.; Häussinger, D. Enhanced Expression of Basolateral Multidrug Resistance Protein Isoforms Mrp3 and Mrp5 in Rat Liver by LPS. *Biol. Chem.* **2004**, *385*, 331–339. [CrossRef] [PubMed]
13. Saib, S.; Delavenne, X. Inflammation Induces Changes in the Functional Expression of P-Gp, BCRP, and MRP2: An Overview of Different Models and Consequences for Drug Disposition. *Pharmaceutics* **2021**, *13*, 1544. [CrossRef] [PubMed]
14. Corsini, A.; Bortolini, M. Drug-Induced Liver Injury: The Role of Drug Metabolism and Transport. *J. Clin. Pharmacol.* **2013**, *53*, 463–474. [CrossRef] [PubMed]
15. Giacomini, K.M.; Huang, S.-M.; Tweedie, D.J.; Benet, L.Z.; Brouwer, K.L.R.; Chu, X.; Dahlin, A.; Evers, R.; Fischer, V.; Hillgren, K.M.; et al. Membrane Transporters in Drug Development. *Nat. Rev. Drug Discov.* **2010**, *9*, 215–236. [PubMed]
16. Karlgren, M.; Vildhede, A.; Norinder, U.; Wisniewski, J.; Kimoto, E.; Lai, Y.; Haglund, U.; Artursson, P. Classification of Inhibitors of Hepatic Organic Anion Transporting Polypeptides (OATPs): Influence of Protein Expression on Drug–Drug Interactions. *J. Med. Chem.* **2012**, *55*, 4740–4763. [CrossRef]
17. Matsson, P.; Pedersen, J.; Norinder, U.; Bergström, C.; Artursson, P. Identification of Novel Specific and General Inhibitors of the Three Major Human ATP-Binding Cassette Transporters P-Gp, BCRP and MRP2 Among Registered Drugs. *Pharm. Res.* **2009**, *26*, 1816–1831. [CrossRef]
18. Lengyel, G.; Veres, Z.; Tugyi, R.; Vereczkey, L.; Molnár, T.; Glavinas, H.; Krajcsi, P.; Jemnitz, K. Modulation of Sinusoidal and Canalicular Elimination of Bilirubin-Glucuronides by Rifampicin and Other Cholestatic Drugs in a Sandwich Culture of Rat Hepatocytes. *Hepatol. Res.* **2008**, *38*, 300–309. [CrossRef]
19. Ishida, K.; Ullah, M.; Tóth, B.; Juhasz, V.; Unadkat, J. Transport Kinetics, Selective Inhibition, and Successful Prediction of In Vivo Inhibition of Rat Hepatic Organic Anion Transporting Polypeptides. *Drug Metab. Dispos.* **2018**, *46*, 1251–1258. [CrossRef]
20. Takashima, T.; Kitamura, S.; Wada, Y.; Tanaka, M.; Shigihara, Y.; Ishii, H.; Ijuin, R.; Shiomi, S.; Nakae, T.; Watanabe, Y.; et al. PET Imaging–Based Evaluation of Hepatobiliary Transport in Humans with (15R)-11C-TIC-Me. *J. Nucl. Med.* **2012**, *53*, 741–748. [CrossRef]
21. Ghibellini, G.; Leslie, E.; Brouwer, K. Methods to Evaluate Biliary Excretion of Drugs in Humans: An Updated Review. *Mol. Pharm.* **2006**, *3*, 198–211. [CrossRef] [PubMed]
22. Gandhi, A.; Moorthy, B.; Ghose, R. Drug Disposition in Pathophysiological Conditions. *Curr. Drug Metab.* **2012**, *13*, 1327–1344. [PubMed]

23. Marie, S.; Hernandez-Lozano, I.; Langer, O.; Tournier, N. Repurposing ^{99m}Tc-Mebrofenin as a Probe for Molecular Imaging of Hepatocyte Transporters. *J. Nucl. Med.* **2021**, *62*, 1043–1047. [CrossRef] [PubMed]

24. Neyt, S.; Huisman, M.; Vanhove, C.; Man, H.; Vliegen, M.; Moerman, L.; Dumolyn, C.; Mannens, G.; Vos, F. In Vivo Visualization and Quantification of (Disturbed) Oatp-Mediated Hepatic Uptake and Mrp2-Mediated Biliary Excretion of ^{99m}Tc-Mebrofenin in Mice. *J. Nucl. Med.* **2013**, *54*, 624–630. [CrossRef] [PubMed]

25. Bhargava, K.; Joseph, B.; Ananthanarayanan, M.; Balasubramaniyan, N.; Tronco, G.; Palestro, C.; Gupta, S. Adenosine Triphosphate–Binding Cassette Subfamily C Member 2 Is the Major Transporter of the Hepatobiliary Imaging Agent ^{99m}Tc-Mebrofenin. *J. Nucl. Med.* **2009**, *50*, 1140–1146. [CrossRef]

26. Marie, S.; Hernández-Lozano, I.; Breuil, L.; Truillet, C.; Hu, S.; Sparreboom, A.; Tournier, N.; Langer, O. Imaging-Based Characterization of a Slco2b1(-/-) Mouse Model Using [11C]Erlotinib and [^{99m}Tc]Mebrofenin as Probe Substrates. *Pharmaceutics* **2021**, *13*, 918. [CrossRef]

27. Pfeifer, N.; Goss, S.; Swift, B.; Ghibellini, G.; Ivanovic, M.; Heizer, W.; Gangarosa, L.; Brouwer, K. Effect of Ritonavir on ^{99m}Technetium–Mebrofenin Disposition in Humans: A Semi-PBPK Modeling and In Vitro Approach to Predict Transporter-Mediated DDIs. *CPT Pharmacomet. Syst. Pharmacol.* **2013**, *2*, e20. [CrossRef]

28. Ali, I.; Slizgi, J.; Kaullen, J.; Ivanovic, M.; Niemi, M.; Stewart, P.; Barritt, A.; Brouwer, K. Transporter-Mediated Alterations in Patients with NASH Increase Systemic and Hepatic Exposure to an OATP and MRP2 Substrate. *Clin. Pharmacol. Ther.* **2017**, *104*, 749–756. [CrossRef]

29. Marie, S.; Hernández-Lozano, I.; Breuil, L.; Saba, W.; Novell, A.; Gennisson, J.; Langer, O.; Truillet, C.; Tournier, N. Validation of Pharmacological Protocols for Targeted Inhibition of Canalicular MRP2 Activity in Hepatocytes Using [^{99m}Tc]Mebrofenin Imaging in Rats. *Pharmaceutics* **2020**, *12*, 486. [CrossRef]

30. Elferink, M.; Olinga, P.; Draaisma, A.; Merema, M.; Faber, K.; Slooff, M.; Meijer, D.; Groothuis, G. LPS-Induced Downregulation of MRP2 and BSEP in Human Liver Is Due to a Posttranscriptional Process. *Am. J. Physiol.-Gastrointest. Liver Physiol.* **2004**, *287*, G1008–G1016. [CrossRef]

31. Tournier, N.; Stieger, B.; Langer, O. Imaging Techniques to Study Drug Transporter Function In Vivo. *Pharmacol. Ther.* **2018**, *189*, 104–122. [CrossRef] [PubMed]

32. Lv, C.; Huang, L. Xenobiotic Receptors in Mediating the Effect of Sepsis on Drug Metabolism. *Acta Pharm. Sin. B* **2020**, *10*, 33–41. [CrossRef] [PubMed]

33. Ramappa, V.; Aithal, G. Hepatotoxicity Related to Anti-Tuberculosis Drugs: Mechanisms and Management. *J. Clin. Exp. Hepatol.* **2013**, *3*, 37–49. [CrossRef] [PubMed]

34. Segovia-Zafra, A.; Di Zeo-Sánchez, D.; López-Gómez, C.; Pérez-Valdés, Z.; García-Fuentes, E.; Andrade, R.; Lucena, M.; Villanueva-Paz, M. Preclinical Models of Idiosyncratic Drug-Induced Liver Injury (IDILI): Moving towards Prediction. *Acta Pharm. Sin. B* **2021**, *11*, 3685–3726. [CrossRef] [PubMed]

35. Roughneen, P.T.; Kumar, S.C.; Pellis, N.R.; Rowlands, B.J. Endotoxemia and Cholestasis. *Surg. Gynecol. Obstet.* **1988**, *167*, 205–210.

36. Chand, N.; Sanyal, A. Sepsis-Induced Cholestasis. *Hepatology* **2007**, *45*, 230–241. [CrossRef]

37. Schmith, V.; Foss, J. Effects of Inflammation on Pharmacokinetics/Pharmacodynamics: Increasing Recognition of Its Contribution to Variability in Response. *Clin. Pharmacol. Ther.* **2008**, *83*, 809–811. [CrossRef]

38. Le Vee, M.; Jouan, E.; Noel, G.; Stieger, B.; Fardel, O. Polarized Location of SLC and ABC Drug Transporters in Monolayer-Cultured Human Hepatocytes. *Toxicol. In Vitro* **2015**, *29*, 938–946. [CrossRef]

39. Hernández Lozano, I.; Karch, R.; Bauer, M.; Blaickner, M.; Matsuda, A.; Wulkersdorfer, B.; Hacker, M.; Zeitlinger, M.; Langer, O. Towards Improved Pharmacokinetic Models for the Analysis of Transporter-Mediated Hepatic Disposition of Drug Molecules with Positron Emission Tomography. *AAPS J.* **2019**, *21*, 61. [CrossRef]

pharmaceuticals

MDPI

Article

In Vivo Imaging of Rat Vascularity with FDG-Labeled Erythrocytes

Shaowei Wang [1], Mikalai Budzevich [2], Mahmoud A. Abdalah [3], Yoganand Balagurunathan [4] and Jung W. Choi [3,*]

1   Department of Medical Engineering, University of South Florida, Tampa, FL 33620, USA; shaowei.wang@usf.edu
2   Small Animal Imaging Laboratory, Department of Cancer Physiology, H. Lee Moffitt Cancer Center and Research Institute, Tampa, FL 33612, USA; mikalai.budzevich@moffitt.org
3   Department of Diagnostic Imaging and Interventional Radiology, H. Lee Moffitt Cancer Center and Research Institute, Tampa, FL 33612, USA; mahmoud.abdalah@moffitt.org
4   Division of Machine Learning, Department of Cancer Imaging and Metabolism, H. Lee Moffitt Cancer Center and Research Institute, Tampa, FL 33612, USA; yoganand.balagurunathan@moffitt.org
*   Correspondence: jung.choi@moffitt.org

Abstract: Microvascular disease is frequently found in major pathologies affecting vital organs, such as the brain, heart, and kidneys. While imaging modalities, such as ultrasound, computed tomography, single photon emission computed tomography, and magnetic resonance imaging, are widely used to visualize vascular abnormalities, the ability to non-invasively assess an organ's total vasculature, including microvasculature, is often limited or cumbersome. Previously, we have demonstrated proof of concept that non-invasive imaging of the total mouse vasculature can be achieved with 18F-fluorodeoxyglucose (18F-FDG)-labeled human erythrocytes and positron emission tomography/computerized tomography (PET/CT). In this work, we demonstrate that changes in the total vascular volume of the brain and left ventricular myocardium of normal rats can be seen after pharmacological vasodilation using 18F-FDG-labeled rat red blood cells (FDG RBCs) and microPET/CT imaging. FDG RBC PET imaging was also used to approximate the location of myocardial injury in a surgical myocardial infarction rat model. Finally, we show that FDG RBC PET imaging can detect relative differences in the degree of drug-induced intra-myocardial vasodilation between diabetic rats and normal controls. This FDG-labeled RBC PET imaging technique may thus be useful for assessing microvascular disease pathologies and characterizing pharmacological responses in the vascular bed of interest.

Keywords: vascular imaging; FDG; PET/CT; microvasculature imaging

Citation: Wang, S.; Budzevich, M.; Abdalah, M.A.; Balagurunathan, Y.; Choi, J.W. In Vivo Imaging of Rat Vascularity with FDG-Labeled Erythrocytes. *Pharmaceuticals* **2022**, *15*, 292. https://doi.org/10.3390/ph15030292

Academic Editor: Xuyi Yue

Received: 9 January 2022
Accepted: 24 February 2022
Published: 27 February 2022

Publisher's Note: MDPI stays neutral with regard to jurisdictional claims in published maps and institutional affiliations.

Copyright: © 2022 by the authors. Licensee MDPI, Basel, Switzerland. This article is an open access article distributed under the terms and conditions of the Creative Commons Attribution (CC BY) license (https://creativecommons.org/licenses/by/4.0/).

1. Introduction

Microvascular disease (MVD) is a type of vascular disorder affecting the arterioles, venules, and capillaries of the vasculature [1,2]. Moreover, MVD can be found in vital organs, such as the brain, heart, and kidneys, in various diseases, including atherosclerosis, dementia, and stroke [3–7]. Unfortunately, non-invasive MVD-specific detection methods are limited, making the early diagnosis, characterization, and treatment of MVD challenging. For example, in the heart, coronary microvascular dysfunction (CVD) can occur with or without obstructive epicardial coronary artery disease, leading to ischemia and angina [8–11]. Imaging techniques, such as coronary flow reserve (CFR), fractional flow reserve (FFR), and index of microcirculatory resistance (IMR), are accepted as measures of myocardial blood flow and even microvascular disease; but are essentially limited to invasive coronary angiography [12–14]. Ultrasound microbubble imaging has been used to assess blood flow and velocity in multiple organs but is dependent on adequate acoustic window access to vessels of appropriate size and proximity and is limited by inherent physical microbubble properties [15–17]. Measures of blood volume and blood flow in organs using iodine and gadolinium-based extracellular contrast agents in computed tomography (CT) or magnetic resonance imaging (MRI) are either semi-quantitatively performed or require compartmental kinetic modeling involving a sufficiently sized vessel in close proximity to the vascular bed of interest [18,19]. The intravascular MRI imaging agent gadofosveset (Vasovist®/Ablavar®) had been successfully used to image the whole organ vasculature in patients but was later withdrawn from the market and is no longer available for commercial use [20]. Arterial spin label MRI has also been successfully used to quantify cerebral blood flow but is technically challenging to perform and is sensitive to differences in MRI technique and parameter settings [21]. As a non-invasive imaging technology with high tracer sensitivity, positron emission tomography (PET) has been shown to accurately provide quantitative measurements of total myocardial perfusion using PET perfusion tracers ^{82}Rb, ^{13}N-NH$_3$, and ^{15}O-H$_2$O [22,23]. Unfortunately, these PET imaging agents currently have minimal clinical implementation due to their restricted applicability owing either to very short tracer half-lives (^{13}N, ^{15}O) or requiring access from vendor-supplied tracer generators (^{82}Rb) [24–26]. More recently, the PET perfusion agent F18-flurpiridaz has recently shown much promise in a phase III clinical trial but currently remains limited to myocardial perfusion imaging [27]. As such, facile methods for robust, non-invasive assessment of the total vascularity of an organ of interest are lacking. The availability of such methods would be expected to both improve the clinical diagnosis and management of patients with MVD, as well as facilitate the development of pharmaceutical agents targeting MVD.

We have previously shown that human red blood cells (RBCs) rapidly incorporate the inexpensive and widely available PET tracer 18F-fluorodeoxyglucose (FDG) due to the inherently dense cell surface expression of glucose transporters, resulting in "trapped" FDG metabolites in human RBCs [28]. We then used FDG-labeled human RBCs as an intravascular PET imaging agent to visualize the entire body vasculature of an immunodeficient mouse model with a microPET/CT scanner. Given the high tracer detection sensitivity of PET imaging relative to contrast agents used in other readily available clinical imaging modalities, we sought to determine whether FDG-labeled RBCs can be used to non-invasively detect changes in the total vascular volume, including the microvasculature, of an organ of interest in rats. As proof of concept, we evaluate the ability of FDG RBC PET imaging to characterize the total vasculature of the brain and left ventricular (LV) myocardium of rats under both rest conditions and pharmacologically induced vasodilatory conditions. In addition, we evaluate the use of FDG RBC PET imaging to characterize myocardial perfusion defects in a surgically induced myocardial infarction rat model. Finally, we show that FDG RBC PET imaging can detect differences in the degree of drug-induced vasodilation of the total LV myocardial vasculature between diabetic rats and normal rat controls.

2. Results

2.1. Detecting Changes in Total Rat Cerebrovascular Volume after Pharmacological Vasodilation

We have previously shown that FDG-labeled human red blood cells (RBCs) can be used to visualize the total body vasculature of immunocompromised mice with a small animal microPET/CT scanner [28]. Similar to results shown in our prior publication, we found that after intravenous injection of FDG-labeled rat RBCs, there is diffuse, intense tracer activity throughout the rat brain volume on PET imaging (Figure 1). To determine the feasibility of using FDG RBC PET imaging to characterize the total in vivo vascular volume of the rat brain, we imaged the rat brain before and after administration of the pharmacological cerebral vasodilator acetazolamide (Figure 2). After administration of acetazolamide, there is a significant vasodilator-induced increase in the total cerebrovascular volume of 72.2 ± 14.7% ($n = 6$). The difference in the rat cerebrovascular volume between vasodilatory and rest conditions was statistically significant (two-tailed Wilcoxon signed rank test: U = 0 ≤ 2, α = 0.05, *n* = 6).

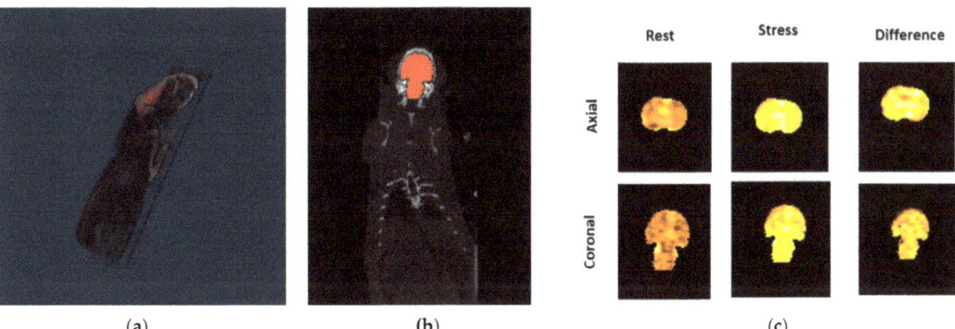

Figure 1. Sagittal (**a**) and coronal (**b**) PET/CT images of a representative rat brain segmentation. (**c**) FDG RBC tracer activity in the segmented rat brain under rest and pharmacological vasodilatory ("stress") conditions, as well as the difference between the two after digital subtraction imaging ("Difference"). Total rats = 6.

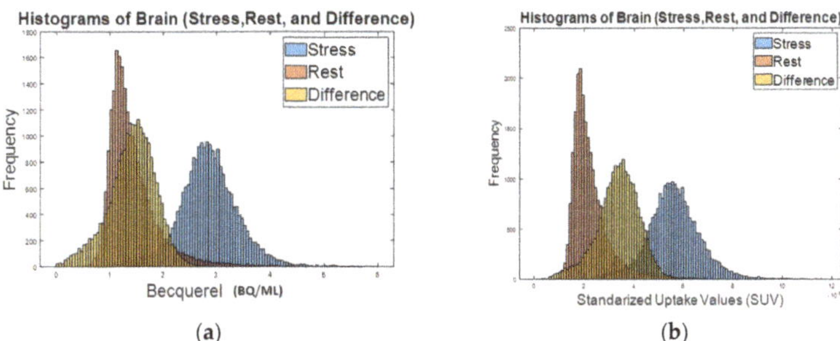

Figure 2. Histograms of FDG RBC tracer activity in a representative rat brain (pharmacological vasodilatory "stress", rest, and the difference between stress and rest): (**a**) intensity activity (Bq/mL) in the brain; (**b**) standardized uptake value (SUV) in the brain.

2.2. Detecting Changes in Rat LV Myocardial Volume after Pharmacological Vasodilation

To further validate the utility of FDG RBC PET imaging to characterize organ vascularity, we showed that FDG RBC tracer activity can be directly visualized within the wall of the left ventricle (LV) in rats after manual image segmentation of the LV myocardium from the chamber blood pool activity on electrocardiogram (ECG)-gated PET images (Figure 3). In addition, we imaged the total LV intramyocardial vascular volume in rats under pharmacological stress conditions after injection of the coronary artery vasodilator regadenoson. Immediately after stress imaging was completed, the vasodilatory effects of regadenoson were pharmacologically reversed by administering a known antagonist, aminophylline, and the rat heart was imaged again (Figure 4). We were able to detect a relative pharmacologically induced difference in LV intramyocardial vessel volume with a mean volume difference of 52.3 ± 11.3% (Bq/mL) (*n* = 5). The volume difference was statistically significant (two-tailed Wilcoxon signed rank test: U = 0 ≤ 2, α = 0.05, *n* = 5). All myocardial image analysis was performed on ECG-gated binned images of the rat heart in the diastolic phase.

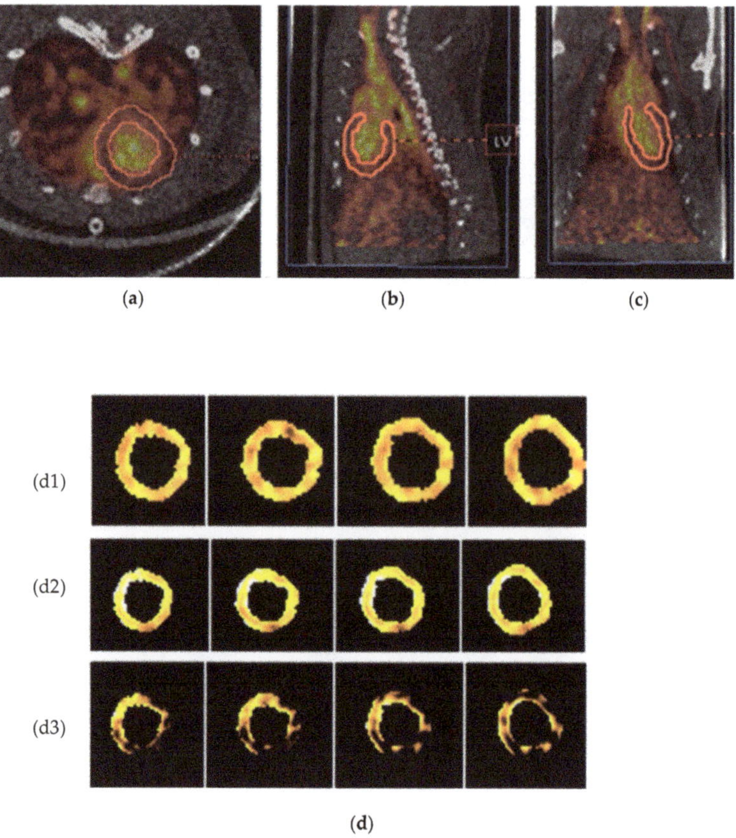

Figure 3. Volume segmentation of the left ventricular (LV) myocardium of a representative rat on FDG RBC PET/CT images. (**a–c**) Axial, coronal, and sagittal (respectively) segmentation of normal rat myocardium. Manual segmentation of the left ventricle is depicted by a red outline. (**d**) Four consecutive cardiac short axis views of a rat left ventricular wall under rest conditions (**d1**), vasodilatory stress conditions (**d2**), and the net activity difference between stress and rest conditions after digital subtraction imaging (**d3**). Total rats = 5.

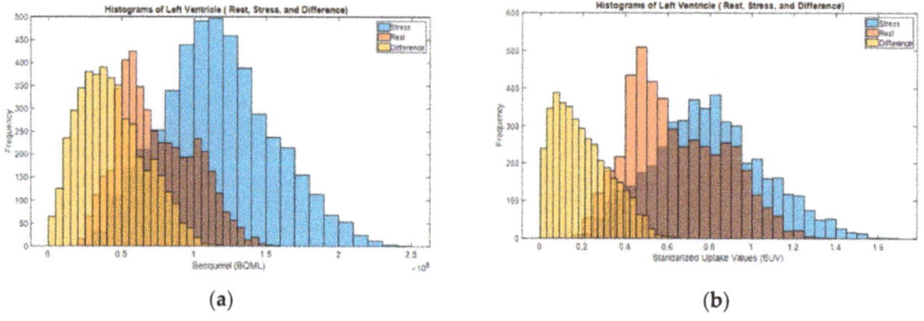

Figure 4. Histograms of FDG RBC tracer activity in the LV myocardial wall of a representative rat (Stress, after-stress = "rest" and the net difference). (**a**) Tracer activity (Bq/mL) in the LV myocardium. (**b**) Standardized uptake value (SUV) in the LV myocardium.

2.3. Imaging LV Intramyocardial Vascularity with FDG RBC PET in a Rat Myocardial Infarction Model

We then sought to determine if FDG RBC PET imaging could detect and approximate the location of any myocardial perfusion defects in rats after surgical ligation of the left coronary artery. We also imaged the same rats after intravenous administration of pure FDG to identify any infarction-induced defects in myocardial FDG metabolism. The rats were sacrificed, the rat hearts were removed, and tissue viability staining was also performed. A representative short axis slice image of the left ventricle of each rat was then evaluated for infarct size based on each method. We found that the myocardial perfusion defect on FDG RBC PET images appeared to correlate in terms of approximate location with both the metabolic myocardial defect on intravenous FDG PET images as well as the infarct location on ex vivo myocardial tissue staining in these rats, when accounting for variability in myocardial infarct size between rats (Figures 5 and 6). In some rats, the margins of the perfusion defect appeared to over-estimate the infarct size as measured on metabolic FDG images or with tissue staining, possibly reflecting areas of relatively decreased perfusion in otherwise viable peri-infarct myocardium (Table 1). The mean percent LV infarct size calculated from the representative slice image for each method was 26.9 ± 0.8% for FDG PET, 33.9 ± 1.6% for FDG RBC PET, and 26.0 ± 1.5% for TTC staining (mean ± S.E.) (Table 1). Further arbitrary segmentation of myocardial perfusion levels on imaging into a low, medium, or high range of tracer activity did not consistently improve visual interpretation of perfusion imaging in these rats, likely in part reflecting technical limitations in the tracer activity acquisition/resolution of our microPET/CT scanner.

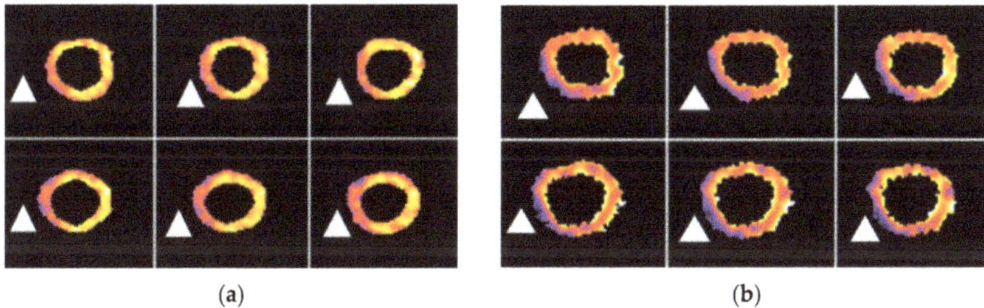

(a) (b)

Figure 5. FDG PET and FDG RBC PET images of the LV myocardium (short axis view) from a representative myocardial infarction rat. (**a**) Metabolic FDG PET images. White triangles indicate the approximate location of decreased metabolism in the left ventricular wall. (**b**) FDG RBC PET images. White triangles indicate the approximate location of relative decreased myocardial perfusion/vascularity. Total rats = 6.

Table 1. Results of mean infarction size measurement based on FDG RBC PET images, FDG PET images, and TTC pathology staining.

Rat	FDG%	RBC%	TTC%
1	24.38%	37.19%	25.97%
2	29.19%	29.49%	26.28%
3	25.46%	32.85%	23.58%
4	29.26%	29.81%	21.35%
5	27.46%	35.39%	31.90%
6	25.62%	38.64%	27.04%
Mean ± S.E.	26.93 ± 0.83%	33.89 ± 1.56%	26.02 ± 1.45%

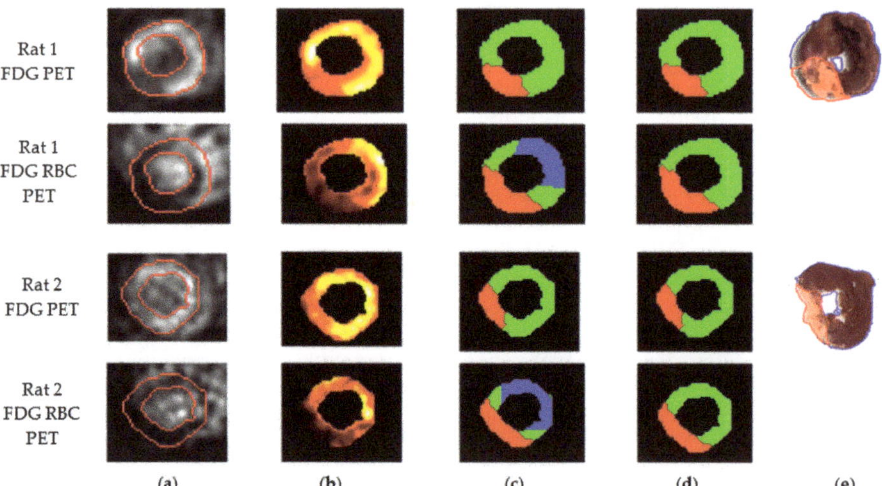

Figure 6. Two consecutive short axis views of the left ventricle from two representative myocardial infarction rats on metabolic FDG PET images and FDG RBC PET images. (**a**) LV myocardial segmentation on PET images. (**b**) Color heat map PET images of segmented LV. (**c**) Three-color stratified FDG RBC PET images (low = red, medium = green, high = blue) of tracer activity. (**d**) Two-color stratified PET image (red = infarct; green = viable). (**e**) TTC tissue viability staining of transverse slices of the LV myocardium from the same infarction rats. There is significant scar tissue and/or atrophy in the infarcted portion of the LV wall (red outline) of some of the rats compared to the uninjured myocardium (blue outline). Dark red tissue = viable myocardium. Tan tissue = infarcted myocardium. Total rats = 6.

2.4. Detecting Differences in Pharmacologically Induced Vasodilation in the Total Rat LV Intramyocardial Vasculature between Normal and Diabetic Rats

Diabetes is a prevalent disease in many countries associated with injury to both the macrovasculature and microvasculature of various organs, including the heart, brain, kidneys, and skeletal muscles [29]. Diabetic rats were created by intravenous injection of the drug streptozotocin, an antibiotic that is used to induce pancreatic islet β-cell destruction in rats [30]. Blood glucose was monitored weekly after the injection. Diabetic rats routinely showed blood glucose levels >400 mg/dL after a few weeks as well as loss of body weight, indicative of progressive diabetic pathophysiology. The mean blood glucose measurement (mg/dL) of the diabetic rats at the time of FDG RBC PET imaging was 394.0 $\pm$ 38.7 compared to 80.8 $\pm$ 12.7 for control rats (mean $\pm$ S.E.). The mean body weight (g) of the diabetic rats at that time was 306.8 $\pm$ 17.6 compared to 419.2 $\pm$ 8.3 for normal controls. Stress and rest myocardial PET imaging of diabetic and normal age-matched control rats was performed after injection of FDG-labeled RBCs. The diabetic rats were found to have a significantly lower degree of regadenoson-related LV myocardial vasodilation when compared to the control group. The mean percentage increase in LV intra-myocardial vascular volume after regadenoson injection in the diabetic rats was 18.8 $\pm$ 2.0% compared to 35.1 $\pm$ 1.9% in the control group (mean $\pm$ S.E.) (Tables 2 and 3). The impairment in vasodilatory response in the diabetic group was statistically significant (two-tailed Wilcoxon rank-sum test: U = 0 $\leq$ 2, $p < 0.05$; diabetic rats, $n = 5$; normal controls, $n = 5$).

Table 2. Tracer activity (Bq/mL) in the segmented LV intra-myocardial vascular volume in stress and rest conditions and relative percentage stress-related increase in LV intra-myocardial vascular volume ("stress–rest difference") after regadenoson injection in the control rats. Stress–rest difference is calculated on a pixelwise level, as detailed in the Materials and Methods section.

Control Rat	LV Stress Activity	LV Rest Activity	Stress–Rest Difference	Increased Ratio
1	2496.72	2008.44	635.04	31.62
2	14,927.50	9340.60	6061.95	32.49
3	12,271.33	9568.55	3108.66	41.00
4	20,315.93	14,905.81	5483.7	36.79
5	46,501.94	34,927.14	11,760.07	33.70
Mean (± S.E.)				35.12 ± 1.91

Table 3. Tracer activity (Bq/mL) in the segmented LV intra-myocardial vascular volume in stress and rest conditions and relative percentage stress-related increase in LV intra-myocardial vascular volume ("stress–rest difference") after regadenoson injection in the diabetic rats. Stress–rest difference is calculated on a pixelwise level, as detailed in the Materials and Methods section.

Diabetic Rat	LV Stress Activity	LV Rest Activity	Stress–Rest Difference	Increased Ratio
1	3386.6	3042.22	410.54	13.40
2	14,225.28	14,026.88	2195.59	15.56
3	18,781.97	15,579.75	3796.07	24.37
4	6124.33	5829.02	1084.05	18.59
5	7063.27	5778.81	1284.46	22.20
Mean (±S.E.)				18.82 ± 2.03

3. Discussion

In this work, we show that FDG RBC PET imaging can detect differences in the total vascular volume of the LV myocardial wall and whole brain of normal rats after pharmacologically induced vasodilation within these organs. In a surgical myocardial infarction rat model, we are also able to spatially correlate the approximate location of myocardial perfusion defects with both the myocardial abnormalities identified with metabolic FDG PET imaging and areas of infarcted myocardium determined by viability tissue staining. We further show in a diabetic rat model that FDG RBC PET imaging can detect impairments in drug-induced coronary vasodilation in the LV intramyocardial vascular bed compared to normal rats. As the microvasculature represents the vast majority of the total vessel volume of major organs, such as the heart ($\geq$90% of the total intra-myocardial vasculature) [31], FDG RBC PET imaging offers the potential for non-invasive detection and serial assessment of the small vessel pathology in whole organs or large body parts seen in diabetic vasculopathy and other diseases.

Unlike other previously explored PET radiotracers, FDG RBC PET imaging offers the unique advantage of utilizing the relatively affordable and readily accessible PET tracer FDG. As described in our previous work, the cell labeling technique can be achieved in a relatively short time period and utilizes a straightforward cell incubation and washing protocol [28]. The cell labeling technique is also similar in the degree of complexity of 111-indium-labeled white blood cell imaging [32].

There is progressive development and ongoing adoption of time-of-flight (TOF) capabilities in modern PET/CT and PET/MR scanners which offer improvements in signal-to-noise ratios, image quality, and patient throughput. Recent studies combining TOF PET imaging with traditional myocardial perfusion agents are being used to quantify myocardial blood flow and improve detection of tracer uptake in lesions of interest [33]. Estimation of tumor flow has also been performed by dynamic PET imaging of the first vascular pass of FDG; however, this estimation requires flow modeling with dynamic measures of arterial tracer concentration. It also requires accurate estimation of the presumed FDG extraction rate parameter by the target tissue and may be susceptible to mild underestimation of blood flow [34]. It remains to be seen whether first-pass dynamic FDG PET imaging or TOF PET imaging can reliably quantify the total vascular volume of an organ or tissue of interest. Unlike first pass FDG PET imaging or TOF PET imaging, FDG RBC PET imaging of the blood pool should allow for vascularity assessment of multiple organs or tissues of interest across different fields of view.

Quantitative measures of total organ vasculature may be possible with clinical PET/CT scanners using FDG RBC PET/CT, based on CT segmentation-based organ volume calculations normalized to measure in vivo FDG RBC concentrations within a large, segmented vessel (aorta, vena cava). Our current goals include exploring the potential of FDG RBC PET imaging to assess in vivo tumor vascularity response to anti-angiogenic drugs for our oncological patient population. FDG RBC PET imaging may also be useful for the translational researcher seeking to non-invasively assess the efficacy of promising pharmaceutical agents targeting other microvascular disease in both small animal models and human subjects.

4. Materials and Methods

All experimental procedures were approved by the University of South Florida (USF) Institutional Animal Care and Use Committee (IACUC). All experiments were performed in accordance with U.S.A. federal regulations and USF IACUC principles and procedures. All of the chemicals were obtained from Sigma-Aldrich, St. Louis, MO, USA, unless specified. A total of ten 4–8-week-old male Sprague-Dawley (SD) rats were used, including four healthy male rats and six myocardial infarction rats. The vendor-supplied myocardial infarction rats were created using a standardized surgical ligation technique of the proximal left coronary artery with an estimated ~35–45% LV infarct size (Envigo, Indianapolis, IN, USA). For drug-induced increases in vascular volume experiments, four 4–8-week-old SD rats were used for cerebrovascular imaging, and five 4–8-week-old SD rats were used for intra-myocardial vascular imaging. For the creation of a diabetic rat model, five 2–4-month-old SD male rats were intravenously injected once or twice with streptozotocin (40 mg/kg weight), and blood glucose was monitored weekly. Diabetic rats underwent myocardial microPET/CT imaging approximately 5 or more weeks after streptozotocin injection. Diabetic rats were compared to five normal age-matched control male SD rats.

4.1. Myocardial and Cerebral Vascular Value Measurement

The total myocardial and cerebral vascular tracer activity was measured in both rest and vasodilatory stress conditions with microPET/CT imaging. Stress condition myocardial PET imaging was performed after intravenous injection of the pharmaceutical coronary vasodilator regadenoson (25 µg/kg). Coronary artery vasodilation was then pharmacologically reversed by intravenous administration of aminophylline (40 mg/kg), and cardiac imaging was repeated. "Stress" condition cerebrovascular PET imaging was performed after intravenous injection of the pharmaceutical cerebral venous vasodilator acetazolamide (100 mg/kg). To segment the left ventricle (LV), the PET and CT images were loaded in Mirada Medical software (Mirada DBx 2.1.0, Denver, CO, USA) to align both image series together. CT images were used to identify the outer margins of the left ventricle, while PET images were used to define the boundaries that separate the left ventricular wall from the intraluminal blood chamber and other soft tissues. Using the built-in segmentation tools offered by Mirada, the left ventricular myocardium was manually delineated voxel by voxel in each short-axis slice (~25 slices, depending on the animal). The segmentation was further refined by adjusting the segmented LV boundaries in the coronal and sagittal views. Finally, any papillary muscles or surrounding soft tissue were excluded from the segmentation. Both normal SD rats and diabetic SD rats underwent stress and rest myocardial PET imaging. The same general approach was also used for the rat brain segmentation for normal rats. The histograms of the LV myocardial becquerel

(Bq) activity and standardized uptake values (SUV) (stress, after-stress, and difference) were quantified.

4.2. FDG-Labeled RBC Preparation

Rat red blood cells were labeled with FDG using a previously published protocol [28], with minor differences, described below. About 500–1000 μL rat blood was collected in a heparin phlebotomy tube through the rat saphenous vein and stored in a 37 °C tissue culture incubator (Sanyo Scientific, New York, NY, USA) for 1–2 h to increase erythrocyte and plasma glucose depletion. After that, rat erythrocytes were centrifuged $1000\times g$ for 10 min, and the plasma and buffer coat were gently aspirated. The remaining red blood cell pellets were gently resuspended in 4X volume of filter-sterilized "1X EDTA" solution (140 mM NaCl, 4 mM KCl, 2.5 mM ethylenediaminetetraacetic acid dipotassium salt dihydrate (K_2EDTA dihydrate)), and centrifuged $1000\times g$ for 10 min. The wash buffer was gently aspirated, and 100 μL 5X EDTA solution and 50 μL deionized water were then added to the 250 μL washed erythrocytes. Finally, 100 μL (37–74 MBq) USP grade 2-deoxy-2[18F]-fluoro-2-D-glucose (FDG) (Cardinal Health, Tampa Fl, USA) was added to a final volume of 500 μL. Samples were incubated at 37 °C for 30 min, centrifuged 3 times and washed 3 times with 12X volume of 1x EDTA/5 mM glucose solution. First, 18F-FDG-labeled RBC PET/CT imaging was performed. For the myocardial infarction rats, myocardial PET imaging was then performed in the same rat after intravenous injection of pure 18F-FDG within the following week. After image processing, the rat was sacrificed. The rat heart was excised intact, saline flushed, and then placed in a −20 °C freezer for 30 min. The semi-frozen rat heart was cut transversely into 2 mm thick sections within a 3D printed rat heart mold for uniform transverse sectioning. The heart tissue was then stained in 1% 2,3,5-triphenyl tetrazolium chloride (TTC) buffer solution at 37 °C for 20–30 min [35]. Stained myocardial slices were subsequently treated with 10% formalin for 20–30 min [36]. Digital images of the formalin-treated heart slices were obtained.

4.3. Small Animal PET/CT Imaging

A more detailed description of our imaging protocol has been previously published [30]. The rat was anesthetized via a nosecone manifold under 2–4% inhalational isoflurane. A tail vein micro-catheter was inserted into one of the dilated rat tail veins after warming the rat tail. The rat was then secured onto the micro-PET/CT scanner bed under anesthesia. Then, 250–500 μL of FDG-labeled RBC suspension (3.7×10^7–1.01×10^8 Bq) was injected slowly through the tail vein microcatheter, after which CT calibration imaging was performed. Electrocardiogram (ECG) leads were placed on two front limbs and one hind limb of the rat (ground lead on a rear leg) for ECG-gated PET imaging. The signals detected by these electrodes were recorded during the 20 min time period using a BioVet® physiological monitoring and heating system (m2m Imaging, Richmond Heights, OH, USA). The threshold for TTL cardiac gating signals was set in a rising mode of R-wave peak. PET list-mode data were reconstructed using 3D-OSEM (ordered-subset expectation maximization) iterative algorithm with four iterations and eight subsets, with a final image volume of $256 \times 256 \times 256$ voxels. Effective voxel dimensions were set at 1.4 mm $\times$ 1.4 mm $\times$ 1.4 mm. For each animal, there are three data sets: standard three-dimensional (3D) PET reconstruction, resulting in a motion–time average 3D PET image; dynamic 3D PET reconstruction with 30 frames; and the phase-based four-dimensional (regular 3D plus time, 4D) PET cardiac reconstruction, with four cardiac gate binning.

4.4. PET Image Analysis

The whole-body PET images of the rats were acquired using Siemens Inveon Workstation Software (Siemens Medical Inc., Knoxville, TN, USA). For 3D PET and 4D PET data sets, multiple volumes of interest (VOI) were selected based on corresponding CT as needed. Voxel activities were represented in standardized uptake values (SUVs). Dynamic activity curves for VOIs were obtained using the dynamic 3D PET data set for each animal.

The 4D PET data were used to define cardiac function. ECG-gated binning of images of the heart in diastolic phase was performed. Heart segmentation on the CT images was based on anatomical features, and the segmented volume was transferred for image-co-registration (cardiac PET VOI). After segmentation of the left ventricular myocardium, the segmentation was then used to obtain the myocardial tracer activity inside the LV muscle. The activity values were converted to SUV units using the following formula: SUV (g/mL) = voxel value × (C × (weight (kg)/(dose (Bq))) × 1000 (g/kg), where C is the correction factor for 18-fluorine tracer decay. Images were represented as maximum intensity projection (MIP) reconstructions of the source data.

The difference in tracer activity within the rat LV myocardium between the vasodilatory "stress" state and the "rest" state were obtained by co-registration of the segmented rat brain or rat left ventricular (LV) myocardium. The difference in activity of the co-registered images was calculated on a pixelwise basis. Pixel differences that yielded negative values are thought to largely reflect volume averaging artifacts related to the abutting high tracer activity of the blood within the LV chamber, and were thus set to zero.

4.5. Myocardial Infarction Size Measurement

The rat heart slices were stained with triphenyl tetrazolium chloride (TTC) to identify the myocardial infarct region. The digitized images were analyzed with ImageJ software (National Institutes of Health, Bethesda, MD, USA) using the "color threshold" mode to manually delineate and measure infarcted vs. viable myocardial tissue. Left ventricular short axis images were obtained from both FDG images and FDG RBC images and then compared to TTC stained images.

Regions of decreased FDG RBC tracer activity on FDG RBC PET images and decreased FDG metabolism on FDG PET images were presumed to correspond to infarcted myocardium and delineated from areas of uninjured myocardium. The infarct size was estimated as a percentage of the left ventricular myocardial cross-sectional area on a given short axis slice. In addition, the distribution of activity inside the myocardial muscle was also arbitrarily further divided into three subregions of tracer activity: "low," "medium," and "high" activity, by first using the OTSU thresholding algorithm [37]. The OTSU algorithm evaluates the activity distribution and divides it into three sub-divisions based on the shape of the activity histogram. The three regions defined by OTSU represent low, medium, and high perfusion subregions inside the myocardium. The medium and high perfused subregions were considered viable myocardium and excluded from the final infarct estimate. Further fine adjustment of the boundaries of the three subregions was performed manually with advice from an experienced radiologist (JWC).

The relative infarct percentage was defined as infarct percentage = (area of the infarcted muscle (low))/(area of all muscle (low + medium + high)). The other two regions, high and medium, were considered uninjured heart muscle.

5. Conclusions

We present data that FDG-labeled erythrocytes can be used to characterize pharmacologic induced changes in the total vascularity of the rat myocardium and rat brain with PET/CT imaging. We also present data that FDG-labeled erythrocyte PET (FDG RBC PET) imaging can detect abnormalities in the left ventricular myocardium of both a surgical myocardial infarction rat model and a diabetic rat model. FDG RBC PET imaging may thus be useful for non-invasively assessing microvascular disease in various clinical settings. It may also be useful for evaluating potential drug candidates targeting microvascular disease.

6. Patents

A provisional patent application regarding the use of FDG RBC PET imaging for imaging microvascular disease was filed by the H. Lee Moffitt Cancer Center & Research Institute, Tampa, Florida, U.S.A., in July 2021.

Author Contributions: Conceptualization, S.W. and J.W.C.; methodology, S.W., M.B. and M.A.A.; software, S.W., M.B. and M.A.A.; validation, S.W., J.W.C., M.B. and M.A.A.; formal analysis, S.W., M.B., M.A.A., Y.B. and J.W.C.; investigation, S.W., M.B., M.A.A. and J.W.C.; resources, M.B., M.A.A. and J.W.C.; data curation, S.W., M.B. and M.A.A.; writing—original draft preparation, S.W.; writing—review and editing, S.W., J.W.C. and Y.B.; visualization, S.W., M.B., M.A.A. and J.W.C.; supervision, M.B. and J.W.C.; project administration, J.W.C.; funding acquisition, J.W.C. All authors have read and agreed to the published version of the manuscript.

Funding: This research received no external funding. Funding was provided through intramural support from the Department of Diagnostic Imaging, H. Lee Moffitt Cancer Center & Research Institute.

Institutional Review Board Statement: All experimental procedures were approved by the University of South Florida (USF) Institutional Animal Care and Use Committee (IACUC). All experiments were performed in accordance with United States federal regulations and USF IACUC principles and procedures (OLAW PHS # A-4100-01; USDA # 58-R-0015; AAALAC # 000434). The USF IACUC protocol number for this body of work is R IS00006926, approved on September 9, 2019. The original USF IACUC approval statement for this protocol is as follows: "The Institutional Animal Care and Use Committee (IACUC) reviewed your application requesting the use of animals in research for the above-entitled study. The IACUC APPROVED your request to use the following animals in your protocol for a one-year period beginning 9/9/2019."

Informed Consent Statement: Not applicable.

Data Availability Statement: The data presented in this study are available on request from the corresponding author. The data are not publicly available due to the filing of a provisional patent application by the H. Lee Moffitt Cancer Center and Research Institute covering this body of research.

Acknowledgments: This research was internally funded with intramural grants from the Department of Diagnostic Imaging and Interventional Radiology at the H. Lee Moffitt Cancer Center & Research Institute. We would like to acknowledge Robert Gatenby for his continued support for this research project.

Conflicts of Interest: The authors declare no conflict of interest. The funders had no role in the design of the study; in the collection, analyses, or interpretation of data; in the writing of the manuscript, or in the decision to publish the results.

References

1. Taqueti, V.R.; Di Carli, M.F. Coronary Microvascular Disease Pathogenic Mechanisms and Therapeutic Options JACC State-of-the-Art Review. *J. Am. Coll. Cardiol.* **2018**, *72*, 2625–2641. [PubMed]
2. Berry, C.; Sidik, N.; Pereira, A.C.; Ford, T.J.; Touyz, R.M.; Kaski, J.C.; Hainsworth, A.H. Small-Vessel Disease in the Heart and Brain: Current Knowledge, Unmet Therapeutic Need, and Future Directions. *J. Am. Heart Assoc.* **2019**, *8*, e011104. [CrossRef]
3. Benjamin, E.J.; Muntner, P.; Alonso, A.; Bittencourt, M.S.; Callaway, C.W.; Carson, A.P.; Chamberlain, A.M.; Chang, A.R.; Cheng, S.; Das, S.R.; et al. Heart Disease and Stroke Statistics—2019 Update: A Report From the American Heart Association. *Circulation* **2019**, *139*, e56–e528. [CrossRef]
4. Reis, S.E.; Holubkov, R.; Conrad Smith, A.J.; Kelsey, S.F.; Sharaf, B.L.; Reichek, N.; Rogers, W.J.; Merz, C.N.; Sopko, G.; Pepine, C.J. WISE Investigators. Coronary microvascular dysfunction is highly prevalent in women with chest pain in the absence of coronary artery disease: Results from the NHLBI WISE study. *Am. Heart J.* **2001**, *141*, 735–741. [CrossRef] [PubMed]
5. Wardlaw, J.M.; Smith, E.E.; Biessels, G.J.; Cordonnier, C.; Fazekas, F.; Frayne, R.; Lindley, R.I.; O'Brien, J.T.; Barkhof, F.; Benavente, O.R.; et al. Standards for Reporting Vascular changes on neuroimaging (STRIVE v1). Neuroimaging standards for research into small vessel disease and its contribution to ageing and neurodegeneration. *Lancet Neurol.* **2013**, *12*, 822–838. [CrossRef]
6. Nowroozpoor, A.; Gutterman, D.; Safdar, B. Is microvascular dysfunction a systemic disorder with common biomarkers found in the heart, brain, and kidneys?—A scoping review. *Microvasc. Res.* **2021**, *134*, 104123. [CrossRef] [PubMed]
7. Tonelli, M.; Wiebe, N.; Culleton, B.; House, A.; Rabbat, C.; Fok, M.; McAlister, F.; Garg, A.X. Chronic kidney disease and mortality risk: A systematic review. *J. Am. Soc. Nephrol.* **2006**, *17*, 2034–2047. [CrossRef] [PubMed]
8. Rahman, H.; Demir, O.M.; Khan, F.; Ryan, M.; Ellis, H.; Mills, M.T.; Chiribiri, A.; Webb, A.; Perera, D. Physiological Stratification of Patients With Angina Due to Coronary Microvascular Dysfunction. *J. Am. Coll. Cardiol.* **2020**, *75*, 2538–2549. [CrossRef] [PubMed]
9. Ford, T.J.; Rocchiccioli, P.; Good, R.; McEntegart, M.; Eteiba, H.; Watkins, S.; Shaukat, A.; Lindsay, M.; Robertson, K.; Hood, S.; et al. Systemic microvascular dysfunction in microvascular and vasospastic angina. *Eur. Heart J.* **2018**, *39*, 4086–4097. [CrossRef]

10. Sinha, A.; Rahman, H.; Perera, D. Ischaemia without obstructive coronary artery disease: The pathophysiology of microvascular dysfunction. *Curr. Opin. Cardiol.* **2020**, *6*, 720–725. [CrossRef]

11. Sucato, V.; Novo, G.; Saladino, A.; Rubino, M.; Caronna, N.; Luparelli, M.; D'Agostino, A.; Novo, S.; Evola, S.; Galassi, A.R. Ischemia in patients with no obstructive coronary artery disease: Classification, diagnosis and treatment of coronary microvascular dysfunction. *Coron. Artery Dis.* **2020**, *31*, 472–476. [CrossRef] [PubMed]

12. Fearon, W.F.; Yuhei, K. Invasive Assessment of the Coronary Microvasculature: The Index of Microcirculatory Resistance. *Circ. Cardiovasc. Interv.* **2017**, *10*, e005361. [CrossRef] [PubMed]

13. Berry, C.; Corcoran, D.; Hennigan, B.; Watkins, S.; Layland, J.; Oldroyd, K.G. Fractional flow reserve-guided management in stable coronary disease and acute myocardial infarction: Recent developments. *Eur. Heart J.* **2015**, *36*, 3155–3164. [CrossRef] [PubMed]

14. Mathew, R.C.; Bourque, J.M.; Salerno, M.; Kramer, C.M. Cardiovascular Imaging Techniques to Assess Microvascular Dysfunction. *JACC Cardiovasc. Imaging* **2020**, *13*, 1577–1590. [CrossRef] [PubMed]

15. Firschke, C.; Lindner, J.R.; Wei, K.; Goodman, N.C.; Skyba, D.M.; Kaul, S. Myocardial perfusion imaging in the setting of coronary artery stenosis and acute myocardial infarction using venous injection of a second-generation echocardiographic contrast agent. *Circulation* **1997**, *96*, 959–967.

16. Porter, T.R.; Xie, F.; Kricsfeld, D.; Armbruster, R.W. Improved myocardial contrast with second harmonic transient ultrasound response imaging in humans using intravenous perfluorocarbon-exposed sonicated dextrose albumin. *J. Am. Coll Cardiol.* **1996**, *27*, 1497–1501. [CrossRef]

17. Lee, H.; Kim, H.; Han, H.; Lee, M.; Lee, S.; Yoo, H.; Chang, J.H.; Kim, H. Microbubbles used for contrast enhanced ultrasound and theragnosis. *Biomed. Eng. Lett.* **2017**, *7*, 59–69. [CrossRef]

18. Rossi, A.; Wragg, A.; Klotz, E.; Pirro, F.; Moon, J.C.; Nieman, K.; Pugliese, F. Dynamic Computed Tomography Myocardial Perfusion Imaging: Comparison of Clinical Analysis Methods for the Detection of Vessel-Specific Ischemia. *Circ. Cardiovasc. Imaging* **2017**, *10*, e005505. [CrossRef]

19. Morton, G.; Chiribiri, A.; Ishida, M.; Hussain, S.T.; Schuster, A.; Indermuehle, A.; Perera, D.; Knuuti, J.; Baker, S.; Hedström, E.; et al. Quantification of absolute myocardial perfusion in patients with coronary artery disease: Comparison between cardiovascular magnetic resonance and positron emission tomography. *J. Am. Coll. Cardiol.* **2012**, *60*, 1546–1555. [CrossRef]

20. Oliveira, I.S.; Hedgire, S.S.; Li, W.; Ganguli, S.; Prabhakar, A.M. Blood pool contrast agents for venous magnetic resonance imaging. *Cardiovasc. Diagn. Ther.* **2016**, *6*, 508–518. [CrossRef]

21. Alsaedi, A.; Thomas, D.; Bisdas, S.; Golay, X. Overview and Critical Appraisal of Artereial Spin labelling technique in Brain Perfusion Imaging. *Contrast. Media Mol. Imaging* **2018**, *8*, e5360376.

22. Maddahi, J.; Packard, R.R. Cardiac PET perfusion tracers: Current status and future directions. *Semin. Nucl. Med.* **2014**, *44*, 333–343. [CrossRef] [PubMed]

23. Driessen, R.S.; Raijmakers, P.G.; Stuijfzand, W.J.; Knaapen, P. Myocardial perfusion imaging with PET. *Int. J. Cardiovasc. Imaging* **2017**, *33*, 1021–1031. [CrossRef]

24. Feher, A.; Srivastava, A.; Quail, M.A.; Boutagy, N.E.; Khanna, P.; Wilson, L.; Miller, E.J.; Liu, Y.H.; Lee, F.; Sinusas, A.J. Serial Assessment of Coronary Flow Reserve by Rubidium-82 Positron Emission Tomography Predicts Mortality in Heart Transplant Recipients. *JACC Cardiovasc. Imaging* **2020**, *13*, 109–120. [CrossRef] [PubMed]

25. Munk, O.L.; Bass, L.; Feng, H.; Keiding, S. Determination of regional flow by use of intravascular PET tracers: Microvascular theory and experimental validation for pig livers. *J. Nucl. Med.* **2003**, *44*, 1862–1870.

26. Salerno, M.; Beller, G.A. Noninvasive Assessment of Myocardial Perfusion. *Circ. Cardiovasc. Imaging* **2009**, *2*, 412–424. [CrossRef] [PubMed]

27. Maddahi, J.; Lazewatsky, J.; Udelson, J.E.; Berman, D.S.; Beanlands, R.S.B.; Heller, G.V.; Bateman, T.M.; Knuuti, J.; Orlandi, C. Phase-III Clinical Trial of Fluorine-18 Flurpiridaz Positron Emission Tomography for Evaluation of Coronary Artery Disease. *J. Am. Coll. Cardiol.* **2020**, *76*, 391–401. [CrossRef]

28. Choi, J.W.; Budzevich, M.; Wang, S.; Gage, K.; Estrella, V.; Gillies, R.J. In vivo positron emission tomographic blood pool imaging in an immunodeficient mouse model using 18F-fluorodeoxyglucose labeled human erythrocytes. *PLoS ONE* **2019**, *14*, e0211012. [CrossRef]

29. Barrett, E.J.; Liu, Z.; Khamaisi, M.; King, G.L.; Klein, R.; Klein, B.E.K.; Hughes, T.M.; Craft, S.; Freedman, B.I.; Bowden, D.W.; et al. Diabetic Microvascular Disease: An Endocrine Society Scientific Statement. *J. Clin. Endocrinol. Metab.* **2017**, *102*, 4343–4410. [CrossRef]

30. Furman, B.L. Streptozotocin-Induced Diabetic Models in Mice and Rats. *Curr. Protoc. Pharmacol.* **2015**, *70*, 5.47.1–5.47.20. [CrossRef]

31. Feher, A.; Sinusas, A.J. Quantitative Assessment of Coronary Microvascular Function. *Circ. Cardiovas. Imaging* **2017**, *10*, e006427.

32. Herron, T.; Gossman, W. 111 Indium White Blood Cell Scan. In *StatPearls*; StatPearls Publishing: Treasure Island, FL, USA, 2022. Available online: https://www.ncbi.nlm.nih.gov/books/NBK554556/ (accessed on 30 January 2022).

33. Surti, S. Update on time-of-flight PET imaging. *J. Nucl. Med.* **2015**, *56*, 98–105. [CrossRef] [PubMed]

34. Mullani, N.A.; Herbst, R.S.; O'Neil, R.G.; Gould, K.L.; Barron, B.J.; Abbruzzese, J.L. Tumor blood flow measured by PET dynamic imaging of first-pass 18F-FDG uptake: A comparison with 15O-labeled water-measured blood flow. *J. Nucl. Med.* **2008**, *49*, 517–523. [CrossRef] [PubMed]

35. Bohl, S.; Medway, D.J.; Schulz-Menger, J.; Schneider, J.E.; Neubauer, S.; Lygate, C.A. Refined approach for quantification of in vivo ischemia-reperfusion injury in the mouse heart. *Am. J. Physiol. Heart Circ. Physiol.* **2009**, *297*, H2054–H2058. [CrossRef] [PubMed]
36. Redfors, B.; Shao, Y.; Omerovic, E. Myocardial infarct size and area at risk assessment in mice. *Exp. Clinical. Cardiol.* **2012**, *17*, 268–272.
37. Otsu, N. A Threshold Selection Method from Gray-Level Histograms. *IEEE Trans. Syst. Man Cybern. Syst. Man Cybern.* **1979**, *9*, 62–66. [CrossRef]

 *pharmaceuticals*

Article

Biodistribution of Intra-Arterial and Intravenous Delivery of Human Umbilical Cord Mesenchymal Stem Cell-Derived Extracellular Vesicles in a Rat Model to Guide Delivery Strategies for Diabetes Therapies

Junfeng Li [1,*], Hirotake Komatsu [1], Erasmus K. Poku [2], Tove Olafsen [3], Kelly X. Huang [1], Lina A. Huang [1], Junie Chea [2], Nicole Bowles [2], Betty Chang [3], Jeffrey Rawson [1], Jiangling Peng [1], Anna M. Wu [3], John E. Shively [3] and Fouad R. Kandeel [1,*]

[1] Department of Translational Research & Cellular Therapeutics, Beckman Research Institute of the City of Hope, Duarte, CA 91010, USA; hkomatsu@coh.org (H.K.); kellyhuang1220@g.ucla.edu (K.X.H.); lahuang@exeter.edu (L.A.H.); jrawson@coh.org (J.R.); jianglingpeng@gdut.edu.cn (J.P.)
[2] Department of Radiopharmacy, Beckman Research Institute of the City of Hope, Duarte, CA 91010, USA; kpoku@coh.org (E.K.P.); jchea@coh.org (J.C.); nbowles@coh.org (N.B.)
[3] Department of Cancer Molecular Imaging and Therapy, Beckman Research Institute of the City of Hope, Duarte, CA 91010, USA; tolafsen@coh.org (T.O.); betchang@coh.org (B.C.); awu@coh.org (A.M.W.); jshively@coh.org (J.E.S.)
* Correspondence: juli@coh.org (J.L.); FKandeel@coh.org (F.R.K.); Tel.: +1-626-218-4507 (J.L.); +1-626-218-2251 (F.R.K.)

Citation: Li, J.; Komatsu, H.; Poku, E.K.; Olafsen, T.; Huang, K.X.; Huang, L.A.; Chea, J.; Bowles, N.; Chang, B.; Rawson, J.; et al. Biodistribution of Intra-Arterial and Intravenous Delivery of Human Umbilical Cord Mesenchymal Stem Cell-Derived Extracellular Vesicles in a Rat Model to Guide Delivery Strategies for Diabetes Therapies. *Pharmaceuticals* **2022**, *15*, 595. https://doi.org/10.3390/ph15050595

Academic Editor: Christos Liolios

Received: 17 March 2022
Accepted: 10 May 2022
Published: 12 May 2022

Publisher's Note: MDPI stays neutral with regard to jurisdictional claims in published maps and institutional affiliations.

Copyright: © 2022 by the authors. Licensee MDPI, Basel, Switzerland. This article is an open access article distributed under the terms and conditions of the Creative Commons Attribution (CC BY) license (https://creativecommons.org/licenses/by/4.0/).

Abstract: Umbilical cord mesenchymal stem cell-derived extracellular vesicles (UC-MSC-EVs) have become an emerging strategy for treating various autoimmune and metabolic disorders, particularly diabetes. Delivery of UC-MSC-EVs is essential to ensure optimal efficacy of UC-MSC-EVs. To develop safe and superior EVs-based delivery strategies, we explored nuclear techniques including positron emission tomography (PET) to evaluate the delivery of UC-MSC-EVs in vivo. In this study, human UC-MSC-EVs were first successfully tagged with I-124 to permit PET determination. Intravenous (I.V.) and intra-arterial (I.A.) administration routes of [^{124}I]I-UC-MSC-EVs were compared and evaluated by in vivo PET-CT imaging and ex vivo biodistribution in a non-diabetic Lewis (LEW) rat model. For I.A. administration, [^{124}I]I-UC-MSC-EVs were directly infused into the pancreatic parenchyma via the celiac artery. PET imaging revealed that the predominant uptake occurred in the liver for both injection routes, and further imaging characterized clearance patterns of [^{124}I]I-UC-MSC-EVs. For biodistribution, the uptake (%ID/gram) in the spleen was significantly higher for I.V. administration compared to I.A. administration (1.95 ± 0.03 and 0.43 ± 0.07, respectively). Importantly, the pancreas displayed similar uptake levels between the two modalities (0.20 ± 0.06 for I.V. and 0.24 ± 0.03 for I.A.). Therefore, our initial data revealed that both routes had similar delivery efficiency for [^{124}I]I-UC-MSC-EVs except in the spleen and liver, considering that higher spleen uptake could enhance immunomodulatory application of UC-MSC-EVs. These findings could guide the development of safe and efficacious delivery strategies for UC-MSC-EVs in diabetes therapies, in which a minimally invasive I.V. approach would serve as a better delivery strategy. Further confirmation studies are ongoing.

Keywords: extracellular vesicles (EVs); umbilical cord mesenchymal stem cell (UCMSC); diabetes; I-124; positron emission tomography (PET); intravenous (I.V.) administration; intra-arterial (I.A.) administration; biodistribution

1. Introduction

Extracellular vesicles (EVs) have recently demonstrated tremendous potential as a therapeutic alternative to cell-based therapy for a wide range of diseases. These nanosized, membrane-bound vesicles are integral modulators of intercellular communication and contain a variety of cellular components, such as cytokines, growth factors, signaling lipids, mRNA, and regulatory miRNAs, depending on the cell origin [1,2]. Compared to cell-based therapy, EVs provide significant clinical advantages including low immunogenic and tumorigenic properties, efficient cellular uptake, homing capabilities, easier quantification and storage, and the ability to escape degradation to provide long-term releasing effects [3]. With such broad regulatory functions, EVs have emerged as a promising treatment strategy for various clinical applications.

A growing number of studies have focused on harnessing the capacity of EVs to recapitulate the therapeutic effects of mesenchymal stem cells (MSC). MSC-derived EVs (MSC-EVs) efficiently transfer therapeutic agents to enhance proliferation, attenuate apoptosis, activate autophagy, and regulate immune reactivity [4–9]. In preclinical studies, MSC-EVs are capable of accelerating tissue regeneration and inducing angiogenesis for osteochondral defects [10], skeletal muscle injury [11], and myocardial ischemia and reperfusion [5,12]. Their therapeutic capacity has also extended to repairing liver fibrosis [13], promoting the recovery of acute renal injury [14], enhancing cutaneous wound healing [15], improving pulmonary hypertension [16], reducing amyloid-beta deposition for Alzheimer's disease [8], and suppressing cell growth and migration for cancers [17,18].

Due to their immunomodulatory and metabolic functions, MSC-EVs have become an emerging strategy for treating various autoimmune and metabolic disorders, particularly type 1 diabetes mellitus (T1DM) and type 2 diabetes mellitus (T2DM). T1DM is characterized by beta-cell dysfunction and death from autoreactive immune cells, reduced insulin production, and hyperglycemia. Meanwhile, T2DM involves peripheral insulin resistance and a reduction in pancreatic beta-cell mass. Diabetes mellitus often leads to severe complications, such as chronic refractory wounds, diabetic nephropathy, diabetic neuropathy, retinopathy, and increased risk of stroke. Preclinical studies have demonstrated MSC-EVs are capable of effectively regenerating pancreatic beta-cell mass, ameliorating autoimmune reaction, restoring insulin production, and preventing disease onset in T1DM murine models [19–21]. Additionally, MSC-EVs promoted hepatic glucose and lipid metabolism, reversed insulin resistance, and reduced beta-cell destruction in rat models of T2DM [22,23]. These studies shed light on the potential of MSC-EVs to modulate and remedy diabetes pathogenesis. MSC-EVs have also shown to ameliorate comorbidities by suppressing renal cell apoptosis for diabetic nephropathy [24], improving functional recovery for diabetic peripheral neuropathy [25], promoting neurorestoration following stroke in T2DM [26], reducing inflammatory reaction for diabetic retinopathy [27], and accelerating diabetic wound healing [28–30]. Although MSC-EVs have never been clinically studied for diabetic patients in the United States to our knowledge, preclinical evaluation of the therapeutic potential of MSC-EVs would greatly accelerate the translation to clinical therapies.

Umbilical cord-derived MSCs are attracting substantial research attention as a promising source of MSC-derived EVs (UC-MSC-EVs) for diabetes therapy. In vivo studies have revealed that various types of cell-derived phospholipid bilayer enclosed vesicles, namely EVs, displayed different distributions. Currently, it is unclear whether intravenous (I.V.) or intra-arterial (I.A.) administration of UC-MSC-EVs provides more efficient delivery. Therefore, comparing the administration routes is a crucial step to translate UC-MSC-EVs into practice. Positron emission tomography (PET) is well established as an imaging modality, offering high sensitivity to monitor the target radiolabeled PET isotopes in various tissues in vivo. In this study, we utilized PET imaging strategies to directly compare delivery routes for human UC-MSC-EVs. By administering radioiodinated human UC-MSC-EVs ([^{124}I]I-UC-MSC-EVs) in a rat model, we aim to elucidate whether I.V. or I.A. administration is better to enhance EV-based delivery. Considering the incredible potential of human UC-MSC-EVs as a versatile therapeutic strategy, these findings may guide the development of safe and efficacious delivering strategies for clinical treatments.

2. Results

2.1. Radioiodination and Stability of UC-MSC-EVs

^{124}I possesses an intermediate half-life of 4.18 days, making it one of the optimal radioisotopes for the evaluation of biomolecules, including EVs. Radioiodination of human UC-MSC-EVs with ^{124}I-sodium iodide was successfully performed using the direct IODOGEN methodology. Purification of [^{124}I]I-UC-MSC-EVs was carried out via Superdex 200 chromatography, achieving high radiochemical purity (>99%) by ITLC. Due to the lengthy process required for I.A. injections, radiolysis of [^{124}I]I-UC-MSC-EVs was of concern, but further analysis revealed that [^{124}I]I-UC-MSC-EVs displayed high stability (>95%) even after 4 h (Figure 1). Only two minor peaks were found (total <5%). These findings indicate that the same batch of [^{124}I]I-UC-MSC-EVs may service 2–4 rats for PET-CT scanning.

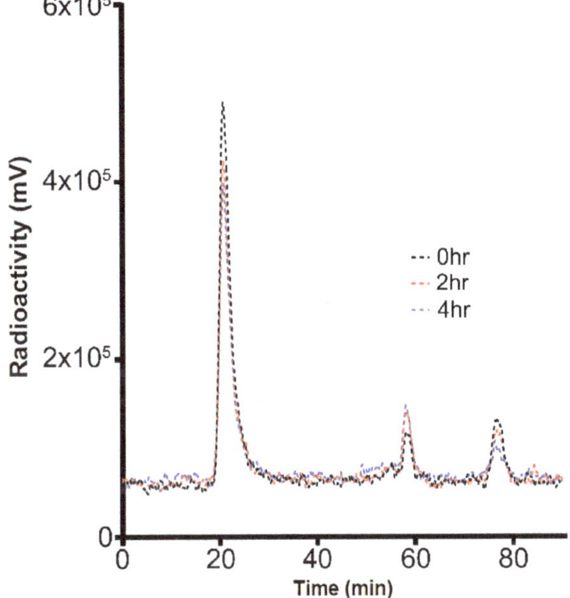

Figure 1. Radiochemical stability test for [^{124}I]I-UC-MSC-EVs. The samples were incubated for 0, 2, and 4 h to assess radiochemical stability. The results indicated that the radiochemical purity of [^{124}I]I-UC-MSC-EVs was >95% (T = 0, 2, and 4 h), and only two minor peaks were observed.

2.2. I.V. Administration and PET-CT Small Animal Imaging

MicroPET imaging of [^{124}I]I-UC-MSC-EVs was conducted for rats that underwent the I.V. injection route. Briefly, male rats (350–500 g) were anesthetized with 2–4% isoflurane in oxygen and received approximately 8 MBq of [^{124}I]I-UC-MSC-EVs in 1% HSA with PBS via the tail for I.V. injections. 0–90 min dynamic body PET scan (48 frames: 15 s × 16; 30 s × 6; 60 s × 8; 300 s × 1; 60 s × 10; 300 s × 2; 600 s × 5) followed by 1 min computed tomography (CT) scan were performed. The representative dynamic images from 0 to 90 min and regions of interest (ROIs) are shown in Figure 2. MicroPET imaging provided clear visualizations of the accumulation and clearance of [^{124}I]I-UC-MSC-EVs in the organs/tissues.

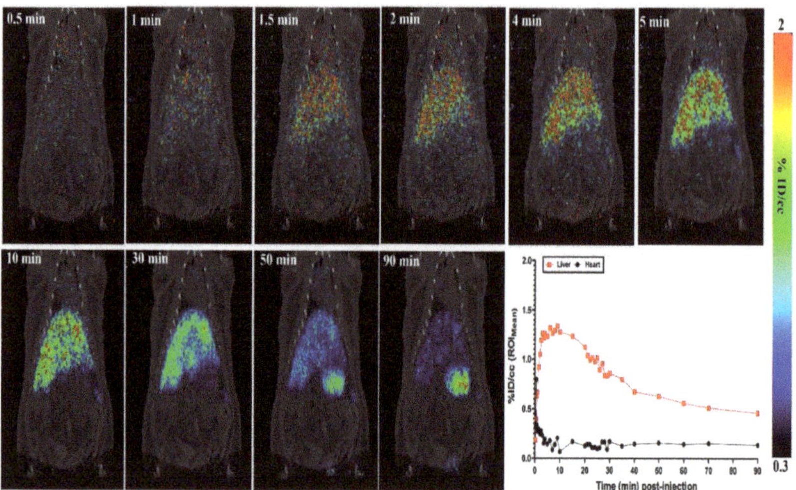

Figure 2. MicroPET imaging of I.V. administered [^{124}I]I-UC-MSC-EVs in male rats. The representative MRP PET/CT serial images were shown (**top and left bottom**) at 0.5 min, 1 min, 1.5 min, 2 min, 4 min, 5 min, 10 min, 30 min, 50 min, and 90 min post injection. MicroPET imaging provided clear visualizations of [^{124}I]I-UC-MSC-EVs accumulation and clearance in the organs from 0 to 90 min. Tissue time-activity curve of the liver and heart were also shown (**right bottom**).

Early dynamic PET images did not display high uptake in all organs within 1 min after injection. Heart uptake ROIs reached the highest levels (%ID/cc of ~0.45) at 1 min post-injection, followed by rapid clearance. The highest liver uptake was observed at 10 min post-injection (%ID/cc of ~1.40), and gradually decreased by ~60% at the end of the scan. Although clearance of [^{124}I]I-UC-MSC-EVs was observed, it was difficult to characterize clearance patterns through PET imaging for the stomach, spleen, pancreas, and small intestine, due to the proximity of the organs to each other in the rat. As a result, we utilized ex vivo biodistribution analysis to validate uptake levels from PET imaging. No significant uptake in the lung and pancreas was observed.

We conducted a separate independent microPET-CT study (~8 MBq [^{124}I]I-UC-MSC-EVs) to evaluate uptake levels in the male rat head. The findings indicated that low uptake was observed in the whole brain throughout the 90 min dynamic PET scan post-injection (microPET results not shown).

2.3. I.A. Administration and PET-CT Small Animal Imaging

I.A. administration of ~ 8 MBq [^{124}I]I-UC-MSC-EVs was successfully achieved in one surgical procedure. [^{124}I]I-UC-MSC-EVs were infused into the pancreatic parenchyma of the rat body and tail through the celiac artery after blockage of the splenic artery, common hepatic artery, and left gastric artery. After the injection was completed, all clamps on the arteries were released, and the abdominal incision was closed. The rats subsequently underwent a 70 min dynamic body PET scan using 17 frames (60 s × 10; 300 s × 2; 600 s × 5) followed by a 1 min CT scan. For the I.A. injection, it was necessary to wait until hemostasis was achieved and to suture the wound closed. In order to compare the same time points for biodistribution analyses between the I.A. and I.V. injection routes, rats in the I.A. injection group underwent 70 min PET scans after a 20 min delay post-injection.

The microPET images and ROIs were shown in Figure 3. PET scans for rats in the I.A. group displayed similar patterns as PET images for rats in the I.V. group. Both I.V and I.A. administration methods showed predominant liver distribution, with the I.A. group having greatest uptake at around 25–30 min post-injection (5–10 min after the initial PET scan). The time-activity curves (TACs) showed that liver uptake for the I.A. group was ~1.7 times higher compared to corresponding time points for the I.V. group. Importantly, the pancreas did not display high accumulation early in PET scanning. Other organs displayed uptake levels similar to the uptake levels for the I.V. injection group. Furthermore, clearance of [^{124}I]I-UC-MSC-EVs from the liver was observed during the middle and end of the PET scan, but it again proved difficult to clarify clearance patterns. We therefore analyzed the biodistribution of [^{124}I]I-UC-MSC-EVs to validate accumulation in organs.

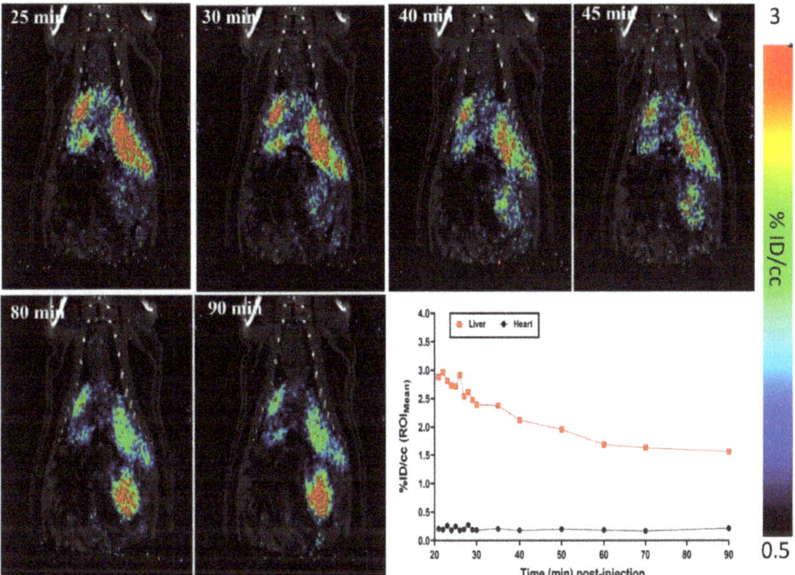

Figure 3. MicroPET imaging of I.A. administered [^{124}I]I-UC-MSC-EVs in male rats. The representative MRP PET/CT serial images were taken (**top and left bottom**) at 25 min, 30 min, 40 min, 45 min, 80 min, and 90 min post-injection. Tissue time-activity curve of the liver and heart were also shown (**right bottom**).

2.4. Biodistribution Analysis of the I.V. Injection Group and the I.A. Injection Group

Biodistribution analyses were performed following PET-CT imaging for both administration routes (Figure 4 and Table 1). Results from both groups indicated low overall uptake levels (%ID/g) in the blood, brain, heart, lung, small intestine, large intestine, muscle, kidneys, and bone. In the pancreas, both I.V. and I.A. administration routes demonstrated comparable accumulation levels of 0.20 ± 0.06 %ID/g and 0.24 ± 0.03 %ID/g, respectively.

The liver, which was the organ of predominant uptake, displayed accumulation levels consistent with the TAC results in both groups: 0.63 ± 0.21 %ID/g (I.V.) and 1.25 ± 0.16 %ID/g (I.A.). Uptake levels in the stomach were similar in the I.V. injection route (1.05 ± 0.11 %ID/g) and the I.A. administration route (0.95 ± 0.07 %ID/g), which was due to radio-deiodination of [^{124}I]I-UC-MSC-EVs. The spleen displayed significantly higher uptake levels in the I.V. group compared to the I.A. group (1.95 ± 0.03 %ID/g vs. 0.43 ± 0.07 %ID/g, respectively). Furthermore, biodistribution results clarified that the region of high uptake towards the left lower area of the liver shown in the PET images was from the spleen (1.62 ± 0.07 %ID/organ) and stomach (2.43 ± 0.07 %ID/organ) in the I.V. group. For the I.A. group, the area of high uptake was from the stomach (~2.86%ID/organ), as the spleen had low accumulation levels (~0.95 %ID/gram).

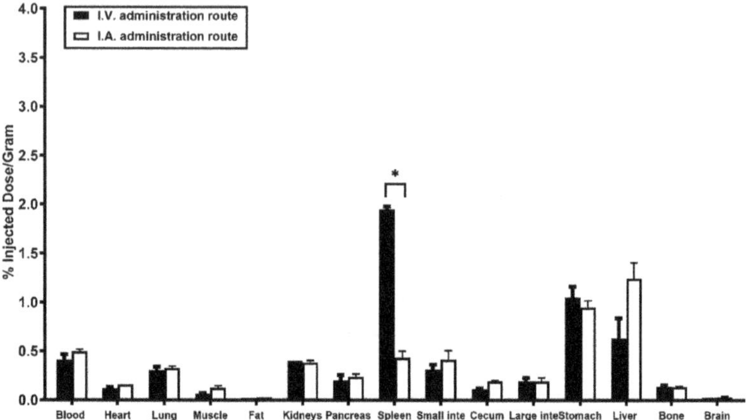

Figure 4. Comparative biodistribution of [^{124}I]I-UC-MSC-EVs for the I.V. group vs. the I.A. group after PET-CT scanning (96–98 min post injection). Rats were injected with ~8 MBq of [^{124}I]I-UC-MSC-EVs, and tissue biodistribution analyses were performed. * p <0.01.

Table 1. Biodistribution of [^{124}I]I-UC-MSC-EVs uptake (%ID/gram) [a] in the I.V. and I.A. group.

Organ/Tissue	I.V. Group	I.A. Group	*p*-Value
blood	0.41 ± 0.06	0.50 ± 0.02	0.18
heart	0.12 ± 0.01	0.16 ± 0.00	0.06
lung	0.31 ± 0.04	0.33 ± 0.02	0.56
muscle	0.07 ± 0.01	0.13 ± 0.02	0.06
fat	0.02 ± 0.00	0.03 ± 0.00	
kidneys	0.40 ± 0.00	0.39 ± 0.02	0.42
pancreas	0.20 ± 0.06	0.24 ± 0.03	0.47
spleen	1.95 ± 0.03	0.43 ± 0.07	<0.01
small intestine	0.32 ± 0.05	0.42 ± 0.09	0.31
cecum	0.12 ± 0.01	0.20 ± 0.01	0.01
large intestine	0.20 ± 0.04	0.20 ± 0.04	>1.00
stomach	1.05 ± 0.11	0.95 ± 0.07	0.40
liver	0.63 ± 0.21	1.25 ± 0.16	0.08
bone	0.14 ± 0.01	0.14 ± 0.01	0.70
brain	0.03 ± 0.00	0.04 ± 0.01	0.42

[a] Data is shown as Mean ± SD.

3. Discussion

Although the method of delivery for biological molecules is important for optimizing the efficacy of clinical therapies, characterizing and quantifying delivery methods has been challenging. PET-CT is an optimal approach that combines the high sensitivity and quantification of PET with computed tomography (CT) scanning to provide anatomic data for co-registration. By allowing for non-invasive tracking of administered molecules. PET-CT imaging plays a key role in evaluating drug delivery in preclinical in vivo optimization cycles prior to validation in humans using the same techniques.

Due to their therapeutic potential, EVs have been characterized in preclinical trials. Although both I.V. and I.A. administration routes are efficient strategies in clinical applications, there have been no direct comparisons of these administration routes for UC-MSC-EVs. Theoretically, I.A. administration has the advantage of selective delivery to the pancreas but may increase the risk of occlusion or embolization. Meanwhile, I.V. administration is less invasive but may lead to sequestration in other organs, such as the spleen and lung, and thereby reduce EV delivery to the pancreas and cause off-target effects [31,32]. A previous comparison of I.V. and I.A. delivery of bone marrow mononuclear cells for acute ischemic stroke found no differences between the two modalities [33]. However, different EVs derived from cell types could exhibit varying characteristics, cellular targets, and therapeutic outcomes, so it is important to evaluate the potential of I.V. versus I.A. administration of UC-MSC-EVs. In the present study, human UC-MSC-EVs were tagged with a radioisotope (I-124) and underwent combined PET-CT imaging and biodistribution, which allowed for direct and sensitive monitoring of UC-MSC-EVs in vivo. This approach may guide further development of safe and efficacious delivery strategies for future clinical trials.

3.1. Surgical Procedures for I.A. Injection of [^{124}I]I-UC-MSC-EVs

[^{124}I]I-UC-MSC-EVs were directly infused into the pancreatic parenchyma of the rat body and tail through arterial flow after blocking nearby arteries in the pancreas (Figure 5A, see Materials and Methods for more details). There were several major challenges in performing I.A. injections and subsequent imaging. Firstly, to perform precise surgical procedures, we used large rats (350–500 g). However, the large size of the rats precluded the ability to evaluate the tracer kinetic in the whole rat body within a single image, due to the small transaxial field of view (FOV) of 12 cm of the microPET imaging system. Instead, a separate scan was required to obtain whole body PET imaging. Therefore, the pilot studies separated imaging of the rats into two independent studies, including the head PET scan and the body PET scan. The low uptake of [^{124}I]I-UC-MSC-EVs in the head observed in the PET images was confirmed by biodistribution analysis, and our subsequent PET studies were therefore focused on the rat body. Secondly, the surgical procedures required technical expertise. In preliminary trials, we confirmed the feasibility of the surgical procedure and distribution of the injected solution using methylene blue dye injection for non-survival surgeries (Figure 5B, see Materials and Methods for more details). Thirdly, avoiding radio-contamination of the tissues near the I.A. injection site proved challenging, particularly due to compound leakage from the high pressure-arterial flow. Accordingly, a small gauze adjacent to the injection site was pre-placed to absorb the leaked radioactive solution. After the injection was completed and hemostasis was achieved, radioactivity of the gauze was measured, and only non-radio-contaminated rats were employed for PET-CT imaging. Fourthly, the procedure was quite invasive since it involved a large laparotomy in a survival surgery. Therefore, health status during the 70 min imaging period was another concern. To determine health status, we performed a pilot study, which confirmed that breathing rates were stable under general anesthesia for 2 h after the I.A. surgery procedure, demonstrating the feasibility of the surgical procedure being followed by image acquisition. In fact, the complexity of the surgical procedures was attributed to the small animal experimental model. In the clinical setting, I.A. is safely performed as a catheter-assisted procedure, in which the catheter is generally inserted through the femoral artery of the patient to the splenic artery for pancreas-selective distribution of the solution.

3.2. Radio-Deiodination In Vivo

The low stability of the carbon-iodine bond in [^{124}I] analogues may cause substantial radio-deiodination in vivo, thus leading to the formation of free radioiodine that will rapidly accumulation in the thyroid and stomach. Therefore, all rats were pretreated with 10 drops of saturated KI per 100 mL of drinking water for 24 h before injection of [^{124}I]I-UC-MSC-EVs. PET imaging results indicate that rat thyroid uptake of radioiodine was successfully blocked. Additionally, uptake of radioiodine in the stomach is typically blocked by gastric lavage with potassium perchlorate in PBS 30 min before injection. However, to avoid complicating the surgery procedure of the I.A. group, we did not pre-treat the stomach. As a result, accumulation of radioiodine was observed in the stomach during the middle and later stages of PET imaging.

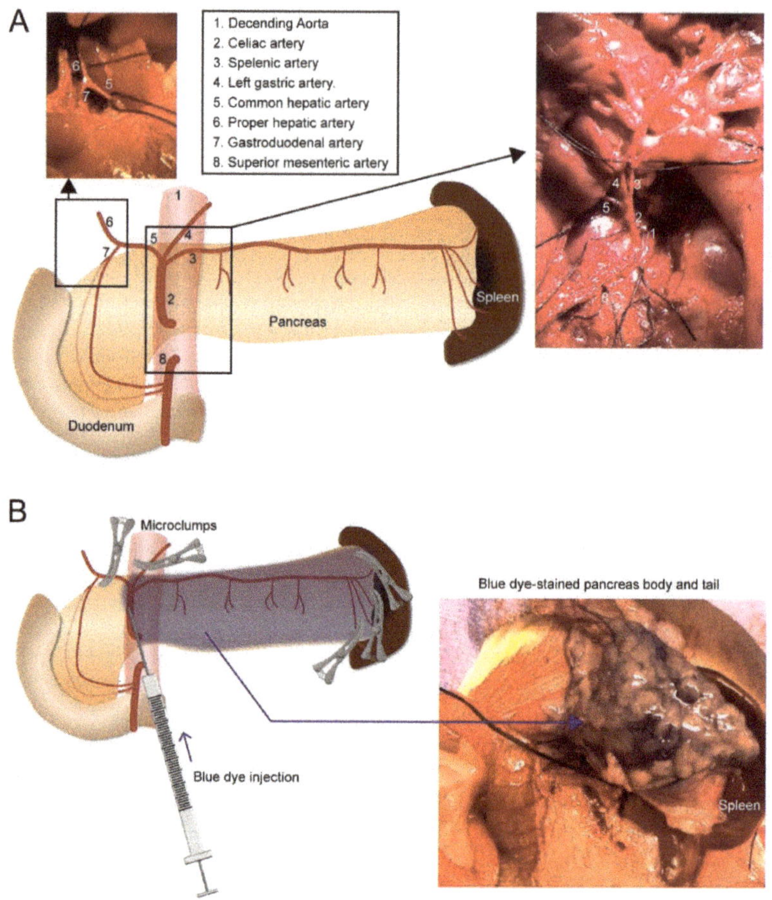

Figure 5. Intra-arterial injection procedure. (**A**) Anatomy of arteries around pancreas. The schema illustrates the description of the arteries, and two photographs show the dissected key arteries. Arteries were looped with black silk sutures. (**B**) Injection procedure. To specifically distribute the injected solution (blue dye in the preliminary tests or [^{124}I]I-UC-MSC-EVs for PET imaging), common hepatic artery, left gastric artery, and terminal branches of splenic arteries were clumped. A needle was cannulated into the Celiac artery to inject the solution. Microclumps were released after the injection was completed.

3.3. PET-CT Small Animal Imaging and Biodistribution

In PET imaging, the predominant uptake of [^{124}I]I-UC-MSC-EVs occurred in the liver for both injection routes during the early and middle stages. Clearance of [^{124}I]I-UC-MSC-EVs was observed from the liver during the middle and late stages. However, it was difficult to characterize clearance patterns for the stomach, spleen, pancreas, and small intestine, since these organs were located close together in the rat. Therefore, we utilized biodistribution analysis to validate uptake levels from PET imaging. Splenic uptake for both modalities displayed significant differences, which may be explained by the pathways of the administration routes. For the I.V. injection route, [^{124}I]I-UC-MSC-EVs were likely filtered out once they reached the spleen. However, splenic uptake levels following I.A. injection depended on which arteries were blocked during the surgical procedure. After releasing the vessel clamps, the [^{124}I]I-UC-MSC-EVs were likely filtered by the liver, so higher uptake levels were observed in the spleen for I.V. administration compared to I.A. administration, and vice versa for the liver. Uptake levels (%ID/gram) in the spleen were significantly different between the two injection routes, whereas uptake levels in the liver were not, due to the extremely small amount of [^{124}I]I-UC-MSC-EVs compared to the entirety of the liver and the spleen, and the greater weight of the liver compared to the spleen (~20–24 g for liver vs. 0.8–1.2 g for spleen) in a big rat. Overall, the high sensitivity of the nuclear imaging technique provided a reliable and quantitative evaluation of pancreas uptake in both administration routes, and demonstrated, importantly, that both I.V. and I.A. injection resulted in similar uptake levels in the pancreas. As such, both the I.A. injection and I.V. injection routes displayed similar accumulation levels of [^{124}I]I-UC-MSC-EVs in the organs, except for the spleen and liver. The higher spleen uptake could importantly provide significant immunomodulatory benefits to UC-MSC-EVs applications for diabetes therapies. These findings may guide the development of safe and efficacious delivery strategies for UC-MSC-EVs.

4. Materials and Methods

4.1. Chemicals

All reagents were purchased from commercial sources as reagent grade and used without further purification unless otherwise stated. ^{124}I-sodium iodide was purchased from 3D Imaging (Little Rock, AR, USA).

4.2. Human UC-MSC-EVs Extraction and Purification

Human UC-MSC-EVs were provided from the EVs team at City of Hope using the following steps: (1) isolation of MSCs from human umbilical cord; (2) culture of MSCs in flasks; (3) starvation of MSCs to maximize EV production; and (4) isolation and purification of EVs to obtain UC-MSC-EVs.

4.3. Animal

Male Lewis (LEW) rats (Charles River Laboratories, Wilmington, MA, USA) weighing 350–500 g were used to study I.V. and I.A. administration of [^{124}I]I-UC-MSC-EVs. Rat thyroid uptake of radioiodine was blocked by pretreatment using 10 drops of saturated KI per 100 mL of drinking water for 24 h before injection of [^{124}I]I-UC-MSC-EVs. The use of animals and animal procedures performed in this study were approved by the City of Hope/Beckman Research Institute Institutional Animal Care and Use Committee.

4.4. Radioiodination of UC-MSC-EVs and Radiochemical Stability Assessment

Radioiodination of human UC-MSC-EVs with ^{124}I-sodium iodide was performed using the direct IODOGEN methodology. Approximately 23 μL of UC-MSC-EVs (2.28 $\times 10^{11}$ particles/mL) was added to a tube pre-coated with 150 μg IODOGEN (Pierce, Rockford, IL), followed by 137 MBq [^{124}I] NaI diluted to 44 μL in 0.1 M phosphate buffer (pH 7.5) and incubated at room temperature for 3 min. At the end of the incubation, the radioiodinated UC-MSC-EVs were purified by Superdex 200 chromatography (GE Healthcare). The radiochemical purity post-purification was >99% by ITLC. [^{124}I]I-UC-MSC-EVs were serviced for animal studies for in vivo evaluation.

The stability of [^{124}I]I-UC-MSC-EVs was evaluated for 0, 2, and 4 h. Samples post-incubation were passed to Superoser® 6 10/300 GL column (Running Buffer: 1 $\times$ PBS + 0.05% NaN$_3$ at 0.4 mL/min) to determine stability.

4.5. Procedures for I.V. Administration and Animal PET-CT Imaging

The needle catheter was inserted into a lateral rat tail vein after rats were anesthetized with 2–4% isoflurane in oxygen. Placement of the needle inside the vein was confirmed by infusing a small volume of saline. Rats were transferred into the PET-CT scanner before injection of ~8 MBq [^{124}I]I-UC-MSC-EVs. Dynamic microPET scans of 0–90 min were conducted, followed by 1-min CT scans using the small-animal GNEXT PET/CT imaging system (SOFIE, Dulles, VA, USA). The images were reconstructed by three-dimensional ordered subsets expectation-maximization (3D-OSEM) using the integrated GNEXT Acquisition Engine software. Separately, one group of rats underwent I.V. injection as described above, a 90 min dynamic PET scan of the head, and then a 1 min CT scan.

4.6. Procedures for I.A. Administration and Animal PET-CT Imaging

All surgical procedures were performed under general anesthesia. To deliver [^{124}I]I-UC-MSC-EVs selectively to the pancreatic parenchyma of the body and tail through arterial flow, the following procedures were performed [34]. The abdominal cavity was exposed via a midline abdominal incision through the linea alba of rectus sheath. The descending aorta, celiac trunk, and its main branches of the splenic artery, common hepatic artery, and left gastric artery were identified and exposed (Figure 5A). In addition, small branches of splenic arteries between the pancreatic tail and the spleen were also identified and exposed, since the splenic artery is the dominant feeding artery to pancreatic body and tail as well as spleen. In order to direct the injected solution to the pancreatic body and tail, the arterial flow of the common hepatic artery, left gastric artery, and small branches of splenic arteries between the pancreatic tail and spleen were temporarily blocked using microvascular clamps. (Figure 5A). Approximately 8 MBq [^{124}I]I-UC-MSC-EVs were prepared in a 1-mL conventional insulin syringe (using a 25 gauge-needle; BD, Franklin Lakes, NJ, USA) to inject through the celiac trunk. Injection flow was manually controlled at 50 µL/min. After the injection was completed, all clamps were released (duration of arterial clamps were within 5 min), the needle was removed from the celiac trunk, and compression hemostasis was performed with a cotton swab for 5 min. After hemostasis was confirmed, the abdominal incision was closed, and the rat subsequently underwent a 70 min PET-CT scan at 20 min post-injection and then a 1 min CT scan.

For the technical proof of the I.A. surgical procedure, we conducted preliminary tests using 200 µL of Methylene blue dye (diluted with saline at 1:1 dilution [v/v] and 0.22 µm-filtered for sterilization) in terminal surgeries. The distribution of the solution in the designated region was confirmed (Figure 5B).

4.7. Biodistribution of [^{124}I]I-UC-MSC-EVs in Rats

The biodistribution of [^{124}I]I-UC-MSC-EVs was investigated at the end of PET-CT scans (96–98 min post injection) for all rats following I.A. and I.V. administration. Rats were anesthetized and euthanized. Blood, heart, lung, liver, spleen, stomach, kidneys, pancreas, small intestine, cecum, large intestine, muscle, fat, bone, and brain were collected and weighed, and the uptake of radioactivity was measured using Hidex AMG Automatic Gamma Counter (HIDEX, Turku, Finland) and its decay was corrected. Results were reported as percentage injected dose per gram (%ID/g).

5. Conclusions

In this study, we explored a nuclear imaging technique to monitor and evaluate different delivery modalities for human UC-MSC-EVs in vivo. UC-MSC-EVs were successfully tagged with I-124 via the IODOGEN methodology. Our initial results indicated that UC-MSC-EVs displayed similar delivery efficacy, except in the spleen and liver, for both I.V. and I.A. administration routes in a non-diabetic Lewis (LEW) rat model. Higher uptake in the spleen could, importantly, provide more advantageous immunomodulatory properties for diabetes therapies. As such, these results can guide optimization of UC-MSC-EV delivery strategies for clinical therapies for diabetes. Studies for further confirmation are ongoing.

Author Contributions: Conceptualization, J.L., A.M.W., J.E.S. and F.R.K.; methodology, J.L., H.K., E.K.P., T.O., J.C., N.B., B.C., J.R. and J.P.; analyses, J.L., T.O., K.X.H., L.A.H. and B.C.; investigation, J.L., H.K., E.K.P., T.O., K.X.H., L.A.H. and J.R.; writing—original draft preparation, J.L., H.K., E.K.P., T.O., K.X.H. and L.A.H.; writing—review and editing, J.L., H.K., K.X.H. and L.A.H.; supervision, J.L., A.M.W., J.E.S. and F.R.K.; project administration, J.L., A.M.W., J.E.S. and F.R.K.; funding acquisition, J.L. and H.K.; All authors have read and agreed to the published version of the manuscript.

Funding: The animal study was supported with funding by the Nora Eccles Treadwell Foundation (50214-2008261-Y02); Radiosynthesis study was supported with funding by the Arthur Riggs Diabetes and Metabolism Research Institute of City of Hope Polit Project Award (60074-2010638-ENDOW).

Institutional Review Board Statement: All animal experiments were carried out under protocols approved by the City of Hope/Beckman Research Institute Institutional Animal Care and Use Committee (approve code 13036, and date of approval, 8 January 2020).

Informed Consent Statement: Not applicable.

Data Availability Statement: All data is contained within article.

Acknowledgments: We would like to thank Wei Tang, Kim Nguyen, Angel Gu, and Keith Rodriguez for providing EVs.

Conflicts of Interest: The authors declare no conflict of interest.

References

1. Valadi, H.; Ekstrom, K.; Bossios, A.; Sjostrand, M.; Lee, J.J.; Lotvall, J.O. Exosome-mediated transfer of mRNAs and microRNAs is a novel mechanism of genetic exchange between cells. *Nat. Cell Biol.* **2007**, *9*, 654–659. [CrossRef] [PubMed]
2. Wang, K.; Jiang, Z.; Webster, K.A.; Chen, J.; Hu, H.; Zhou, Y.; Zhao, J.; Wang, L.; Wang, Y.; Zhong, Z.; et al. Enhanced Cardioprotection by Human Endometrium Mesenchymal Stem Cells Driven by Exosomal MicroRNA-21. *Stem Cells Transl. Med.* **2017**, *6*, 209–222. [CrossRef] [PubMed]
3. Perets, N.; Betzer, O.; Shapira, R.; Brenstein, S.; Angel, A.; Sadan, T.; Ashery, U.; Popovtzer, R.; Offen, D. Golden Exosomes Selectively Target Brain Pathologies in Neurodegenerative and Neurodevelopmental Disorders. *Nano Lett.* **2019**, *19*, 3422–3431. [CrossRef] [PubMed]
4. Zhang, S.; Chuah, S.J.; Lai, R.C.; Hui, J.H.P.; Lim, S.K.; Toh, W.S. MSC exosomes mediate cartilage repair by enhancing proliferation, attenuating apoptosis and modulating immune reactivity. *Biomaterials* **2018**, *156*, 16–27. [CrossRef]
5. Zhao, J.; Li, X.; Hu, J.; Chen, F.; Qiao, S.; Sun, X.; Gao, L.; Xie, J.; Xu, B. Mesenchymal stromal cell-derived exosomes attenuate myocardial ischaemia-reperfusion injury through miR-182-regulated macrophage polarization. *Cardiovasc. Res.* **2019**, *115*, 1205–1216. [CrossRef]
6. Zhang, B.; Yeo, R.W.Y.; Lai, R.C.; Sim, E.W.K.; Chin, K.C.; Lim, S.K. Mesenchymal stromal cell exosome-enhanced regulatory T-cell production through an antigen-presenting cell-mediated pathway. *Cytotherapy* **2018**, *20*, 687–696. [CrossRef]
7. Shahir, M.; Mahmoud Hashemi, S.; Asadirad, A.; Varahram, M.; Kazempour-Dizaji, M.; Folkerts, G.; Garssen, J.; Adcock, I.; Mortaz, E. Effect of mesenchymal stem cell-derived exosomes on the induction of mouse tolerogenic dendritic cells. *J. Cell Physiol.* **2020**, *235*, 7043–7055. [CrossRef]
8. Ding, M.; Shen, Y.; Wang, P.; Xie, Z.; Xu, S.; Zhu, Z.; Wang, Y.; Lyu, Y.; Wang, D.; Xu, L.; et al. Exosomes Isolated from Human Umbilical Cord Mesenchymal Stem Cells Alleviate Neuroinflammation and Reduce Amyloid-Beta Deposition by Modulating Microglial Activation in Alzheimer's Disease. *Neurochem. Res.* **2018**, *43*, 2165–2177. [CrossRef]
9. Qu, Y.; Zhang, Q.; Cai, X.; Li, F.; Ma, Z.; Xu, M.; Lu, L. Exosomes derived from miR-181-5p-modified adipose-derived mesenchymal stem cells prevent liver fibrosis via autophagy activation. *J. Cell Mol. Med.* **2017**, *21*, 2491–2502. [CrossRef]
10. Zhang, S.; Chu, W.C.; Lai, R.C.; Lim, S.K.; Hui, J.H.; Toh, W.S. Exosomes derived from human embryonic mesenchymal stem cells promote osteochondral regeneration. *Osteoarthr. Cartil.* **2016**, *24*, 2135–2140. [CrossRef]
11. Nakamura, Y.; Miyaki, S.; Ishitobi, H.; Matsuyama, S.; Nakasa, T.; Kamei, N.; Akimoto, T.; Higashi, Y.; Ochi, M. Mesenchymal-stem-cell-derived exosomes accelerate skeletal muscle regeneration. *FEBS Lett.* **2015**, *589*, 1257–1265. [CrossRef]
12. Teng, X.; Chen, L.; Chen, W.; Yang, J.; Yang, Z.; Shen, Z. Mesenchymal Stem Cell-Derived Exosomes Improve the Microenvironment of Infarcted Myocardium Contributing to Angiogenesis and Anti-Inflammation. *Cell Physiol. Biochem.* **2015**, *37*, 2415–2424. [CrossRef]
13. Li, T.; Yan, Y.; Wang, B.; Qian, H.; Zhang, X.; Shen, L.; Wang, M.; Zhou, Y.; Zhu, W.; Li, W.; et al. Exosomes derived from human umbilical cord mesenchymal stem cells alleviate liver fibrosis. *Stem Cells Dev.* **2013**, *22*, 845–854. [CrossRef]
14. Yan, Y.; Jiang, W.; Tan, Y.; Zou, S.; Zhang, H.; Mao, F.; Gong, A.; Qian, H.; Xu, W. hucMSC Exosome-Derived GPX1 Is Required for the Recovery of Hepatic Oxidant Injury. *Mol. Ther.* **2017**, *25*, 465–479. [CrossRef]
15. Zhang, B.; Shi, Y.; Gong, A.; Pan, Z.; Shi, H.; Yang, H.; Fu, H.; Yan, Y.; Zhang, X.; Wang, M.; et al. HucMSC Exosome-Delivered 14-3-3zeta Orchestrates Self-Control of the Wnt Response via Modulation of YAP During Cutaneous Regeneration. *Stem Cells* **2016**, *34*, 2485–2500. [CrossRef]
16. Aliotta, J.M.; Pereira, M.; Wen, S.; Dooner, M.S.; Del Tatto, M.; Papa, E.; Goldberg, L.R.; Baird, G.L.; Ventetuolo, C.E.; Quesenberry, P.J.; et al. Exosomes induce and reverse monocrotaline-induced pulmonary hypertension in mice. *Cardiovasc. Res.* **2016**, *110*, 319–330. [CrossRef]
17. Wang, B.; Xu, Y.; Wei, Y.; Lv, L.; Liu, N.; Lin, R.; Wang, X.; Shi, B. Human Mesenchymal Stem Cell-Derived Exosomal microRNA-143 Promotes Apoptosis and Suppresses Cell Growth in Pancreatic Cancer via Target Gene Regulation. *Front. Genet.* **2021**, *12*, 581694. [CrossRef]

18. Du, L.; Tao, X.; Shen, X. Human umbilical cord mesenchymal stem cell-derived exosomes inhibit migration and invasion of breast cancer cells via miR-21-5p/ZNF367 pathway. *Breast Cancer* **2021**, *28*, 829–837. [CrossRef]

19. Nojehdehi, S.; Soudi, S.; Hesampour, A.; Rasouli, S.; Soleimani, M.; Hashemi, S.M. Immunomodulatory effects of mesenchymal stem cell-derived exosomes on experimental type-1 autoimmune diabetes. *J. Cell Biochem.* **2018**, *119*, 9433–9443. [CrossRef]

20. Mahdipour, E.; Salmasi, Z.; Sabeti, N. Potential of stem cell-derived exosomes to regenerate beta islets through Pdx-1 dependent mechanism in a rat model of type 1 diabetes. *J. Cell Physiol.* **2019**, *234*, 20310–20321. [CrossRef]

21. Shigemoto-Kuroda, T.; Oh, J.Y.; Kim, D.K.; Jeong, H.J.; Park, S.Y.; Lee, H.J.; Park, J.W.; Kim, T.W.; An, S.Y.; Prockop, D.J.; et al. MSC-derived Extracellular Vesicles Attenuate Immune Responses in Two Autoimmune Murine Models: Type 1 Diabetes and Uveoretinitis. *Stem Cell Rep.* **2017**, *8*, 1214–1225. [CrossRef]

22. He, Q.; Wang, L.; Zhao, R.; Yan, F.; Sha, S.; Cui, C.; Song, J.; Hu, H.; Guo, X.; Yang, M.; et al. Mesenchymal stem cell-derived exosomes exert ameliorative effects in type 2 diabetes by improving hepatic glucose and lipid metabolism via enhancing autophagy. *Stem Cell Res. Ther.* **2020**, *11*, 223. [CrossRef]

23. Sun, Y.; Shi, H.; Yin, S.; Ji, C.; Zhang, X.; Zhang, B.; Wu, P.; Shi, Y.; Mao, F.; Yan, Y.; et al. Human Mesenchymal Stem Cell Derived Exosomes Alleviate Type 2 Diabetes Mellitus by Reversing Peripheral Insulin Resistance and Relieving beta-Cell Destruction. *Acs Nano* **2018**, *12*, 7613–7628. [CrossRef]

24. Mao, R.; Shen, J.; Hu, X. BMSCs-derived exosomal microRNA-let-7a plays a protective role in diabetic nephropathy via inhibition of USP22 expression. *Life Sci.* **2021**, *268*, 118937. [CrossRef]

25. Fan, B.; Li, C.; Szalad, A.; Wang, L.; Pan, W.; Zhang, R.; Chopp, M.; Zhang, Z.G.; Liu, X.S. Mesenchymal stromal cell-derived exosomes ameliorate peripheral neuropathy in a mouse model of diabetes. *Diabetologia* **2020**, *63*, 431–443. [CrossRef]

26. Venkat, P.; Zacharek, A.; Landschoot-Ward, J.; Wang, F.; Culmone, L.; Chen, Z.; Chopp, M.; Chen, J. Exosomes derived from bone marrow mesenchymal stem cells harvested from type two diabetes rats promotes neurorestorative effects after stroke in type two diabetes rats. *Exp. Neurol.* **2020**, *334*, 113456. [CrossRef]

27. Li, W.; Jin, L.Y.; Cui, Y.B.; Xie, N. Human umbilical cord mesenchymal stem cells-derived exosomal microRNA-17-3p ameliorates inflammatory reaction and antioxidant injury of mice with diabetic retinopathy via targeting STAT1. *Int. Immunopharmacol.* **2021**, *90*, 107010. [CrossRef]

28. Li, B.; Luan, S.; Chen, J.; Zhou, Y.; Wang, T.; Li, Z.; Fu, Y.; Zhai, A.; Bi, C. The MSC-Derived Exosomal lncRNA H19 Promotes Wound Healing in Diabetic Foot Ulcers by Upregulating PTEN via MicroRNA-152-3p. *Mol. Ther. Nucleic Acids* **2020**, *19*, 814–826. [CrossRef]

29. Yu, M.; Liu, W.; Li, J.; Lu, J.; Lu, H.; Jia, W.; Liu, F. Exosomes derived from atorvastatin-pretreated MSC accelerate diabetic wound repair by enhancing angiogenesis via AKT/eNOS pathway. *Stem Cell Res. Ther.* **2020**, *11*, 350. [CrossRef]

30. Liu, W.; Yu, M.; Xie, D.; Wang, L.; Ye, C.; Zhu, Q.; Liu, F.; Yang, L. Melatonin-stimulated MSC-derived exosomes improve diabetic wound healing through regulating macrophage M1 and M2 polarization by targeting the PTEN/AKT pathway. *Stem Cell Res. Ther.* **2020**, *11*, 259. [CrossRef]

31. Grange, C.; Tapparo, M.; Bruno, S.; Chatterjee, D.; Quesenberry, P.J.; Tetta, C.; Camussi, G. Biodistribution of mesenchymal stem cell-derived extracellular vesicles in a model of acute kidney injury monitored by optical imaging. *Int. J. Mol. Med.* **2014**, *33*, 1055–1063. [CrossRef] [PubMed]

32. Wen, S.; Dooner, M.; Papa, E.; Del Tatto, M.; Pereira, M.; Borgovan, T.; Cheng, Y.; Goldberg, L.; Liang, O.; Camussi, G.; et al. Biodistribution of Mesenchymal Stem Cell-Derived Extracellular Vesicles in a Radiation Injury Bone Marrow Murine Model. *Int. J. Mol. Sci.* **2019**, *20*, 5468. [CrossRef] [PubMed]

33. Yang, B.; Migliati, E.; Parsha, K.; Schaar, K.; Xi, X.; Aronowski, J.; Savitz, S.I. Intra-arterial delivery is not superior to intravenous delivery of autologous bone marrow mononuclear cells in acute ischemic stroke. *Stroke* **2013**, *44*, 3463–3472. [CrossRef] [PubMed]

34. Choi, J.; Wang, J.; Ren, G.; Thakor, A.S. A Novel Approach for Therapeutic Delivery to the Rodent Pancreas Via Its Arterial Blood Supply. *Pancreas* **2018**, *47*, 910–915. [CrossRef] [PubMed]

pharmaceuticals

Article

Thin Layer-Protected Gold Nanoparticles for Targeted Multimodal Imaging with Photoacoustic and CT

Jing Chen [1], Van Phuc Nguyen [2], Sangeeta Jaiswal [1], Xiaoyu Kang [1], Miki Lee [1], Yannis M. Paulus [2,3] and Thomas D. Wang [1,3,4,*]

1   Division of Gastroenterology, Department of Internal Medicine, University of Michigan, Ann Arbor, MI 48109, USA; chenjingsimm@gmail.com (J.C.); jaiswals@umich.edu (S.J.); kangx@umich.edu (X.K.); leemiki@umich.edu (M.L.)
2   Kellogg Eye Center, Department of Ophthalmology and Visual Sciences, University of Michigan, Ann Arbor, MI 48109, USA; vanphucn@umich.edu (V.P.N.); ypaulus@umich.edu (Y.M.P.)
3   Department of Biomedical Engineering, University of Michigan, Ann Arbor, MI 48109, USA
4   Department of Mechanical Engineering, University of Michigan, Ann Arbor, MI 48109, USA
*   Correspondence: thomaswa@umich.edu

Abstract: The large size of nanoparticles prevents rapid extravasation from blood vessels and diffusion into tumors. Multimodal imaging uses the physical properties of one modality to validate the results of another. We aim to demonstrate the use of a targeted thin layer-protected ultra-small gold nanoparticles (Au-NPs) to detect cancer in vivo using multimodal imaging with photoacoustic and computed tomography (CT). The thin layer was produced using a mixed thiol-containing short ligands, including MUA, CVVVT-ol, and HS-(CH2)$_{11}$-PEG$_4$-OH. The gold nanoparticle was labeled with a heterobivalent (HB) peptide ligand that targets overexpression of epidermal growth factor receptors (EGFR) and ErbB2, hereafter HB-Au-NPs. A human xenograft model of esophageal cancer was used for imaging. HB-Au-NPs show spherical morphology, a core diameter of 4.47 ± 0.8 nm on transmission electron microscopy, and a hydrodynamic diameter of 6.41 ± 0.73 nm on dynamic light scattering. Uptake of HB-Au-NPs was observed only in cancer cells that overexpressed EGFR and ErbB2 using photoacoustic microscopy. Photoacoustic images of tumors in vivo showed peak HB-Au-NPs uptake at 8 h post-injection with systemic clearance by ~48 h. Whole-body images using CT validated specific tumor uptake of HB-Au-NPs in vivo. HB-Au-NPs showed good stability and biocompatibility with fast clearance and contrast-enhancing capability for both photoacoustic and CT imaging. A targeted thin layer-protected gold nanoprobe represents a new platform for molecular imaging and shows promise for early detection and staging of cancer.

Keywords: nanoparticle; multimodal imaging; photoacoustic; heterobivalent peptide

Citation: Chen, J.; Nguyen, V.P.; Jaiswal, S.; Kang, X.; Lee, M.; Paulus, Y.M.; Wang, T.D. Thin Layer-Protected Gold Nanoparticles for Targeted Multimodal Imaging with Photoacoustic and CT. *Pharmaceuticals* **2021**, *14*, 1075. https://doi.org/10.3390/ph14111075

Academic Editors: Xuyi Yue and Klaus Kopka

Received: 6 September 2021
Accepted: 20 October 2021
Published: 25 October 2021

Publisher's Note: MDPI stays neutral with regard to jurisdictional claims in published maps and institutional affiliations.

Copyright: © 2021 by the authors. Licensee MDPI, Basel, Switzerland. This article is an open access article distributed under the terms and conditions of the Creative Commons Attribution (CC BY) license (https://creativecommons.org/licenses/by/4.0/).

1. Introduction

Esophageal adenocarcinoma (EAC) is an aggressive disease with a poor 5-year survival rate of between 15–25% [1]. This disease is associated with high morbidity and mortality, thus accurate staging is important to determine the best therapeutic options for patients. EAC often develops in patients without symptoms, such as acid reflux or dysphagia, and many are not enrolled in an endoscopic surveillance program. Magnetic resonance imaging (MRI), computed tomography (CT), and endoscopic ultrasound are frequently used for cancer staging. These imaging modalities detect grossly visible anatomic abnormalities, such as the presence of a mass, to detect cancer, and are not sensitive to small or subtle lesions. By comparison, molecular imaging methods can be developed to detect cancer by visualizing the functional behavior of tumors based on known cellular and molecular signaling pathways. Epidermal growth factor receptors (EGFR) and ErbB2 are transmembrane tyrosine kinase receptors that stimulate epithelial cell growth, proliferation, and differentiation [2,3]. Overexpression of these targets reflects an increase in biological aggressiveness and a higher risk for progression to cancer [4–6]. These cell surface targets can be developed for molecular imaging to improve methods of cancer diagnosis and staging.

Multimodal imaging methods are an emerging approach that uses the physical properties of one modality to validate the results of another. Photoacoustic imaging is a hybrid technology that combines optical excitation with acoustic detection to provide deep tissue penetration with high sensitivity [7]. Light is absorbed by tissue contrast agents and results in thermoelastic expansion that generates acoustic waves. Various types of nanostructures are being developed as contrast agents for photoacoustic imaging [8–10]. CT is commonly used for diagnostic imaging and requires ionizing radiation to create cross-sectional images [11,12]. Functional CT is performed with use of contrast agents [13], such as barium suspensions [14] and iodinated small molecules [15]. However, these agents can be limited in use by potential nephrotoxicity, patient hypersensitivity, and short circulation times [16,17]. Therefore, the development of a targeted contrast agent that can be used for multimodal imaging with photoacoustic and CT methods may be synergistic for early detection and monitoring progression of cancer. Here, we aim to demonstrate a targeted nanoprobe specific for EGFR and ErbB2 to detect esophageal cancer using multimodal imaging.

2. Results

2.1. Preparation and Characterization of HB-Au-NPs

Thin layer-protected gold nanoparticles (Au-NPs) were synthesized using a one-step facile process, Figure 1. Mixed capping ligands of CVVVT-ol peptide, HS-$(CH_2)_{11}$-PEG$_4$-OH, and 11-mercaptoundecanoic acid (MUA) were self-assembled on the surface of Au-NPs. The terminal carboxylic acid of MUA was activated by an NHS ester and reacted with DBCO-PEG$_4$-amine. The azide functionalized heterobivalent (HB) peptide consists of monomers QRHKPRE, hereafter QRH*, and KSPNPRF, hereafter KSP*, specific for EGFR and ErbB2, respectively [18–20]. Mass spectra for KSP*-QRH*-E_3-K-N_3 is shown, Figure S1A,B. The scheme for synthesis of CVVVT-ol is shown, Figure S1C. The peptides were reacted with alkyne via a strain-promoted azide-alkyne cycloaddition (SPAAC) to modify the surface of the Au-NPs, hereafter HB-Au-NPs.

2.2. Nanoparticle Characterization

The nanoparticle properties were characterized using the following methods. Transmission electron microscopy (TEM) showed spherical morphology for Au-NP and HB-Au-NP, Figure 2A,B. The mean diameters of HB-Au-NPs and Au-NPs were 4.24 ± 0.83 and 4.47 ± 0.88 nm, respectively, Figure 2C,D. In the UV absorption spectra, conjugation of the heterobivalent peptide ligand showed negligible change in the peak at 518 nm, Figure 2E. The HB-Au-NPs in PBS solution showed excellent photostability after laser eradiation, Figure 2F, and exhibited good biocompatibility and stability against various endogenous bioactive thiol-containing molecules, Figure S2A–C. An X-ray diffraction (XRD) pattern shows 4 peaks that correspond to standard Bragg reflections from the center faces of a cubic lattice, Figure 2G. The peak at 38.1 deg represents preferential growth in the (111) direction. These results are consistent with a typical purity for crystallinity of the Au nanocrystals. Dynamic light scattering (DLS) showed a mean hydrodynamic size of 5.50 ± 0.63 and 6.41 ± 0.73 nm for Au-NPs and HB-Au-NPs, respectively, Figure 2H, and zeta potential of -12.6 and -9.33 mV, respectively, Figure S3.

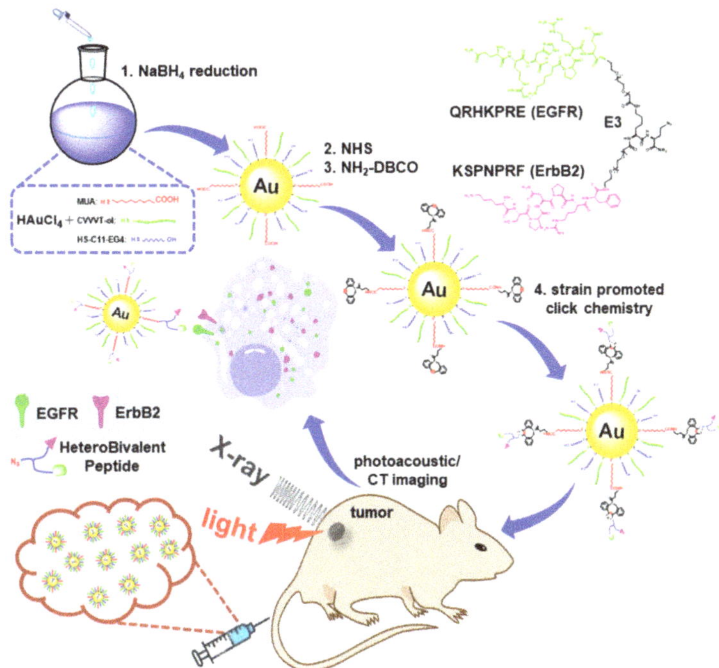

Figure 1. Schematic is shown for labeling the thin layer-protected gold nanoparticles with the heterobivalent (HB) peptide-specific for epidermal growth factor receptors (EGFR) and ErbB2 (HB-Au-NPs) developed for dual photoacoustic/CT imaging.

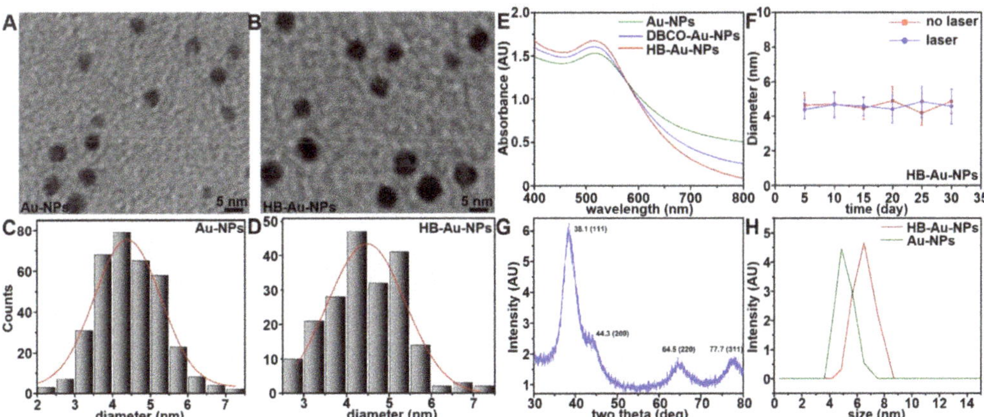

Figure 2. Nanoparticle characterization. TEM images of (**A**) Au-NPs and (**B**) HB-Au-NPs show spherical morphology. (**C, D**) A mean ($\pm$SD) diameter of 4.24 ± 0.83 and 4.47 ± 0.88 nm, was measured for Au-NPs and HB-Au-NPs, respectively. (**E**) The absorbance spectra for Au-NPs, DBCO-Au-NPs, and HB-Au-NPs show no shift in the peak at 518 nm. (**F**) No change is seen in the mean ($\pm$SD) diameter of HB-Au-NPs in PBS as a function of storage time over 30 days at RT with and without 5 min of laser irradiation (100 μJ/cm^2). Error bars represent standard deviations of n = 3 independent measurements. (**G**) The XRD pattern shows crystalline nanoparticles represented by 4 peaks corresponding to the standard Bragg reflections (111), (200), (220), and (311) of the center faces in a cubic lattice. The peak at 38.1 deg represents preferential growth in the (111) direction. (**H**) Dynamic light scattering (DLS) measurements show mean ($\pm$SD) diameter of 6.41 ± 0.73 and 5.50 ± 0.63 nm, respectively, for HB-Au-NPs and Au-NPs.

2.3. In Vitro Cytotoxicity

Cytotoxicity was evaluated by incubating human OE33 (EGFR+/ErbB2+) cancer and Qh-TERT (EGFR-/ErbB2-) benign esophageal cells and SKBr3 (EGFR+/ErbB2+) human breast cancer cells with HB-Au-NPs for 24 h, Figure S4. HB-Au-NPs were found to be non-toxic to cell proliferation at concentrations ranging from 5–400 µg/mL.

2.4. Photoacoustic Microscopy of Cells

Human OE19 (ErbB2+), OE21 (EGFR+), SKBr3 (EGFR+/ErbB2+), and Qh-TERT (EGFR-/ErbB2-) cells were incubated with 100 µg/mL of HB-Au-NPs. Photoacoustic microscopy images show a strong signal for OE19, OE21, and SKBr3 cells and minimal intensity for Qh-TERT cells, Figure 3A–D. Image intensities for each cell were quantified, and the mean values for OE19, OE21, and SKBr3 were significantly greater than that of Qh-TERT cells. Western blot shows expression levels of EGFR and ErbB2 for each cell type, Figure 3F. Without HB-Au-NPs, we were unable to focus on the cells to capture images.

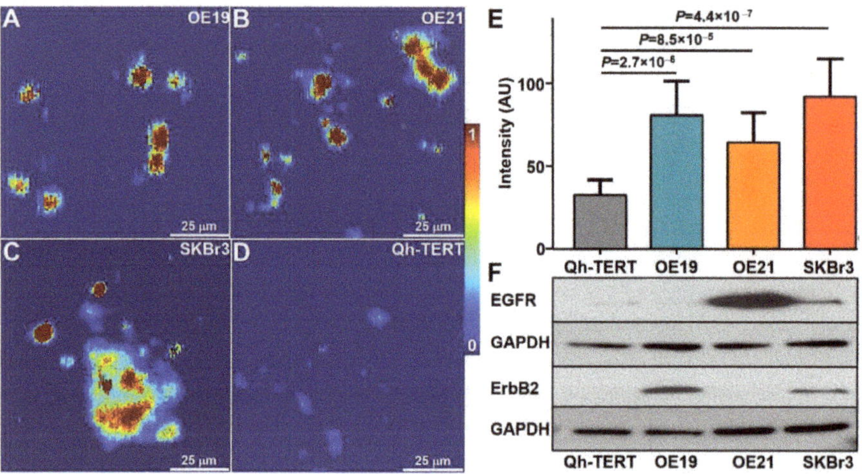

Figure 3. Photoacoustic microscopy. (**A–C**) In vitro images of human OE19 (ErbB2+), OE21 (EGFR+), and SKBr3 cells (EGFR+/ErbB2+) incubated with HB-Au-NPs at a concentration of 100 µg/mL show strong signal. (**D**) Image of Qh-TERT cells (EGFR-/ErbB2-) used for control at same concentration shows minimal signal. (**E**) Quantified image intensities are shown. (**F**) Western blot supports EGFR and ErbB2 expression in each cell.

2.5. Photoacoustic Imaging

Photoacoustic tomography was performed to evaluate nanoparticle uptake by human OE33 xenograft tumors in vivo following intravenous injection, Figure 4A–C. Images were acquired prior to injection (0 h) to assess background, and a weak signal from intrinsic absorption of oxy- and deoxy-hemoglobin was observed. After systemic injection of HB-Au-NPs, Au-NPs, and PBS, images were collected at 2, 4, 8, 12, 24, and 48 h. The maximum intensity projections (MIP) of the tumors are shown in the coronal view. In vitro photoacoustic intensities measured from HB-Au-NPs in vials demonstrate a linear relationship between photoacoustic signal and HB-Au-NPs concentration up to 400 µg/mL, Figure 4D. In vivo photoacoustic signal of HB-Au-NPs and Au-NPs reached a peak value at 8 h post-injection, Figure 4E. The intensities decreased over time to nearly baseline by 48 h. PBS was injected as control and showed no increase in signal. The mean T/B ratio at 8 h was found to be significantly greater for HB-Au-NPs versus Au-NPs and PBS in n = 6 animals, Figure 4F.

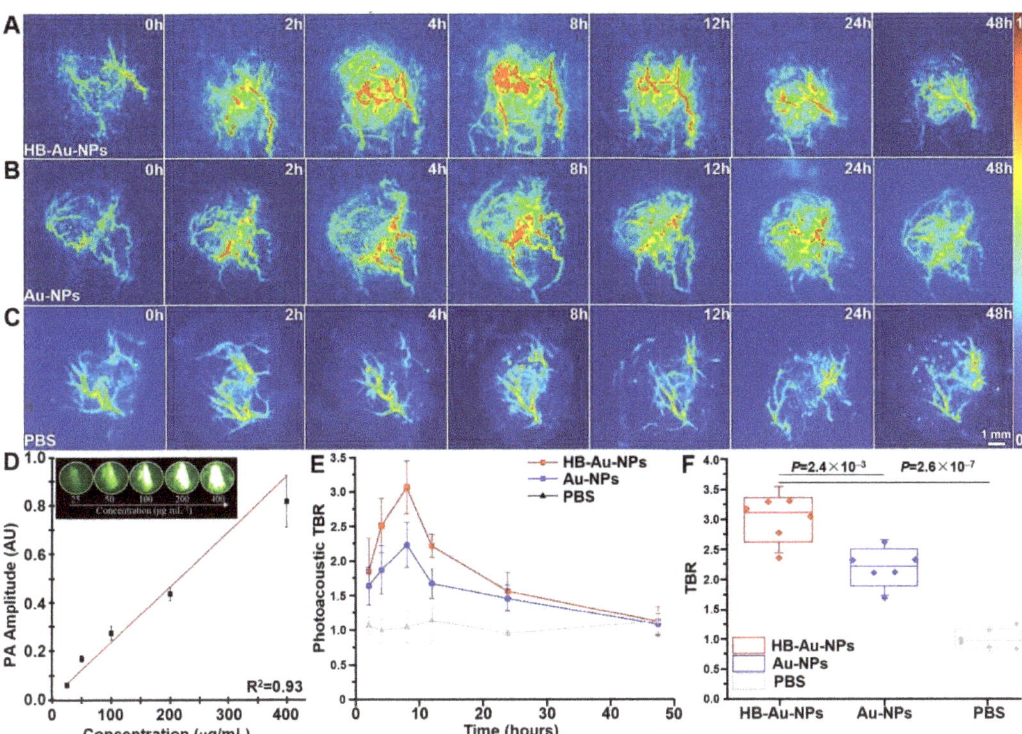

Figure 4. Photoacoustic tomography. In vivo images of human xenograft tumors implanted in mice are shown between 0–48 h post-injection of (**A**) HB-Au-NPs, (**B**) Au-NPs, and (**C**) PBS. (**D**) The in vitro photoacoustic intensity increases linearly with concentration of HB-Au-NPs, R^2 = 0.93. Inset: photoacoustic images are shown of HB-Au-NPs in vials at different concentrations (25, 50, 100, 200, and 400 μg/mL). (**E**) Quantified intensities from the tumors show a peak T/B ratio of 3.08 ± 0.37 and 2.27 ± 0.31 at 8 h post-injection for HB-Au-NPs and Au-NPs, respectively. (**F**) The mean value for HB-Au-NPs is significantly greater than that for either Au-NPs or PBS in n = 6 animals, $p = 2.4 \times 10^{-3}$ and 2.6×10^{-7}, respectively, by unpaired t-test.

2.6. CT Imaging

Whole-body CT images were collected to validate nanoparticle uptake by OE33 human xenograft tumors in vivo seen on photoacoustic tomography. At 24 h post-injection, representative images show the tumor (arrows) in different views for HB-Au-NPs, Figure 5A–D, and Au-NPs, Figure 5E–H. The results were quantified, and the mean value from tumor was significantly greater for HB-Au-NPs versus Au-NPs, Figure 5I. The attenuation of CT signal by HB-Au-NPs in vials of deionized water was evaluated and compared with that for Omnipaque, an FDA-approved iodine-based contrast agent used for clinical imaging, Figure 5J. Both contrast agents showed enhanced attenuation with increasing concentration. The intensity for HB-Au-NPs was over 3-fold greater than that for Omnipaque at a concentration of 25 mg/mL, Figure 5K.

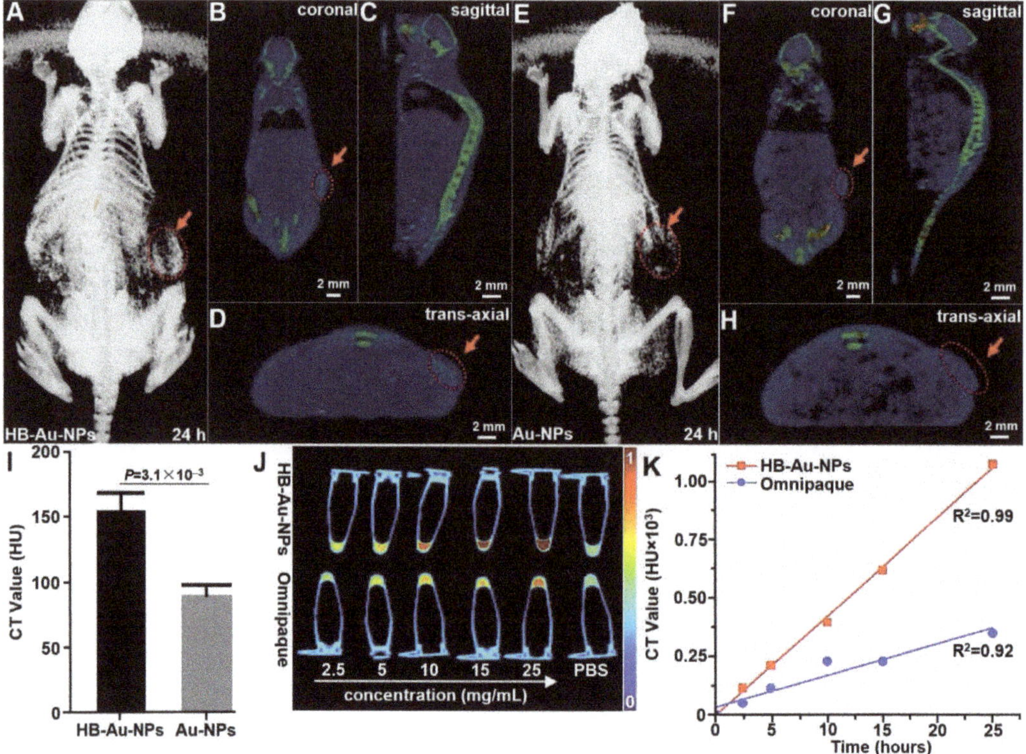

Figure 5. CT images. Representative 3D-reconstructed whole-body CT images at 24 h following intravenous injection of (**A–D**) HB-Au-NPs and (**E–H**) Au-NPs are shown. The locations (arrows) of implanted human OE33 xenograft tumors are shown in the axial, sagittal, and coronal views. (**I**) Quantified results showed a mean value of 155.08 ± 14.94 HU for HB-Au-NPs and 90.47 ± 9.05 HU for Au-NPs from n = 3 animals, p = 3.1 × 10⁻³ by unpaired t-test. (**J**) CT image of HB-Au-NPs and Omnipaque in vials at different concentrations, including 0 (PBS), 2.5, 5, 10, 15, and 25 mg/mL. (**K**) Intensities increase linearly with concentration of HB-Au-NPs, R^2 = 0.99, and Omnipaque, R^2 = 0.92.

2.7. Nanoparticle Biodistribution

The in vivo biodistribution of HB-Au-NPs and Au-NPs in major organs, including liver, spleen, brain, kidney, lung, heart, and tumor was investigated by inductively coupled plasma mass spectrometry (ICP-MS) at 8 h post-injection on OE33-bearing xenograft nude mice, Figure 6. The mean value for uptake in tumor was significantly greater for HB-Au-NPs versus Au-NPs in n = 3 mice. Uptake of both nanoparticles was high in liver and moderate in spleen. These results support nanoparticles being taken up by macrophages and sequestered in the reticuloendothelial system (RES) in part [21].

2.8. Animal Toxicity

Animal necropsy was performed after the completion of imaging, and the main organs, including liver, spleen, lung kidney, and brain, were harvested for necropsy. Histopathology was evaluated to assess toxicity, Figure 7A. No evidence of acute toxicity was observed. Body weight was measured in mice treated with HB-Au-NPs, Au-NPs, and PBS (n = 4 animals per group) every 2 days for up to 30 days. The mice gained weight as expected. No significant differences were observed among the different groups, Figure 7B. Laboratory tests, including hematology and chemistry, were evaluated on day 1 and 30 post-injection of HB-Au-NPs and PBS (control) to assess for potential toxicity, Figure S5A,B. No significant difference was seen in any of the results.

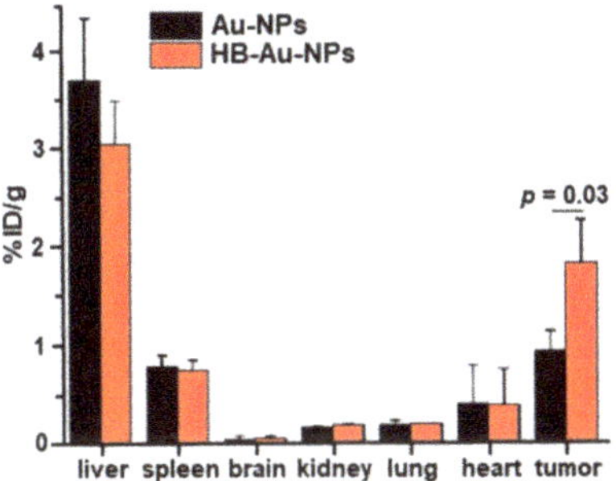

Figure 6. Nanoparticle biodistribution. In vivo uptake of HB-Au-NPs and Au-NPs by major organs, including liver, spleen, brain, kidney, lung, heart, and tumor is shown at 8 h post-injection in mice implanted with human xenograft tumor. The mean value in tumor was significantly higher for HB-Au-NPs than for unlabeled Au-NPs, 1.86 ± 0.43 versus 0.96 ± 0.20 %ID/g, in $n = 3$ animals, $p = 0.03$ by unpaired t-test.

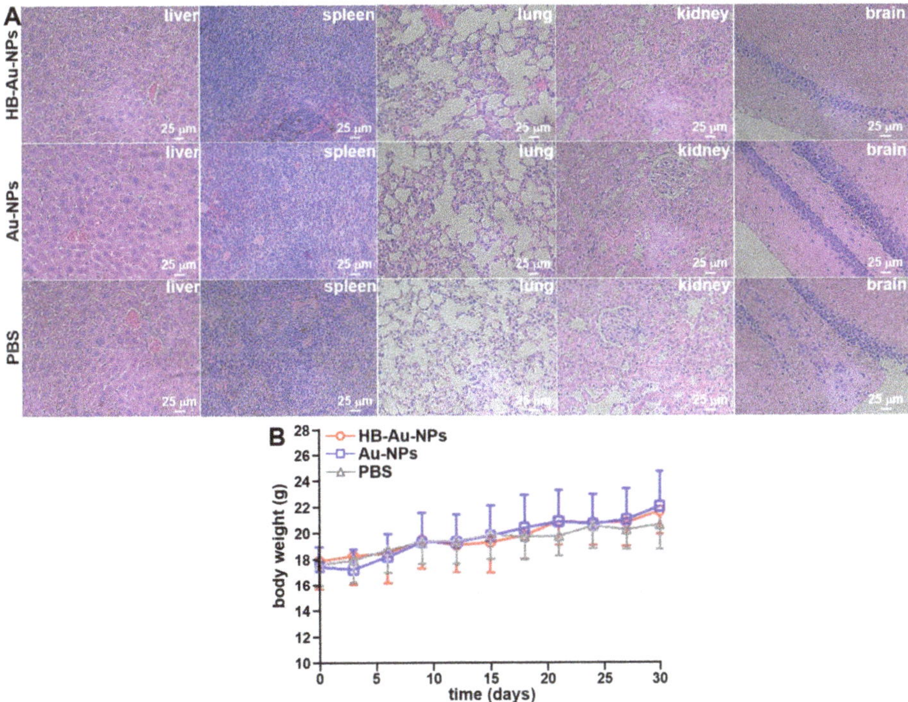

Figure 7. Animal necropsy. Histology (H&E) is shown for major organs, including liver, spleen, lung, kidney, and brain from the mice treated with (**A**) HB-Au-NPs, Au-NPs, and PBS at day 30 following intravenous administration. (**B**) The mean body weight from $n = 4$ animals shows no difference for mice injected with HB-Au-NPs, Au-NPs, and PBS.

3. Discussion

Here, we demonstrate use of a targeted gold nanoprobe HB-Au-NPs to detect a human xenograft model of esophageal cancer in vivo using multimodal imaging with photoacoustic and CT. The thin layer-protected gold nanoparticle represents a new platform for biomedical imaging. The thin layer was produced using a mixed thiol-containing short ligands, including MUA, CVVVT-ol and HS-$(CH_2)_{11}$-PEG$_4$-OH. Assembly on the surface of gold nanospheres resulted in greater compactness than conventional larger PEG-coated contrast agents. This feature improves nanoparticle stability under excitation with light. By labeling with a heterodimeric peptide ligand, the gold nanoparticles specifically target overexpression of EGFR and ErbB2 by the cancer cells. The active targeting ability of these nanoparticles is superior to passive tumor uptake via the enhanced permeability and retention (EPR) effect. Peak uptake was observed at 8 h post-injection with systemic clearance by ~48 h. Moreover, these thin layer-protected gold nanoparticles exhibit strong contrast enhancement for both photoacoustic and CT imaging. We found the heterodimer-labeled gold nanoparticle HB-Au-NPs to have excellent stability and biocompatibility. Strong contrast enhancement with both photoacoustic and CT imaging was observed, and the targeting capability of the heterobivalent peptide specific for EGFR and ErbB2 was demonstrated. This nanoprobe may be used as a new platform for diagnosis and staging of cancer. The distribution of EGFR and ErbB2 expression in the xenograft tumor has been previously reported [18]. Peptide monomers are arranged in the heterodimer configuration and are specific for either EGFR or ErbB2 [19].

The thin layer-protected gold nanoparticle with mixed self-assembled monolayers of small ligands represents a novel platform for nanoparticle-based contrast agents [22–24]. These compounds have been shown to be extremely stable for biological applications [25–28]. The thin layer-protected gold nanoparticles were prepared using a simple one-step sodium borohydride reduction without the limitations of a diluted solution. Moreover, the prepared gold nanoparticles can be lyophilized and reconstituted in water without changing characteristic properties. Their unique properties, including small size, large surface-to-volume ratio, tailored surface modification, and excellent biocompatibility, provide utility as a multifunctional biomaterial. As a contrast agent for photoacoustic imaging, gold nanoparticles have higher extinction coefficients than organic dyes at their plasmonic resonance wavelength for greater contrast [29]. Through a thermoelastic expansion mechanism, the absorbed photons can also produce acoustic waves. For CT imaging, the gold nanoparticles resulted in stronger attenuation than Omnipaque, a clinically used iodinated molecule. This effect results from a higher atomic number and electron density (gold: 79 and 19.23 g/cm^3, respectively, iodine: 53 and 4.9 g/cm^3, respectively) [30]. Furthermore, gold nanoparticles have a small size and higher CT attenuation. This property may reduce the dosage needed to provide contrast as compared with conventional iodinated agents [31]. Therefore, gold nanoparticles show potential for use as a dual-modal contrast agent both photoacoustic and CT imaging [32,33].

We demonstrated a tumor-targeting strategy by arranging monomer peptides in a dimer configuration and labeled Au-NPs for multi-modal imaging. This ligand structure was designed to improve binding affinity, sensitivity, and specificity [34]. Increased binding affinity may result from a multi-valent effect [18,19]. Higher sensitivity can occur from simultaneous detection of two unique targets. Greater specificity may arise from the dimer binding to a larger target surface area compared to that for the monomer. These improvements may be useful for detecting early targets that are expressed at low levels [20]. Previously, the RGD peptide was labeled with Au-NPs to target $\alpha_v\beta_3$ integrins expressed by tumor vasculature [35]. Targeted imaging was performed using CT by taking advantage of the high X-ray attenuation properties of the contrast agent. Additionally, Au-NPs have been used to label monomer peptides, such as conjugated analogs of the peptide bombesin [36]. This gastrin-releasing peptide (GRP) binds specifically to GRP receptors overexpressed in breast, prostate, and lung cancers.

Imaging modalities used to diagnose and stage esophageal cancer include barium esophagograms, endoscopic ultrasonography (EUS), computed tomography (CT), magnetic resonance imaging (MRI), and positron emission tomography (PET) [37]. Each modality has specific capability and strengths as well as limitations in terms of sensitivity and resolution. Use of the targeted gold nanoparticle may result in more accurate cancer staging by visualizing the molecular properties of the tumor. The thin layer-protected gold nanoparticles have been synthesized and demonstrated in vivo using the dual-modality photoacoustic and CT imaging. For CT imaging, the small size of the nanoparticle provides higher CT attenuation than larger-sized particles [38]. For photoacoustic imaging, the gold nanoparticles offer advantages over organic dyes in terms of quantum energy transferring coefficient, easy modification, and stability. Laser excitation at λ_{ex} = 532 nm was used to image cells to achieve the best resolution, while λ_{ex} = 680–950 nm was used to image the tumors to maximize image penetration depth.

4. Materials and Methods

4.1. Materials

Gold (III) chloride trihydrate, sodium borohydride, and 11-mercaptoundecanoic acid (11-MUA) were obtained from Sigma-Aldrich (Burlington, MA, USA). Alkyl PEG (PEGylated-alkanethiol HS-$(CH_2)_{11}$-EG_4-OH) was obtained from Prochimia Surfaces (Gdansk, Poland). All these chemicals were used without any further purification. CVVVT-ol peptide was synthesized with DHP-HM-Resin and purified by RP-HPLC. Deionized water used in all experiments was freshly prepared with Milli-Q water purification system (>18 MU cm). Human SKBr3, OE33, OE19, OE21, and Qh-TERT cell lines were obtained from the American Type Culture Collection (ATCC). Cell culture media was procured from Thermo Scientific, unless specifically stated.

4.2. Preparation and Characterization of HB-Au-NPs

The Au-NPs was synthesized by borohydride reduction of $HAuCl_4 \cdot 3H_2O$ in presence of a mixture of ligands which are CVVVT-ol peptide (T-ol – threoninol), alkyl PEG (PEGy-lated alkanethiol HS-$(CH2)_{11}$-EG_4-OH, sigma-aldrich), and MUA (11-mercaptoundecanoic acid) according to reported protocols [25,27,29]. The concentration of peptides, alkyl PEG and MUA were used in a molar ratio of 40%:40%:20%. Briefly, 0.05 mmol (19.7 mg) of $HAuCl_4 \cdot 3H_2O$ and 0.02 mmol of the ligand mixture containing 0.008 mmol (4 mg) of CVVVT-ol, 0.008 mmol (3.04 mg) of short alkyl PEG, and 0.004 mmol (0.87 mg) of MUA were dissolved in a mixture of solvents, including methanol (3.0 mL) and acetic acid (0.5 mL) by gentle stirring for 5 min, which results in a yellow solution. A freshly pre-pared sodium borohydride solution (30 mg of $NaBH_4$ in 1.5 mL of ice-cooled deionized water) was added dropwise into the above solution under rapid stirring for 4 h at room temperature (RT). Then, 25 µL of 1% Tween-20 was added and gently stirred overnight. The resulting dispersion of functional gold nanoparticles was transferred into a filtration tube (30 KDa MWCO membrane), purified by centrifugation at 3500 rpm, and washed 4 times with methanol containing 0.005% Tween-20 and twice with PBS containing 0.005% Tween 20. The remaining particles were then redispersed in a minimum volume of PBS containing 0.005% Tween-20 and filtered with a syringe filtration unit (0.2 µm) to remove any solid residues. This filtrate was then stored as stock solution. Further dilution of PBS containing 0.005% Tween-20 was carried out to maintain the Au-NPs with OD = 1, and peptide conjugation was conducted at this concentration. For peptide conjugation, about 10 mL of Au-NPs solution was placed into an Eppendorf tube, and 500 µL of EDC and NHS at a concentration of 500 mM each was added. The solution was mixed and purified by centrifugal filtration after 1 h of reaction time and redispersed again in 10 mL of deionized water. DBCO-PEG_4-amine (0.004 mM, 0.21 mg) was added to this solution, and stirring was performed for 2 h. Unreacted active sites were quenched by adding 200 µL of 0.01 M glycine solution. After an additional 30 min, the nanoparticle solution was purified by centrifugal filtration and redispersed in 10 mL of deionized water. The solution of

HB-E$_3$-N$_3$ (0.004 mM, 11.03 mg) was added, and the reaction was allowed to take place for 12 h. Centrifugal filtration was performed to remove excess heterobivalent peptide, and the solution was washed twice with deionized water.

4.3. Nanoparticle Characterization

The samples were dried on carbon-coated copper grids and imaged with a transmission electron microscope (TEM, JEOL JEM-2100) operating at an accelerated voltage of 200 kV. The dimensions of the nanoparticles (n = 200) were measured from the TEM images using ImageJ software. UV-vis absorbance was recorded in the spectral range from 400–800 nm at 5 nm increments using a NanoDrop 2000C spectrophotometer at 20° in 10 mm semi-micro cuvettes (Thermo Fisher Scientific, Waltham, MA, USA). The hydrodynamic diameter and zeta potential of nanoparticles were tested with a Zetasizer Nano ZS instrument (Malvern, UK).

4.4. In Vitro Cytotoxicity

Cytotoxicity of HB-Au-NPs was evaluated using a Cell Counting Kit-8 (CCK-8, Dojindo Molecular Technologies, Inc., Rockville, MD, USA) assay. Human Qh-TERT and OE33 esophageal cancer cells and SKBr3 human breast cancer cells were investigated with this colorimetric assay. Briefly, the cells were seeded in 96-well plates at a density of ~5 × 10^3 cells per well in 100 µL of media and incubated overnight at 37 °C in 5% CO$_2$. The media in each well was replaced with 100 µL of fresh media containing various concentrations of HB-Au-NPs. After incubation for 24 and 48 h, the media was aspirated, and the cells were washed with PBS twice. Then, 10 µL of CCK-8 solution was added to each well, and incubated for another 2 h at 37 °C. The absorbance was measured using a microplate spectrophotometer (Molecular Devices Tunable Microplate Reader VersaMax, SN# BNR06880) at 450 nm. All concentrations were tested in triplicate. The values were normalized and expressed as percent viability.

4.5. Photoacoustic Microscopy of Cells

All cells were maintained at 37 °C and 5% CO$_2$ and were supplemented with 10% FBS and 1% penicillin/streptomycin. Human SKBr3 breast cancer cells were cultured in McCoy's 5A media. Human OE33, OE19, and OE21 esophageal cells were cultured in Roswell Park Memorial Institute (RPMI) 1640 media, and Qh-TERT cells were cultured in keratinocyte-SFM media (Gibco). For photoacoustic microscopy, the cells were fixed in the 4% paraformaldehyde in the bottom of a petri dish filled with PBS. The sample was excited by a pulsed laser (OCT-LK3-BB, Thorlabs, Inc., Newton, NJ, USA) at the excitation wavelength of l$_{ex}$ = 532 nm. An ultrasonic transducer with center frequency of 27 MHz (Optosonic Inc., Arcadia, CA, USA) was immersed into water to detect the photoacoustic signals. The image was obtained by mechanically scanning the objective. The data were analyzed using custom MATLAB (Mathworks) software.

4.6. Photoacoustic Imaging

All experimental procedures were performed in accordance with relevant guidelines and requirements of the University of Michigan. Mouse imaging studies were conducted with approval of the University of Michigan Committee on the Use and Care of Animals (UCUCA) under project identification code PRO00009130 with date of approval 8/21/2019. Animals were housed per guidelines of the Unit for Laboratory Animal Medicine (ULAM). Female nude mice at 4 weeks of age were injected in the flank with ~5 × 10^7 OE33 human esophageal cancer cells in 200 µL of PBS. The mice were imaged when tumor dimensions reached ~5 mm.

Images were collected in vivo using a photoacoustic tomography system (Nexus128, Endra). A tunable pulsed laser (7 ns, 20 Hz, 25 mJ/pulse) provided excitation wavelengths ranging from l$_{ex}$ = 680–950 nm. The photoacoustic signals were acquired by 128 unfocused 3 mm diameter transducers with 5 MHz center frequency arranged in a helical pattern in a hemispherical bowl filled with water. The console provided data acquisition/reconstruction, servo motors for 3D rotation of the bowl, and a temperature monitor for the water bath. The anesthetized animals were placed in a transparent imaging tray located above the transducers.

4.7. CT Imaging

CT images were acquired and reconstructed using an IVIS Spectrum CT instrument (PerkinElmer Imaging Systems) at X-ray voltage of 45 KV and anode current of 500 mA. The images were processed using MicroView (ver 2.5.0) software. Suspensions of HB-Au-NPs and Omnipaque, an iodine-based contrast agent that is FDA-approved for clinical use, containing equivalent concentrations of 2.5, 5, 10, 15, 20, and 25 mg/mL were placed into 0.5 mL Eppendorf tubes. CT images were acquired using an IVIS Spectrum CT. The X-ray voltage was set at 45 kV. A circular ROI was drawn on the coronal view of each tube, and the mean attenuation value for an ROI of 3 slices per tube was recorded and normalized to the value of PBS. The attenuation values for each concentration from $n = 3$ samples were averaged.

4.8. Animal Toxicity

Healthy nude mice were injected with HB-Au-NPs and Au-NPs at a dose of 2.5 mg/mL and volume of 100 μL in PBS. The animals were euthanized 30 days post-injection, and major organs, including liver, spleen, lung kidney, and brain, were harvested. The organs were immobilized in 4% paraformaldehyde at 4 °C for 24 h, paraffin-embedded, and cut into 10 mm sections for evaluation by routine histology (Hand E). A total of $n = 4$ mice were used in each group.

4.9. Statistical Analysis

All statistical analysis was performed using GraphPad Prism, and plots were generated using Origin 8.0 software.

5. Conclusions

We demonstrate use of a heterobivalent peptide labeled with a gold nanoparticle to provide strong contrast enhancement with both photoacoustic and CT imaging. The targeting capability of this multimeric peptide specific for EGFR and ErbB2 was shown in human xenograft tumors in vivo. This nanoprobe shows excellent stability and biocompatibility and has the potential to be used as a new platform for diagnosis and staging of cancer with multi-modal imaging.

6. Patents

Patents resulting from the work reported in this manuscript include Wang TD, Chen J. Heterodimeric Peptide Reagents and Methods, WO2019222450A1, https://patents.google.com/patent/WO2019222450A1 (21 November 2019).

Supplementary Materials: The following are available online at https://www.mdpi.com/article/10.3390/ph14111075/s1, Figure S1: Mass spectra, Figure S2: Nanoparticle stability, Figure S3: Nanoparticle parameters, Figure S4: Nanoparticle cytotoxicity, Figure S5: No nanoparticle serum toxicity.

Author Contributions: Conceptualization, J.C., T.D.W.; methodology, J.C., T.D.W.; validation, J.C., V.P.N., S.J., X.K., and M.L.; formal analysis, J.C.; investigation, J.C., T.D.W.; resources, T.D.W.; data curation, J.C.; writing—original draft preparation, J.C.; writing—review and editing, J.C., T.D.W.; visualization, J.C., V.P.N., S.J., X.K., and M.L.; supervision, Y.M.P., T.D.W.; project administration, T.D.W.; funding acquisition, T.D.W. All authors have read and agreed to the published version of the manuscript.

Funding: This work was supported in part by NIH U54 CA163059 and U01 CA189291 (TDW).

Institutional Review Board Statement: The study was conducted according to the guidelines of the Declaration of Helsinki, and approved by the University of Michigan Committee on the Use and

Care of Animals (UCUCA) under project identification code PRO00009130 with date of approval 21 August 2019.

Informed Consent Statement: Not applicable.

Data Availability Statement: Data is contained within the article.

Conflicts of Interest: The authors report no conflict.

References

1. Arnal, M.J.D.; Arenas, Á.F.; Arbeloa, Á.L. Esophageal cancer: Risk factors, screening and endoscopic treatment in Western and Eastern countries. *World J. Gastroenterol.* **2015**, *21*, 7933–7943. [CrossRef] [PubMed]
2. Cronin, J.; McAdam, E.; Danikas, A.; Tselepis, C.; Griffiths, P.; Baxter, J.; Thomas, L.; Manson, J.; Jenkins, G. Epidermal growth factor receptor (EGFR) is overexpressed in high-grade dysplasia and adenocarcinoma of the esophagus and may represent a biomarker of histological progression in Barrett's esophagus (BE). *Am. J. Gastroenterol.* **2011**, *106*, 46–56. [CrossRef]
3. Hu, Y.; Bandla, S.; Godfrey, T.E.; Tan, D.; Luketich, J.D.; Pennathur, A.; Qiu, X.; Hicks, D.G.; Peters, J.H.; Zhou, Z. HER2 amplification, overexpression and score criteria in esophageal adenocarcinoma. *Mod. Pathol.* **2011**, *24*, 899–907. [CrossRef]
4. Dulak, A.M.; Schumacher, S.E.; van Lieshout, J.; Imamura, Y.; Fox, C.; Shim, B.; Ramos, A.H.; Saksena, G.; Baca, S.C.; Baselga, J.; et al. Gastrointestinal adenocarcinomas of the esophagus, stomach, and colon exhibit distinct patterns of genome instability and oncogenesis. *Cancer Res.* **2012**, *72*, 4383–4393. [CrossRef] [PubMed]
5. Dulak, A.M.; Stojanov, P.; Peng, S.; Lawrence, M.S.; Fox, C.; Stewart, C.; Bandla, S.; Imamura, Y.; Schumacher, S.E.; Shefler, E.; et al. Exome and whole-genome sequencing of esophageal adenocarcinoma identifies recurrent driver events and mutational complexity. *Nat. Genet.* **2013**, *45*, 478–486. [CrossRef] [PubMed]
6. Miller, C.T.; Moy, J.R.; Lin, L.; Schipper, M.; Normolle, D.; Brenner, D.E.; Iannettoni, M.D.; Orringer, M.B.; Beer, D.G. Gene amplification in esophageal adenocarcinomas and Barrett's with high-grade dysplasia. *Clin. Cancer Res.* **2003**, *9*, 4819–4825.
7. Wang, L.V.; Hu, S. Photoacoustic Tomography: In Vivo Imaging from Organelles to Organs. *Science* **2012**, *335*, 1458–1462. [CrossRef] [PubMed]
8. Fu, Q.; Zhu, R.; Song, J.; Yang, H.; Chen, X. Photoacoustic Imaging: Contrast Agents and Their Biomedical Applications. *Adv. Mater.* **2019**, *31*, 1805875. [CrossRef]
9. Weber, J.; Beard, P.C.; Bohndiek, S.E. Contrast agents for molecular photoacoustic imaging. *Nat. Methods* **2016**, *13*, 639–650. [CrossRef] [PubMed]
10. Li, J.; Rao, J.; Pu, K. Recent progress on semiconducting polymer nanoparticles for molecular imaging and cancer phototherapy. *Biomaterials* **2018**, *155*, 217–235. [CrossRef]
11. Lee, N.; Choi, S.H.; Hyeon, T. Nano-Sized CT Contrast Agents. *Adv. Mater.* **2013**, *25*, 2641–2660. [CrossRef]
12. Goodman, L.R. The Beatles, the Nobel Prize, and CT scanning of the chest. *Radiol. Clin. North. Am.* **2010**, *20*, 1–7. [CrossRef] [PubMed]
13. Lusic, H.; Grinstaff, M.W. X-ray-Computed Tomography Contrast Agents. *Chem. Rev.* **2013**, *113*, 1641–1666. [CrossRef]
14. Yu, S.B.; Watson, A.D. Metal-Based X-ray Contrast Media. *Chem. Rev.* **1999**, *99*, 2353–2378. [CrossRef]
15. Hallouard, F.; Anton, N.; Choquet, P.; Constantinesco, A.; Vandamme, T. Iodinated blood pool contrast media for preclinical X-ray imaging applications—A review. *Biomaterials* **2010**, *31*, 6249–6268. [CrossRef] [PubMed]
16. Singh, J.; Daftary, A. Iodinated Contrast Media and Their Adverse Reactions. *J. Nucl. Med. Technol.* **2008**, *36*, 69–74. [CrossRef]
17. Wang, C.L.; Cohan, R.H.; Ellis, J.H.; Adusumilli, S.; Dunnick, N.R. Frequency, Management, and Outcome of Extravasation of Nonionic Iodinated Contrast Medium in 69657 Intravenous Injections. *Radiology* **2007**, *243*, 80–87. [CrossRef] [PubMed]
18. Chen, J.; Gao, Z.; Li, G.; Wang, T.D. Dual-modal in vivo fluorescence and photoacoustic imaging using a heterodimeric peptide. *Chem. Commun.* **2018**, *54*, 13196–13199. [CrossRef]
19. Chen, J.; Zhou, J.; Gao, Z.; Li, X.; Wang, F.; Duan, X.; Li, G.; Joshi, B.P.; Kuick, R.; Appelman, H.D.; et al. Multiplexed Targeting of Barrett's Neoplasia with a Heterobivalent Ligand: Imaging Study on Mouse Xenograft in Vivo and Human Specimens ex Vivo. *J. Med. Chem.* **2018**, *61*, 5323–5331. [CrossRef] [PubMed]
20. Chen, J.; Jiang, Y.; Chang, T.S.; Joshi, B.; Zhou, J.; Rubenstein, J.H.; Wamsteker, E.J.; Kwon, R.S.; Appelman, H.; Beer, D.G.; et al. Multiplexed endoscopic imaging of Barrett's neoplasia using targeted fluorescent heptapeptides in a phase 1 proof-of-concept study. *Gut* **2021**, *70*, 1010–1013. [CrossRef]
21. Khlebtsov, N.; Dykman, L. Biodistribution and toxicity of engineered gold nanoparticles: A review of in vitro and in vivo studies. *Chem. Soc. Rev.* **2011**, *40*, 1647–1671. [CrossRef] [PubMed]
22. Chen, X.; Qoutah, W.W.; Free, P.; Hobley, J.; Fernig, D.G.; Paramelle, D. Features of Thiolated Ligands Promoting Resistance to Ligand Exchange in Self-Assembled Monolayers on Gold Nanoparticles. *Aust. J. Chem.* **2012**, *65*, 266–274. [CrossRef]
23. Zheng, M.; Huang, X. Nanoparticles Comprising a Mixed Monolayer for Specific Bindings with Biomolecules. *J. Am. Chem. Soc.* **2004**, *126*, 12047–12054. [CrossRef] [PubMed]
24. Duchesne, L.; Gentili, D.; Comes-Franchini, M.; Fernig, D.G. Robust Ligand Shells for Biological Applications of Gold Nanoparticles. *Langmuir* **2008**, *24*, 13572–13580. [CrossRef]
25. Zhou, G.; Liu, Y.; Luo, M.; Xu, Q.; Ji, X.; He, Z. Peptide-Capped Gold Nanoparticle for Colorimetric Immunoassay of Conjugated Abscisic Acid. *ACS Appl. Mater. Interfaces* **2012**, *4*, 5010–5015. [CrossRef] [PubMed]

26. Leduc, C.; Si, S.; Gautier, J.; Soto-Ribeiro, M.; Wehrle-Haller, B.; Gautreau, A.; Giannone, G.; Cognet, L.; Lounis, B. A Highly Specific Gold Nanoprobe for Live-Cell Single-Molecule Imaging. *Nano Lett.* **2013**, *13*, 1489–1494. [CrossRef]
27. Wang, Z.; Lévy, R.; Fernig, D.G.; Brust, M. Kinase-Catalyzed Modification of Gold Nanoparticles: A New Approach to Colorimetric Kinase Activity Screening. *J. Am. Chem. Soc.* **2006**, *128*, 2214–2215. [CrossRef]
28. Lévy, R.; Thanh, N.T.K.; Doty, R.C.; Hussain, I.; Nichols, R.J.; Schiffrin, D.J.; Brust, M.; Fernig, D.G. Rational and Combinatorial Design of Peptide Capping Ligands for Gold Nanoparticles. *J. Am. Chem. Soc.* **2004**, *126*, 10076–10084. [CrossRef]
29. Li, W.; Chen, X. Gold nanoparticles for photoacoustic imaging. *Nanomedicine* **2015**, *10*, 299–320. [CrossRef]
30. Popovtzer, R.; Agrawal, A.; Kotov, N.A.; Popovtzer, A.; Balter, J.; Carey, T.E.; Kopelman, R. Targeted Gold Nanoparticles Enable Molecular CT Imaging of Cancer. *Nano Lett.* **2008**, *8*, 4593–4596. [CrossRef]
31. Tsvirkun, D.; Ben-Nun, Y.; Merquiol, E.; Zlotver, I.; Meir, K.; Weiss-Sadan, T.; Matok, I.; Popovtzer, R.; Blum, G. CT Imaging of Enzymatic Activity in Cancer Using Covalent Probes Reveal a Size-Dependent Pattern. *J. Am. Chem. Soc.* **2018**, *140*, 12010–12020. [CrossRef]
32. Jing, L.; Liang, X.; Deng, Z.; Feng, S.; Li, X.; Huang, M.; Li, C.; Dai, Z. Prussian blue coated gold nanoparticles for simultaneous photoacoustic/CT bimodal imaging and photothermal ablation of cancer. *Biomaterials* **2014**, *35*, 5814–5821. [CrossRef]
33. Cheheltani, R.; Ezzibdeh, R.M.; Chhour, P.; Pulaparthi, K.; Kim, J.; Jurcova, M.; Hsu, J.C.; Blundell, C.; Litt, H.I.; Ferrari, V.A.; et al. Tunable, biodegradable gold nanoparticles as contrast agents for computed tomography and photoacoustic imaging. *Biomaterials* **2016**, *102*, 87–97. [CrossRef] [PubMed]
34. Foreman, K.W. A general model for predicting the binding affinity of reversibly and irreversibly dimerized ligands. *PLoS ONE* **2017**, *12*, e0188134. [CrossRef]
35. Zhu, J.; Fu, F.; Xiong, Z.; Shen, M.; Shi, X. Dendrimer-entrapped gold nanoparticles modified with RGD peptide and alpha-tocopheryl succinate enable targeted theranostics of cancer cells. *Colloids Surf. B Biointerfaces* **2015**, *133*, 36–42. [CrossRef]
36. Chanda, N.; Kattumuri, V.; Shukla, R.; Zambre, A.; Katti, K.; Upendran, A.; Kulkarni, R.R.; Kan, P.; Fent, G.M.; Casteel, S.W.; et al. Bombesin functionalized gold nanoparticles show in vitro and in vivo cancer receptor specificity. *Proc. Natl. Acad. Sci. USA* **2010**, *107*, 8760–8765. [CrossRef] [PubMed]
37. Kim, T.J.; Kim, H.Y.; Lee, K.W.; Kim, M.S. Multimodality Assessment of Esophageal Cancer: Preoperative Staging and Monitoring of Response to Therapy. *Radiographics* **2009**, *29*, 403–421. [CrossRef] [PubMed]
38. Xu, C.; Tung, G.A.; Sun, S. Size and Concentration Effect of Gold Nanoparticles on X-ray Attenuation As Measured on Computed Tomography. *Chem. Mater.* **2008**, *20*, 4167–4169. [CrossRef]

pharmaceuticals

Review

Positron Emission Tomography in Animal Models of Alzheimer's Disease Amyloidosis: Translational Implications

Ruiqing Ni [1,2]

1   Institute for Biomedical Engineering, ETH & University of Zurich, 8093 Zurich, Switzerland;
    ni@biomed.ee.ethz.ch
2   Institute for Regenerative Medicine, University of Zurich, 8952 Zurich, Switzerland

Abstract: Animal models of Alzheimer's disease amyloidosis that recapitulate cerebral amyloid-beta pathology have been widely used in preclinical research and have greatly enabled the mechanistic understanding of Alzheimer's disease and the development of therapeutics. Comprehensive deep phenotyping of the pathophysiological and biochemical features in these animal models is essential. Recent advances in positron emission tomography have allowed the non-invasive visualization of the alterations in the brain of animal models and in patients with Alzheimer's disease. These tools have facilitated our understanding of disease mechanisms and provided longitudinal monitoring of treatment effects in animal models of Alzheimer's disease amyloidosis. In this review, we focus on recent positron emission tomography studies of cerebral amyloid-beta accumulation, hypoglucose metabolism, synaptic and neurotransmitter receptor deficits (cholinergic and glutamatergic system), blood–brain barrier impairment, and neuroinflammation (microgliosis and astrocytosis) in animal models of Alzheimer's disease amyloidosis. We further propose the emerging targets and tracers for reflecting the pathophysiological changes and discuss outstanding challenges in disease animal models and future outlook in the on-chip characterization of imaging biomarkers towards clinical translation.

Keywords: Alzheimer's disease; amyloid-beta; animal model; astrocyte; blood–brain barrier; imaging; metabolism; microglia; neuroinflammation; neurotransmitter receptors; positron emission tomography; synaptic density

Citation: Ni, R. Positron Emission Tomography in Animal Models of Alzheimer's Disease Amyloidosis: Translational Implications. *Pharmaceuticals* **2021**, *14*, 1179. https://doi.org/10.3390/ph14111179

Academic Editor: Xuyi Yue

Received: 18 October 2021
Accepted: 15 November 2021
Published: 18 November 2021

Publisher's Note: MDPI stays neutral with regard to jurisdictional claims in published maps and institutional affiliations.

Copyright: © 2021 by the author. Licensee MDPI, Basel, Switzerland. This article is an open access article distributed under the terms and conditions of the Creative Commons Attribution (CC BY) license (https://creativecommons.org/licenses/by/4.0/).

1. Introduction

Alzheimer's disease (AD) is the most common cause of dementia, afflicting 50 million people worldwide [1]. AD is pathologically featured by amyloid-beta(Aβ) plaques and neurofibrillary tangles formed by hyperphosphorylated tau, gliosis, neurotransmitter deficits, and neuronal loss leading to cognitive impairment [2]. The abnormal accumulation of Aβ deposits, especially the neurotoxic oligomeric Aβ plays a crucial role in the disease pathogenesis in animal models and in patients with AD [3–6]. Recent advances in positron emission tomography (PET) using [^{18}F]fluorodeoxyglucose (FDG), tracers for Aβ pathology and tauopathy, structural magnetic resonance imaging, and cerebrospinal fluid biomarkers have provided valuable insights into the time course of the pathophysiology of AD continuum, assisted the early and differential diagnosis, and facilitated the development of therapeutics for AD [7–11]. Disease animal models recapitulating AD amyloidosis have been developed including transgenic APP/PS1, APP23, APPswe, J20, PS2APP, arcAβ, 5 × FAD, 3 × Tg mice, TgF344 and McGill-R-Thy1-APP rats [12–19], second-generation App^{NL-G-F}, App$^{hu/hu}$ knock-in mice [20,21], third-generation mouse models [22,23], as well as non-human primate model [24]. The animal models accumulate cerebral Aβ pathology, develop gliosis, metabolic and synaptic deficits, and cognitive impairment assessed by behavior tests, and facilitate the understanding of disease mechanisms and the development of treatment strategies. In this review, we focused on the recent development in PET imaging for Aβ, alterations in cerebral glucose metabolism, synaptic neurotransmitter receptors, blood–brain barrier, and neuroinflammation in rodent models of AD amyloidosis.

2. Amyloid Imaging

Ex vivo immunohistochemistry in brain tissues from amyloidosis mouse or rat models has revealed that Aβ pathology initiates first in the cortical region and spreads to the limbic region and finally to the cerebellum [25], in an animal line-dependent manner. A more pronounced load of Aβ deposits was observed in 5 × FAD mice, compared with that in APPswe mice [25–27]. In addition to the parenchymal Aβ plaques, cerebral amyloid angiopathy (CAA) is also observed in different amyloidosis animal models, especially in the APPDutch mice, Tg-SwDI, APP/London, APP23, arcAβ, and APPswe mice [28–30]. Several Aβ imaging tracers have been developed and applied in animal models of amyloidosis, including benzothiazole derivatives [^{11}C]PiB, [^{18}F]flutemetamol, [^{18}F]florbetaben, [^{18}F]FIBT, [^{18}F]florbetapir, [^{11}C]AZD2184, [^{18}F]FC119S and [^{18}F]flutafuranol, benzofuran derivatives [^{18}F]FACS and [^{18}F]FPZBF-2, benzoxazole derivatives [^{11}C]BF-227 and [^{18}F]MK3328, benzoselenazole derivative [^{18}F]fluselenamyl. hydroxyquinoline derivative [^{18}F]CABS13, imidazopyridine derivative [^{18}F]DRKXH1, as well as [^{64}Cu]labelled 8a′–8d and HYR-17 [31–47] (Table 1). Higher cortical amyloid PET tracer uptake was observed in various transgenic or knock-in animal models, compared with wild-type littermates, and validated by the ex vivo immunohistochemical stainings. Longitudinal comparative imaging studies across amyloidosis mouse lines have detected distinct Aβ spreading patterns in vivo. Snellman et al. showed a greater Aβ tracer dynamic range in the brain of the APP23 model, compared with that of APPswe and APP/PS1 models by PET imaging using both [^{11}C]PiB and [^{18}F]flutemetamol [38,48]. Brendel et al. compared four amyloidosis mouse strains (PS2APP, APPswe/PS1G384A, APP/PS1, APPswe) and found that PS2APP mice demonstrated greater dynamic changes in the longitudinal [^{18}F]florbetaben imaging study [49] (Figure 1a). Moreover, comparative studies of amyloid imaging tracers have been performed in a head-to-head manner in animal models, such as comparison among [^{11}C]PiB, [^{18}F]florbetaben, and [^{18}F]FIBT [36], and between [^{18}F]florbetaben and [^{18}F]flutemetamol [50]; similar patterns of tracer detection of cerebral Aβ distribution in the animal models have been reported in general.

As the commonly used amyloid tracers cannot differentiate parenchymal Aβ plaques and CAA [51], efforts have been made to develop CAA-specific tracers such as resorufin derivatives [52], [^{3}H]1, 2 [53]. One of the unsolved questions in Aβ imaging is the detection of small forms of Aβ aggregates. Biechele et al. recently indicated that the non-fibrillar Aβ (positive for 3552 antibodies) significantly impacted the [^{18}F]florbetaben PET signal, in addition to the Thiazine red-stained fibrillar Aβ, in App^{NL-G-F} and APP/PS1 mice from 3–12 months of age [54]. In addition to the small chemical dyes, PET using Aβ antibodies conjugated to a transferrin receptor antibody such as [^{124}I]RmAb158-scFv8D3 and [^{124}I]8D3-F(ab′)2-h158 have been developed to detect cerebral accumulation of small forms of Aβ. These tracers harbor an improved blood–brain barrier permeability and have been demonstrated in several transgenic mouse models of amyloidosis. Meier et al. demonstrated that the uptake of [^{124}I]RmAb158-scFv8D3 and [^{124}I]8D3-F(ab′)2-h158 was significantly higher in the cortical regions of transgenic ArcSwe mice, compared with non-transgenic littermates. In addition, the distribution pattern of PET using [^{124}I]8D3-F(ab′)2-h158 differs from that by PET using [^{11}C]PiB in the brain of tg-ArcSwe mice, indicating a preference to different types of Aβ by these two tracers (Figure 1b–d) [55]. Given the quantitativeness of in vivo microPET, non-invasive imaging using [^{18}F]florbetaben and [^{18}F]florbetapir for Aβ load have been applied for longitudinal monitoring of the treatment effect in animal models, such as using γ-secretase modulator and β-secretase 1 inhibitor [56–58]. Xu et al. recently demonstrated using [^{11}C]SGSM-1560 for in vivo detection of an increased level of γ-secretase in 5 × FAD, compared with wild-type mice [59] (Figure 1e–g).

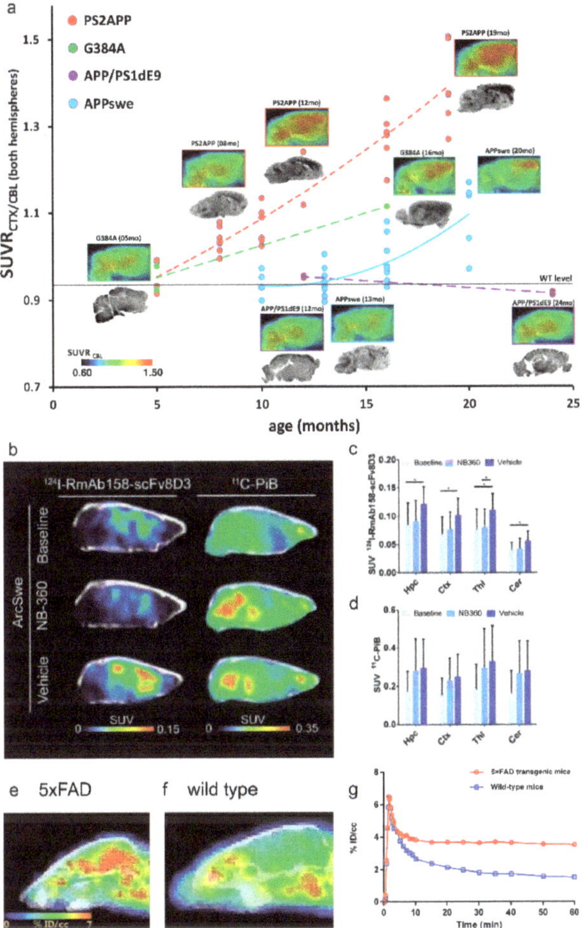

Figure 1. Imaging of amyloid-beta accumulation, and gamma-secretase in amyloidosis animal models of Alzheimer's disease: (**a**) multi-modal analysis of the four AD mouse strains in cross-sectional [^{18}F]florbetaben PET study. Images indicate group averaged sagittal PET slices, normalized to the cerebellum as well as ex vivo autoradiography. Dots indicate PET SUVR cortex/cerebellum in individual mice. Dashed lines express the estimated time-dependent progression in PS2APP, APPswe/PS1G384A, and APP/PS1 mice, fitted with a polynomial function. Reproduced from [49] with permission from PLOS One; (**b–d**) PET images and quantification of [^{11}C]PiB (40–60 min after injection) and [^{124}I]RmAb158-scFv8D3 scans (72 h after injection) expressed as standardized uptake value (SUV): (**b**) comparison of representative [^{124}I]RmAb158-scFv8D3 and [^{11}C]PiB PET images in ArcSwe animals; (**c,d**) quantification of [^{124}I]RmAb158-scFv8D3 and [^{11}C]PiB in hippocampus (Hpc), cortex (Ctx), thalamus (Thl) and cerebellum (Cer). * $p < 0.05$. Reproduced from [55] with permission from the Society of Nuclear Medicine and Molecular Imaging; (**e–g**) PET–CT imaging of γ-secretase in 5 × FAD and wild-type mice; (**e**) PET–CT image of 5 × FAD mice ($n = 2$) and (**f**) wild-type mice ($n = 2$) after i.v. injection of [^{11}C]SGSM-15606; (**g**) time activity curve of whole-brain uptake of [^{11}C]SGSM-15606 in h and i. Data are expressed as the percentage of injected dose per cubic centimeter (% ID/cc). Reproduced from [59] with permission from Rockefellfigureer University Press.

Table 1. Amyloid-beta PET imaging in animal models of Alzheimer's disease amyloidosis.

Tracer	Animal Model	References
[11C]PiB	APPswe mice	[37,48,60]
	5 × FAD mice	[61]
	APP/PS1 mice	[36,48,62–66]
	3 × Tg mice	[67]
	APP23 mice	[33,48,68]
	Aged non-human primates	[69,70]
[18F]florbetapir, AV-45	5 × FAD mice	[61,71]
	TASTPM mice	[72]
	APP/PS1 mice	[58,73]
	PS2APP mice	[49,74]
	APPswe mice	[49,75]
[18F]florbetaben, AV-1	App^{NL-G-F} mice	[54,74,76–78]
	APPswe/PS1G384A mice	[49]
	APP-SL70 mice	[74,79]
	TgF334 rats	[80]
	APP/PS1 mice	[49,54,66,81]
[11C]AZD2184	APPswe mice	[82]
	APP/PS1 mice	[83]
[18F]flutafuranol AZD4694, NAV4694	McGill-R-Thy1-APP rats	[43]
	APPswe mice	[42]
[18F]flutemetamol	APP23, APPswe, APP/PS1 mice	[37,38]
[18F]FIBT	APP/PS1 mice	[36]
[18F]FC119S	5 × FAD, APP/PS1 mice	[34,35]
[18F]FACT, [11C]BF-227	APP/PS1 mice	[84,85]
[18F]fluselenamyl	APP/PS1 mice	[86]
[124I]RmAb158-scFv8D3	Tg-ArcSwe, App^{NL-G-F} mice	[55]
[124I]8D3-F(ab')2-h158	Tg-ArcSwe, APPswe mice	[87]
[18F]CDA-3	5 × FAD mice	[88]
[64Cu]HYR-17	5 × FAD mice	[39]
[64Cu]8a'–8d	5 × FAD mice	[44]
[18F]DRKXH1	APP/PS1 mice	[40]
[18F]CABS13	APP/PS1 mice	[41]

[11C]AZD2184, 2-(6-[11C]methylaminopyridin-3-yl)-1,3-benzothiazol-6-ol; [11C]BF-227, [11C]2-(2-[2-Dimethylaminothiazol-5-yl]ethenyl)-6-(2-[fluoro]ethoxy)benzoxazole; [18F]CABS13, 2-[18F]fluoroquinolin-8-ol; [18F]CDA-3, [18F]croconium dye for amyloid; [18F]DRKXH1, 5-(4-(6-(2-[18F]fluoroethoxy)ethoxy)imidazo[1,2-alpha]pyridin-2-yl)phenyl; Fab, antigen-binding fragment; [18F]FACT, 2-[(2-{(E)-2-[2-(dimethylamino)-1,3-thiazol-5-yl]vinyl}-1,3-benzoxazol-6-yl)oxy]-3-[18F]fluoropropan-1-ol; [18F]FIBT, 2-(p-methylaminophenyl)-7-(2-[18F]fluoroethoxy)imidazo-[2,1-b]benzothiazole; [18F]FC119S, 2-[2-(N-monomethyl)aminopyridine-6-yl]-6-[(S)-3-[18F]fluoro-2-hydroxypropoxy]benzothiazole; [18F]florbetaben, 4-[(E)-2-[4-[2-[2-(2-[18F]fluoranylethoxy)ethoxy]ethoxy]phenyl]ethenyl]-N-methylaniline; [18F]florbetapir, 4-[(E)-2-[6-[2-[2-(2-[18F]fluoranylethoxy)ethoxy]ethoxy]pyridin-3-yl]ethenyl]-Nmethylaniline; [18F]fluselenamyl, (Z)-5-(2-(5-(2-[18F]fluoroethoxy)benzo[d][1,3]selenazol-2-yl)vinyl)-N,N-dimethylpyrimidin-2-amine; [18F]flutafuranol, 2-[2-[18F]fluoro-6-(methylamino)-3-pyridinyl]-1-benzofuran-5-ol; [18F]flutemetamol, 2-[3-[18F]fluoro-4-(methylamino)phenyl]-1,3-benzothiazol-6-ol; [11C]PiB, Pittsburgh compound B, 2-[4-([11C]methylamino)phenyl]-1,3-benzothiazol-6-ol; scFv, single chain fragment variable.

3. Cerebral Glucose Metabolism Imaging

Brain glucose dysregulation plays an important role in AD [89]. Post-mortem studies reported higher levels of brain tissue glucose concentration, lower levels of glucose transporter 3, and glycolytic flux in the brain from patients with AD, compared with controls, associating with the severity of AD pathology [89]. [18F]FDG PETs have been routinely used for detecting the reduced cerebral glucose metabolism (CMRglc) in disease-specific brain regions in patients with AD, Frontotemporal dementia, and Parkinson's disease to improve the diagnostic accuracy [9,90]. In lab settings, [18F]FDG PET have been assessed along with Aβ imaging in various amyloidosis rodent models such as APPswe mice, 5 × FAD, APP/PS1, 3 × Tg, Tg4-42, TASTPM mice, and McGill-R-Thy1-APP rats [43,66,71,91–95] (Table 2) (Figure 2a). However, [18F]FDG uptake is known to be highly sensitive to experimental conditions such as anesthesia and handling, as well as genotype, age, and gender of the animal models [96]. Most of the studies in rodent amyloidosis models reported a global reduction in CMRglc, although few exceptions of increased CMRglc (associating with gliosis) were also reported [61]. A recent study by Xiang et al. further showed that microglial activation states drive glucose uptake and [^{18}F]FDG-PET alterations [97].

Table 2. PET imaging in of neurotransmitter receptors, blood–brain barriers, enzymes, metabolism, and synaptic density in animal models of Alzheimer's disease amyloidosis.

Target	Tracer	Animal Model	References
CMRglc	[^{18}F]FDG	3 × Tg mice	[94,98–102],
		APPswe mice	[92]
		APP/PS1 mice	[58,66,72,103–106]
		Tg4-42 mice	[91,107]
		5 × FAD mice	[61,71,81,108,109]
		3 × Tg rats	[110]
		APP23 mice	[111]
		McGill-R-Thy1-APP rats	[43]
		TASTPM mice	[72,112]
		Aged monkey	[70]
SV2A	[^{11}C]UCB-J	APP/PS1 mice	[113]
		ArcSwe, Tg-L61 mice	[114]
	[^{18}F]SynVesT-1	APP/PS1 mice	[115]
mGluR5	[^{18}F]FPEB	5 × FAD mice	[116,117]
		APP/PS1 mice	[118]
	[^{11}C]ABP688	Tg-ArcSwe mice	[119]
α7nAChR	[^{11}C]MeQAA	Aged monkey	[69]
	[^{18}F]ASEM	TgF344 rats	[80]
AChE	[^{11}C]MP4A	APP23 mice	[120]
BChE	[^{11}C]4	5 × FAD mice	[121]
GABAR	[^{11}C]flumazenil	APP23 mice	[120]
GSM	[^{11}C]SGSM-1560	5 × FAD mice	[59]
IIa HDAC	[^{18}F]TFAHA	3 × Tg mice	[122]
GLP-1R	[^{18}F]FBEM-Cys^{39}-exendin-4	3 × Tg mice	[102]
D$_2$R	[^{18}F]fallypride	3 × Tg, 5 × FAD mice	[102,117]
MC1	[^{18}F]BCPP-EF	Aged monkey, SAMP10 mice	[69,70,123,124]
Copper	[^{64}Cu]GTSM	TASTPM mice	[125]
MT	[^{11}C]MPC-6827	J20 mice	[126]
GSK3β	[^{11}C]OCM-44, [^{3}H]PF-367	APPswe mice	[127]
	[^{11}C]2	3 × Tg mice	[128]
RAGE	[^{11}C]FPS-ZM1	APPswe mice	[129]
ABCC1	[^{11}C]BMP	APP/PS1 mice	[130]
ABCG2	[^{11}C]erlotinib	APP/PS1 mice	[131]
P-GP ABCB1	[^{11}C]tariquidar	APP/PS1 mice	[131]
	[^{11}C]metoclopramide	APP/PS1 mice	[132]
	(R)-[^{11}C]verapamil	APP/PS1 mice	[133]

ABC: ATP-binding cassette transporter; α7 nAChR: α7 icotinic acetylcholine receptor; AChE, acetylcholine esterase; BChE: butyrylcholinesterase; CMRglc: cerebral metabolic rate of glucose; D2: dopamine receptor D2; [^{18}F]FDG: [^{18}F]fluorodeoxyglucose; GABAR: gamma-aminobutyric acid receptor; GLP-1R: glucagon-like peptide-1 receptor; GSK3β: glycogen synthase kinase-3b; GSM: γ-secretase modulator; IIa HDAC: class IIa histone deacetylases; LPS: lipopolysaccharide; MC1: mitochondrial complex 1; mGluR5: metabotropic glutamate receptor type 5; MT: microtubule; NP: nanoparticle; P-GP: P-glycoprotein; SV2A: synaptic vesicle glycoprotein 2A.

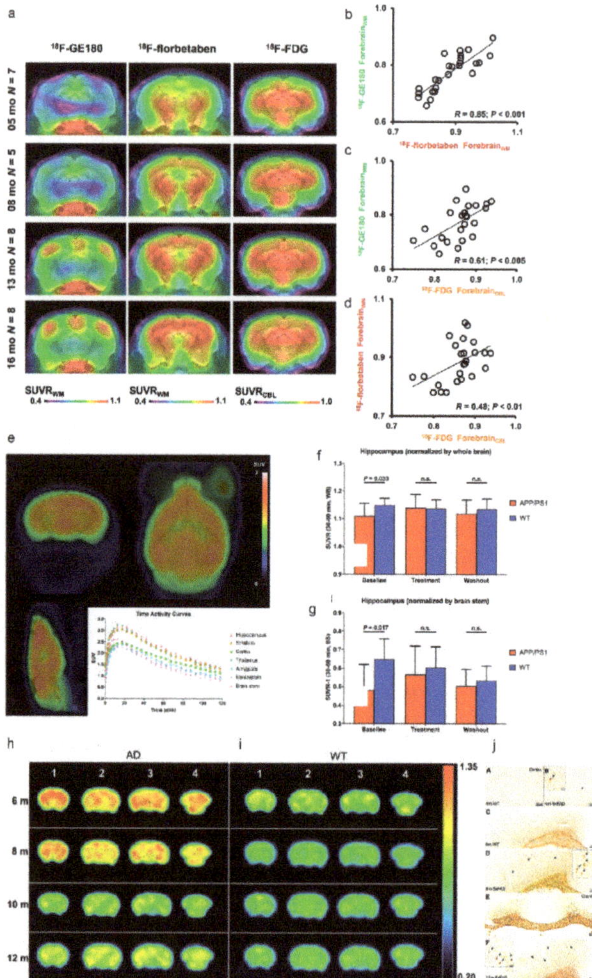

Figure 2. In vivo imaging of translocator protein, cerebral glucose metabolism, synaptic density, and butyrylcholinesterase, in amyloidosis animal models of Alzheimer's disease: (**a–d**) [^{18}F]GE-180, [^{18}F]florbetaben, and [^{18}F]FDG PET imaging at different ages of PS2APP animals; (**a**) coronal planes of mean SUVR maps projected on an MRI mouse atlas (grayscale); (**b–d**) correlations between the different forebrain radiotracer SUVR for all PS2APP mice. Reproduced with permission [134] with permission from the Society of Nuclear Medicine and Molecular Imaging; (**e–g**) representative [^{11}C]UCB-J PET image and time-activity curve in APP/PS1 mice; (**a**) static SUV image (30–60 min after injection) overlaid on atlas brain MR image; (**b**,**c**) hippocampal SUVRs in wild-type and APP/PS1 mice during baseline, treatment, and washout phases: whole-brain SUVR (**b**) and brain stem SUVR (**c**). Reproduced from [113] with permission Society of Nuclear Medicine; (**h–j**) PET images for BChE imaging in 5 × FAD mice; (**c**,**d**) axial view of PET images in 5 × FAD and wild-type mice after i.v. administration of [^{11}C]4 at different ages; (**e**) Staining for BChE enzymatic activity in 4-, 8-, and 12-month-old brains of wild-type (**A**,**C**,**E**) and 5 × FAD mice (**B**,**D**,**F**) using the Karnovsky–Roots method. BChE staining showed an increase in enzyme activity in the cerebral cortex of 5 × FAD at different ages in comparison with wild-type (**A** to **F**) mice. Magnified images show the co-occurrence of plaques with BChE enzyme activity in different regions of the cerebral cortex (**B**,**D**,**F**) Reproduced from [121] with permission from Ivyspring International Publisher.

4. Synaptic and Neurotransmitter Receptor Deficits

4.1. Synaptic Vesicle Glycoprotein 2A

Synapse loss is reported in the post-mortem frontal cortex of patients with AD, correlating with cognitive severity [135]. Synaptic vesicle glycoprotein 2A (SV2A) is located at the synapses across the entire brain and is the binding site for the antiepileptic drug levetiracetam [136]. SV2A involves in vesicle trafficking exocytosis and is crucial for neurotransmission and postnatal brain development [137]. Mendoza-Torreblanca et al. suggested that SV2A either regulates the presynaptic Ca^{2+} levels during repetitive activity or is a target for residual Ca^{2+}. Higher loads of cerebral $A\beta$ deposits have been reported in the brain of SV2A knock-out mice, compared with control littermates [138]. A 40% reduction in SV2A signal by PET using $[^{11}C]$UCB-J was observed in the hippocampus in patients with AD, compared with cognitively normal control cases [139,140]. Kong et al. showed that SV2A over-expression was associated with the downregulation of β-site APP-cleaving enzyme 1 and apolipoprotein E genes, indicating that SV2A impacts $A\beta$ production. However, Nowack et al. showed that overexpression of SV2A increased synaptic levels of the calcium-sensor protein synaptotagmin, resulting in a neurotransmission deficit [141]. Thus, modulation of SV2A as a potential treatment requires careful dosing and close monitoring of the SV2A levels. Several SV2A PET imaging tracers have been developed including $[^{11}C]$UCB-J, $[^{18}F]$UCB-H [142], $[^{18}F]$SynVesT-1 [143], $[^{18}F]$SDM-8 [144], and $[^{18}F]$MNI-1126 [145] (Table 2). PET measures of $A\beta$ deposition were found associated with regional synaptic density measured by $[^{11}C]$UCB-J in patients with early AD [139,146]. Few studies have reported on SV2A imaging in AD animal models. Bertoglio et al. demonstrated that $[^{11}C]$UCB-J is bound specifically to SV2A in mouse brain and that the radioligand binding can be quantified by kinetic modeling using an image-derived input function [147]. Toyonaga et al. showed that in vivo $[^{11}C]$UCB-J detected reduced levels of SV2A in APP/PS1 mice and the treatment effects of tyrosine kinase Fyn inhibitor Saracatinib in mitigating the $[^{11}C]$UCB-J reduction [113] (Figure 2e–g). Xiong et al. recently compared the $[^{11}C]$UCB-J binding in tg-ArcSwe and wild-type mice [114] and did not observe a clear difference between the two groups. $[^{18}F]$SynVesT-1, $[^{18}F]$analog of $[^{11}C]$UCB-J, has demonstrated favorable in vivo brain uptake in non-human primate [148]. Sadasivam et al. showed a lower $[^{18}F]$SynVesT-1 standard uptake value (SUV) across the whole brain of APP/PS1 mice, compared with non-transgenic mice [115]. The results from a static (30–60 min post-injection) $[^{18}F]$SynVesT-1 PET scan were found comparable to kinetic modeling results [115].

4.2. Glutamate Receptors

The glutamate receptors are classified into the N-methyl-D-aspartate receptor (NMDAR), α-amino-3-hydroxy-5-methyl-4-isoxazolepropionate (AMPA)-kainate receptor, and metabotropic glutamate receptors (mGluRs). The glutamate receptors mediate excitatory neurotransmission, involve in multiple second messenger systems, and are essential in learning and memory [149,150]. Glutamate excitotoxicity and disruption of the glutamate receptor-mediated normal signaling are implicated in AD [151,152]. $A\beta$ reduces glutamatergic transmission and inhibits synaptic plasticity [153,154]. Direct interaction between $A\beta$ oligomers and glutamate receptors including NMDAR [155], mGluR subunit mGluR5 [156], AMPA receptor subunit GluA3 [157], and GluA1 [158] have been demonstrated, leading to impaired synaptic plasticity in the animal models [159]. Chronic pharmacological inhibition of mGluR5 has been shown to prevent cognitive impairment and reduce pathological development in APP/PS1 mice [160]. Thus, glutamate receptors have been important targets for AD therapeutics. Several imaging tracers for glutamate receptors have been developed, including $[^{11}C]$K-2 [161] and $[^{11}C]$HMS011 [162] for AMPA receptor, $[^{18}F]$GE-179 [163] and $[^{18}F]$PK-209 for NMDAR [164], $[^{11}C]$Me-NB1 [165] for NMDAR GluN1/GluN2B subunits [166], as well as $[^{18}F]$FPEB, $[^{11}C]$ABP688, and $[^{18}F]$PSS232 for mGluR5 [167–169]. In patients with AD, PET using $[^{18}F]$FPEB [170] and $[^{11}C]$ABP688 [171] revealed consistent reductions in regional mGluR5 binding in the hippocampus and amygdala, compared with non-demented controls. Sofar only mGluR5 imaging has been reported in amyloidosis animal models and showed conflicting results probably due to different animal models utilized (Table 2). Lee et al. demonstrated an age-dependent 35% decrease in the level of [18F]FPEB measures of mGluR5 in the cortical and subcortical brain areas in 5 × FAD mice at 9 months of age, compared with 3 months of age, validated by ex vivo assessment of mGluR5 protein expression levels [116]. However, Varlow et al. showed that [18F]FPEB uptake increased in the brain of 10-month-old APP/PS1 mice, compared with controls [118]. Fang et al. reported similar levels of [18F]FPEB uptake in the brain of Tg-ArcSwe mice, compared with control mice at different ages [119]. However, immunoblotting results indicated that the level of mGluR5 in Tg-ArcSwe mouse brain lysate was higher, compared with control mice, at 12 months of age, not at 8 and 16 months of age [119]. Further studies are needed to elucidate the dynamic alteration in glutamate receptors in AD animal models.

4.3. Cholinergic System

The cholinergic system is essential for learning, memory formation, attention, and regulating inflammation [172]. The cholinergic system includes nicotinic acetylcholine receptors (nAChR), muscarinic acetylcholine receptors (mAChR), acetylcholinesterase (AChE), and butyrylcholinesterase (BChE). $\alpha7$ nAChR and $\alpha4\beta2$ nAChR are the most abundant nAChR subtypes in the brain. The cholinergic system is impaired early in AD associated with the cognitive, behavioral, and global functioning decline [172–174]. Reduced basal forebrain cholinergic neurons, increased levels of $\alpha7$ nAChR [175,176], and reduced levels of M1 mAChR [177] were reported in the cortical regions of post-mortem brain from AD patients, compared with control. Interaction between $\alpha7$ and $\alpha4\beta2$ nAChR and different forms of Aβ aggregates have also been reported [178–181]. Several recent PET tracers, including [11C]NS14492 [182], [11C](R)MeQAA [69], and [18F]ASEM for $\alpha7$ nAChR [183], [11C](+)3-MPB [184] and [18F]fluorobenzyl-dexetimide [185] for mAChR, [11C]LSN3172176 [186] for M1 mAChR, and [11C]MK-6884 for M4 mAChR [187] have been developed (Table 2). PET using [11C]nicotine imaging showed that the cortical nAChR binding correlated with the cognitive function of attention in patients with mild AD [188]. Few in vivo PET studies for the cholinergic system have been performed in AD models. Nishiyama et al. demonstrated higher [11C](R)-MeQAA brain uptake in the thalamus, hippocampus, striatum, and cortical regions, along with increased [11C]PiB detection of Aβ load and impaired [18F]BCPP-EF binding to mitochondrial complex 1 in the brain of aged monkey [69]. Chaney et al. demonstrated lower levels of [18F]ASEM in TgF344 rats, compared with wild-type at 18 months of age [80]. Rejc et al. recently reported increased levels of BChE along with Aβ accumulation using [11C]4 and [18F]florbetaben, respectively, in brain of 5 × FAD mice at 4–12 months of age, compared with wild-type mice [121] (Figure 2h–j). In comparison, comparable levels of AChE were observed in APP23, compared with wild-type mice at 10–13 months of age, assessed by PET using [11C]MP4A [120].

5. Blood–Brain Barrier

Blood–brain barrier (BBB) is impaired at an early disease stage in AD [189,190]. Whether the BBB dysfunction is secondary to Aβ pathology or a causal factor has not been fully elucidated. In amyloidosis animal models of AD, BBB disruption is observed in mouse models such as arcAβ and APP/PS1 but not prevalent in certain mouse lines such as the PS2APP line [191,192]. Several receptors presented in the BBB have been explored as PET imaging targets, such as adenosine triphosphate-binding cassette (ABC) transporter ABCC1, ABCG2, ABCB1 (P-glycoprotein, P-gp), and receptor for advanced glycation end-products (RAGE). P-gp plays an important role in the clearance and efflux of Aβ from the brain into the blood across the brain endothelial luminal membrane [193]. The levels of P-gp expression and activity were found to be decreased in the brains of AD patients, compared with that in control cases, as well as in the APP mouse model, compared with wild-type mice [194]. Several P-gp tracers such as (R)-O-[18F]fluoroethylnorverapamil, (R)-

N-[^{18}F]fluoroethylverapamil, (R)-[^{11}C]verapamil, [^{11}C]tariquidar, [^{11}C]metoclopramide, and [^{18}F]MC225 have been developed [130–133,195–198] (Table 2). Zoufal et al. demonstrated an age-dependent reduction in the cerebral P-gp function in APP/PS1 mice, compared with wild-type mice assessed by PET using (R)-[^{11}C]verapamil [133] (Figure 3a–d) and by using [^{11}C]metoclopramide [132].

However, (R)-[^{11}C]verapamil showed suboptimal brain uptake, and further improvement and evaluation of P-gp function using novel tracers with improved properties are needed. In addition, PET using 6-bromo-7-[^{11}C]methylpurine ([^{11}C]BMP) showed an increased level of ABCC1 along with [^{11}C]PiB detection of an increased level of Aβ pathology in the brain of APP/PS1 mice, compared with wild-type mice [130]. The increase in the ABCC1 level has been assumed to be related to the upregulation of its expression in astrocytes as a protective mechanism. Imaging of ABCG2 by PET using [^{11}C]erlotinib has been reported in APP/PS1 mice: no alteration in the level of ABCG2, compared with wild-type mice, was observed [131].

Receptor for advanced glycation end products (RAGE) is a BBB transporter and a binding site for advanced glycation end products and mediates Aβ transportation across the BBB into the brain [199,200]. The expression level of RAGE was found increased in post-mortem AD brains, compared with that in control cases [199]. RAGE tracers such as [^{11}C]FPS-ZM1 [201], [^{18}F]RAGER [202], [^{18}F]InRAGER [203], and [^{64}Cu]Rho-G4-CML nanoparticle (multimodal) have been developed [204]. The only imaging study conducted in the AD animal model by Luzi et al. showed that [^{11}C]FPS-ZM1 uptake in the brain of APPswe was similar, compared with that of wild-type mice [129]. Further development and studies are needed to evaluate RAGE imaging tracers in AD animal models and in patients with AD.

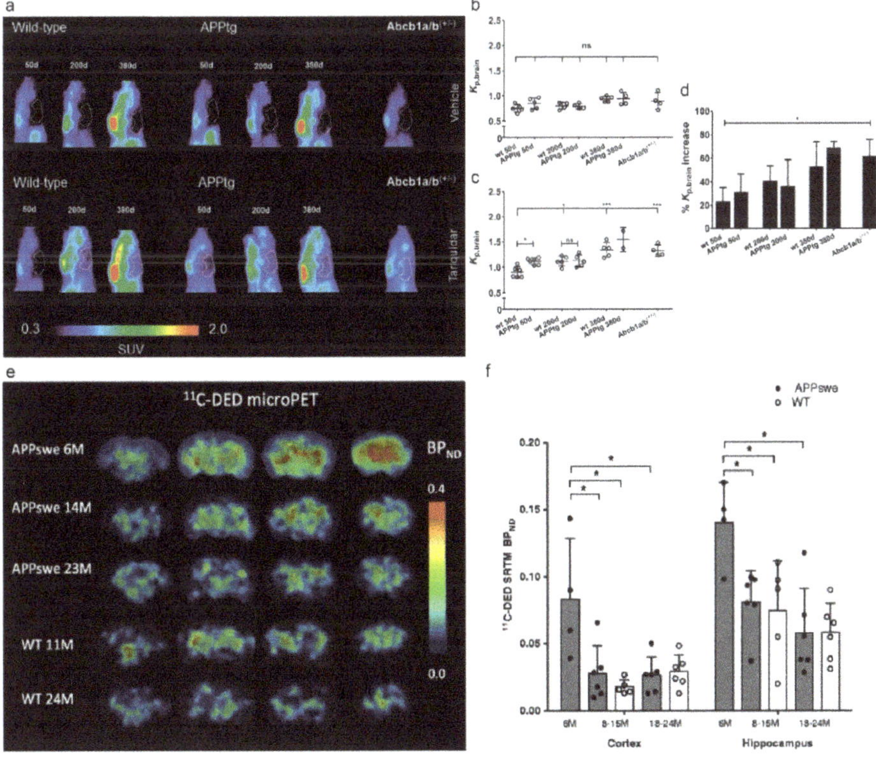

Figure 3. *Cont.*

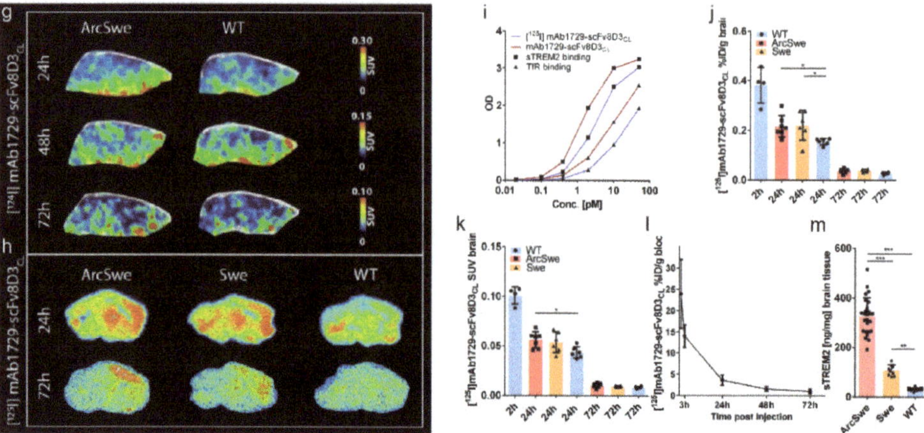

Figure 3. In vivo imaging of blood–brain barrier, astrocytosis, and triggering receptors expressed on myeloid cells (TREM) 2 in amyloidosis animal models of Alzheimer's disease: (**a–d**) imaging of P-glycoprotein (P-gp, ABCB1) using (*R*)-[^{11}C]verapamil; (**a**) sagittal PET summation images (0–60 min) of wild-type and APP/PS1 mice aged 50, 200 and 380 days and Abcb1a/b$^{(+/-)}$ mice pre-treated i.v. with vehicle (**b**) or tariquidar (4 mg/kg) (**c**) at 2 h before start of the PET scan. The whole-brain region is highlighted as a white line. In (**d**), the mean percentage increase in Kp, brain of individual tariquidar-treated animals relative to mean Kp, brain value of vehicle group is shown. Ns: not significant, * $p < 0.05$, *** $p < 0.001$. Reproduced from [133] from Sage Publication; (**e,f**) [^{11}C]deuterium-l-deprenyl ([^{11}C]DED) microPET imaging in APPswe and wild-type (WT) mice: (**e**) [^{11}C]DED microPET coronal parametric BP$_{ND}$ maps images; (**f**) [^{11}C]DED binding in the cortex and hippocampus, expressed as BP$_{ND}$, obtained from simplified reference tissue model of [^{11}C]DED using the cerebellum as a reference region, in three groups of APPswe mice aged 6-, 8–15, and 18–24 months and two groups of wild-type mice aged 8–15 and 18–24 months. Significant differences between groups are indicated by * $p < 0.05$. Reproduced from [82] with permission from Springer Nature; (**g–m**) PET imaging of triggering receptor expressed on myeloid cells 2 (TREM2) level in ArcSwe, Swe, and wild-type mice; (**g**) representative SUV scaled sagittal PET images with [^{124}I]mAb1729-scFv8D3CL at 24 h, 48 h, and 72 h after injection; (**h**) radioligand distribution in brain tissue displayed in sagittal ex vivo autoradiography images in ArcSwe, Swe, and wild-type animals at 24 h and 72 h after injection (**h**); (**i**) binding comparison of [^{125}I]mAb1729-scFv8D3CL and unlabelled mAb1729-scFv8D3CL by using ELISA. Percent of injected dose (**j**) and SUV (**k**) of [^{125}I]mAb1729-scFv8D3CL in brain 2 h, 24 h, and 72 h after injection; (**l**) level of [^{124}I]mAb1729-scFv8D3CL in blood, which was sampled 1 h, 3 h, 24 h, 48 h, and 72 h after injection; (**m**) TREM2 levels in TBS extracted brains of ArcSwe, APPSwe, and wild-type mice at the age of 18 months. Reproduced from [205] with permission from Springer Nature.

6. Neuroinflammation Imaging

Several recent articles have provided thorough reviews on neuroinflammation PET imaging in AD patients and AD animal models [206–211]. Thus, here, we discuss briefly the recent development in neuroinflammation imaging in AD amyloidosis animal models. Neuroinflammation plays an important role in the pathogenesis of AD and appears early in the development of the disease [212–214]. Microglia are the resident macrophages in the central nervous system, engulf Aβ plaques, and are important for maintaining brain homeostasis [214,215]. Recent single-cell sequencing and transcriptomics have demonstrated a transcriptionally distinct and neurodegeneration-specific profile of microglia termed disease-associated microglia (DAM) [216–218]. The 18 kDa translocator protein (TSPO) located on the outer mitochondria membrane of microglia has been the most investigated target for microgliosis PET imaging. Three generations of TSPO tracers have been developed with improved properties: the first-generation (R)-[^{11}C]PK11195 [219]; the second-generation [^{11}C]PBR28 [220], [^{18}F]FEDAA1106 [68], [^{18}F]DPA-714 [105]; the third-generation [^{18}F]GE-180 [134] (Figure 2a) and [^{11}C]ER176 [221]. PET using various 18 kDa translocator protein (TSPO) tracers have demonstrated an early microgliosis preceding the Aβ deposition in several animal models of amyloidosis including APP23, hAPP-J20, APPSL70, App^{NL-G-F}, and PS2APP mice [76,77,215,222–225]. Sacher et al. showed an asymmetry and hemispheric predominance of Aβ accumulation detected by using [^{18}F]florbetaben accompanied by microglial activation assessed by using [^{18}F]GE-180 in five mouse lines, including APP/PS1, PS2APP, APP-SL70, APPswe transgenic mice, and App^{NL-G-F} knock-in mice [74,226]. Due to the diverse cellular location of TSPO expression on astrocytes and endothelial cells, in addition to that on microglia, tracers specific for microglial expression and of the disease-associated profile are of high interest [227–229]. Emerging targets and tracers include [^{11}C]SW125M139 for purinergic P2X7 receptor [230,231], [^{124}I] mAb1729-scFv8D3CL for triggering receptors expressed on myeloid cells (TREM) 2, [^{11}C]AZD1283 for purinergic P2Y12 receptor [232], [^{11}C]CPPC [233] and [^{11}C]GW2580 [234] for colony-stimulating factor 1 receptor, [^{11}C]KTP-Me for cyclooxygenase 1 [235] have been reported in AD animal models. Meier et al. showed a higher expression level of triggering receptor expressed on myeloid cells 2 (TREM2) in the brain from ArcSwe mice, compared with wild-type mice at 24 h, 48 h, and 72 h after injection by autoradiography using [^{124}I] mAb1729-scFv8D3CL [205] (Figure 3g–m). The tracers for purinergic P2Y12 receptor [232] show a more specific microglial localization and thus are of high potential.

7. Discussion

In vivo longitudinal imaging in animal models of AD amyloidosis has provided valuable insights on the spatiotemporal links between different pathophysiology. A range of molecular imaging tracers for neuroinflammation, synaptic density, and neurotransmitter receptor deficits have been developed and provided a comprehensive picture of AD [11,210,236,237]. In addition to the aforementioned targets, many emerging targets show potential as indicators for pathological alterations in AD and are yet to be further investigated in amyloidosis animal models. These include (1) microgliosis; (2) astrocytosis; (3) metal dysregulation and copper trafficking, e.g., using [^{64}Cu]GTSM [125]; (4) reactive oxygen species [238] and pH alterations [239]; (5) microtubule using [^{11}C]MPC-6827, [^{11}C]HD-800, [^{11}C]WX-132-18B [126,240,241]; (6) sigma 1 receptor using [^{11}C]HCC0929, [^{18}F]FTC-146, [^{18}F]IAM6067 and [^{11}C]SA4503 [242–244]; (7) mitochondria imaging using [^{18}F]BCPP-EF [123]; 8) glycogen synthase kinase-3 imaging using [^{11}C]2, [^{11}C]OCM-44, [^{3}H]PF-367 [128,245].

Among the aforementioned emerging microgliosis tracers, the tracers for purinergic P2X7 receptor [230,231], P2Y12 receptor [232] are of high interest due to their specific cellular location on microglia. In addition, astrocytes are essential for maintaining the homeostasis, synaptic plasticity, and inflammatory response in the central nervous system [246] and play key roles in the onset and progression of AD. Reactive astrocytes show disease-associated profiles and exert dynamic functions (neuroprotection and neurotoxicity) in AD [247–251]. Few studies have been reported on PET imaging of astrocytosis in AD animal models. PET using irreversible monoamine oxidase B (MAO-B) inhibitors [^{11}C]deuterium-L-deprenyl (DED) showed an early astrocytosis preceding the Aβ accumulation assessed by using [^{11}C]AZD2184 in the brain of APPswe at 6 months of age, compared with wild-type mice (Figure 3e,f). A similar finding of an early increase in [^{11}C]DED binding was reported in Tg-ArcSwe mice, compared with wild-type littermates [252]. Several novel MAO-B tracers have been developed including [^{11}C]SMBT-1 [253] based on (S)-[^{18}F]THK5117 structure [254] and [^{18}F]6 [255]. In addition, a novel astrocytic tracer [^{11}C]BU99008, which targets imidazoline-2 binding sites (I2BS), has shown specific and high-affinity binding properties in post-mortem characterization [256] and demonstrated promising results in the recent in vivo PET studies in patients with AD [257,258].

Several earlier studies have reported the complicated temporal and spatial association between [^{18}F]FDG, TSPO, and amyloid accumulation: reduced [^{18}F]FDG uptake, increased Aβ deposition using [^{11}C]PiB or [^{18}F]florbetaben [64,134], and increased microglial activation using [^{18}F]GE-180 [134] (Figure 2a–d), and [^{18}F]DPA-714 has been reported in animal models [105]. Tsukada et al. reported reduced [^{18}F]FDG uptake, increased [^{11}C]PiB measures of Aβ deposition, increased [^{11}C]DPA-713 for microglia activation, and reduced [^{18}F]BCPP-EF for mitochondrial complex 1 in the brain of aged monkeys [70]. Given the recent finding of microglial [^{18}F]FDG-PET uptake [97], further studies may potentially use [^{18}F]FDG-PET for monitoring the microglial status in treatment targeting at microglia. In addition, markers that can specifically reflect synaptic and neuronal function are needed. Amyloidosis animal models show cortical, hippocampal atrophy, and enlargement of ventricle assessed by using structural magnetic resonance imaging, although to a less extent, compared with that in tauopathy animal models [259,260]. Multi-modal imaging [261] or multi-tracer imaging studies combining microgliosis, [^{18}F]FDG, and SV2A imaging to provide more comprehensive functional and molecular readouts are thus highly desired [262].

The challenges in bridging the translational gaps of PET imaging in rodent models and in patients with AD may include (1) different rodent models of AD demonstrated divergent time courses and patterns of pathophysiological development. Thus, rational selection of optimal animal models and age for investigation is thus critical in PET imaging studies in tracer evaluation [263]; (2) in addition, species difference in cell types, protein expression level, available binding sites, and post-translational modification of the target added to the complexity [264]. For example, the Aβ deposits formed in the APP mouse models and in aged primates are structurally different from that in the brain from patients with AD [265]. Thus, models that better recapitulate the human AD pathology will greatly boost the AD research, such as the Aβ-KI mouse modeling late-onset AD [23] and the third-generation mouse model [22]; databases of comprehensive deep phenotyping in disease animal models such as "MODEL-AD" by the Alzheimer Consortium Think Tank [266,267] (www.model-ad.org/, accessed on 15 October 2021) are instrumental in facilitating the translational research. Systems biology approaches, including single-cell sequencing, transcriptomics, biochemical characterization, and behavioral assessments, along with in vivo imaging data, will provide accurate interpretation of the readouts [268].

8. Conclusions

We provided an overview of PET imaging in animal models of AD amyloidosis, highlighting recent development in visualizing Aβ, cerebral glucose metabolism, synaptic and neurotransmitter receptor deficits, BBB impairment, and neuroinflammation, and proposed outstanding challenges for future development to increase the translational power of preclinical PET in AD.

Funding: R.N. received funding from Helmut Horten Stiftung, Vontobel Stiftung.

Institutional Review Board Statement: Not applicable.

Informed Consent Statement: Not applicable.

Data Availability Statement: All data are contained within the article.

Conflicts of Interest: The author declares no conflict of interest.

References

1. Bhatt, J.; Comas Herrera, A.; Amico, F.; Farina, N.; Wong, J.; Orange, J.B.; Gaber, S.; Knapp, M.; Salcher-Konrad, M.; Stevens, M.; et al. *The World Alzheimer Report 2019: Attitudes to Dementia*; Alzheimer's Disease International: London, UK, 2019.
2. Scheltens, P.; De Strooper, B.; Kivipelto, M.; Holstege, H.; Chételat, G.; Teunissen, C.E.; Cummings, J.; van der Flier, W.M. Alzheimer's disease. *Lancet* **2021**, *397*, 1577–1590. [CrossRef]
3. Lesné, S.; Koh, M.T.; Kotilinek, L.; Kayed, R.; Glabe, C.G.; Yang, A.; Gallagher, M.; Ashe, K.H. A specific amyloid-β protein assembly in the brain impairs memory. *Nature* **2006**, *440*, 352–357. [CrossRef] [PubMed]
4. Haass, C.; Selkoe, D.J. Soluble protein oligomers in neurodegeneration: Lessons from the Alzheimer's amyloid beta-peptide. *Nat. Rev. Mol. Cell Biol.* **2007**, *8*, 101–112. [CrossRef] [PubMed]
5. Lambert, M.P.; Velasco, P.T.; Chang, L.; Viola, K.L.; Fernandez, S.; Lacor, P.N.; Khuon, D.; Gong, Y.; Bigio, E.H.; Shaw, P.; et al. Monoclonal antibodies that target pathological assemblies of Aβ. *J. Neurochem.* **2007**, *100*, 23–35. [CrossRef] [PubMed]

6. Shankar, G.M.; Li, S.; Mehta, T.H.; Garcia-Munoz, A.; Shepardson, N.E.; Smith, I.; Brett, F.M.; Farrell, M.A.; Rowan, M.J.; Lemere, C.A.; et al. Amyloid-β protein dimers isolated directly from Alzheimer's brains impair synaptic plasticity and memory. *Nat. Med.* **2008**, *14*, 837–842. [CrossRef]
7. Jack, C.R., Jr.; Bennett, D.A.; Blennow, K.; Carrillo, M.C.; Dunn, B.; Haeberlein, S.B.; Holtzman, D.M.; Jagust, W.; Jessen, F.; Karlawish, J.; et al. NIA-AA Research Framework: Toward a biological definition of Alzheimer's disease. *Alzheimers Dement.* **2018**, *14*, 535–562. [CrossRef] [PubMed]
8. Cotta Ramusino, M.; Perini, G.; Altomare, D.; Barbarino, P.; Weidner, W.; Salvini Porro, G.; Barkhof, F.; Rabinovici, G.D.; van der Flier, W.M.; Frisoni, G.B.; et al. Outcomes of clinical utility in amyloid-PET studies: State of art and future perspectives. *Eur. J. Nucl. Med. Mol. Imaging* **2021**, *48*, 2157–2168. [CrossRef]
9. Chételat, G.; Arbizu, J.; Barthel, H.; Garibotto, V.; Law, I.; Morbelli, S.; van de Giessen, E.; Agosta, F.; Barkhof, F.; Brooks, D.J.; et al. Amyloid-PET and ^{18}F-FDG-PET in the diagnostic investigation of Alzheimer's disease and other dementias. *Lancet Neurol.* **2020**, *19*, 951–962. [CrossRef]
10. Dubois, B.; Villain, N.; Frisoni, G.B.; Rabinovici, G.D.; Sabbagh, M.; Cappa, S.; Bejanin, A.; Bombois, S.; Epelbaum, S.; Teichmann, M.; et al. Clinical diagnosis of Alzheimer's disease: Recommendations of the International Working Group. *Lancet Neurol.* **2021**, *20*, 484–496. [CrossRef]
11. Perani, D.; Iaccarino, L.; Lammertsma, A.A.; Windhorst, A.D.; Edison, P.; Boellaard, R.; Hansson, O.; Nordberg, A.; Jacobs, A.H. A new perspective for advanced positron emission tomography-based molecular imaging in neurodegenerative proteinopathies. *Alzheimers Dement.* **2019**, *15*, 1081–1103. [CrossRef]
12. Radde, R.; Bolmont, T.; Kaeser, S.A.; Coomaraswamy, J.; Lindau, D.; Stoltze, L.; Calhoun, M.E.; Jaggi, F.; Wolburg, H.; Gengler, S.; et al. Abeta42-driven cerebral amyloidosis in transgenic mice reveals early and robust pathology. *EMBO Rep.* **2006**, *7*, 940–946. [CrossRef] [PubMed]
13. Hsiao, K.; Chapman, P.; Nilsen, S.; Eckman, C.; Harigaya, Y.; Younkin, S.; Yang, F.; Cole, G. Correlative memory deficits, Abeta elevation, and amyloid plaques in transgenic mice. *Science* **1996**, *274*, 99–102. [CrossRef] [PubMed]
14. Mucke, L.; Masliah, E.; Yu, G.Q.; Mallory, M.; Rockenstein, E.M.; Tatsuno, G.; Hu, K.; Kholodenko, D.; Johnson-Wood, K.; McConlogue, L. High-level neuronal expression of abeta 1-42 in wild-type human amyloid protein precursor transgenic mice: Synaptotoxicity without plaque formation. *J. Neurosci.* **2000**, *20*, 4050–4058. [CrossRef] [PubMed]
15. Richards, J.G.; Higgins, G.A.; Ouagazzal, A.M.; Ozmen, L.; Kew, J.N.; Bohrmann, B.; Malherbe, P.; Brockhaus, M.; Loetscher, H.; Czech, C.; et al. PS2APP transgenic mice, coexpressing hPS2mut and hAPPswe, show age-related cognitive deficits associated with discrete brain amyloid deposition and inflammation. *J. Neurosci.* **2003**, *23*, 8989–9003. [CrossRef] [PubMed]
16. Sturchler-Pierrat, C.; Abramowski, D.; Duke, M.; Wiederhold, K.H.; Mistl, C.; Rothacher, S.; Ledermann, B.; Bürki, K.; Frey, P.; Paganetti, P.A.; et al. Two amyloid precursor protein transgenic mouse models with Alzheimer disease-like pathology. *Proc. Natl. Acad. Sci. USA* **1997**, *94*, 13287–13292. [CrossRef]
17. Oakley, H.; Cole, S.L.; Logan, S.; Maus, E.; Shao, P.; Craft, J.; Guillozet-Bongaarts, A.; Ohno, M.; Disterhoft, J.; Van Eldik, L.; et al. Intraneuronal beta-amyloid aggregates, neurodegeneration, and neuron loss in transgenic mice with five familial Alzheimer's disease mutations: Potential factors in amyloid plaque formation. *J. Neurosci.* **2006**, *26*, 10129–10140. [CrossRef] [PubMed]
18. Oddo, S.; Caccamo, A.; Shepherd, J.D.; Murphy, M.P.; Golde, T.E.; Kayed, R.; Metherate, R.; Mattson, M.P.; Akbari, Y.; LaFerla, F.M. Triple-transgenic model of Alzheimer's disease with plaques and tangles: Intracellular Abeta and synaptic dysfunction. *Neuron* **2003**, *39*, 409–421. [CrossRef]
19. Ni, R.; Dean-Ben, X.L.; Kirschenbaum, D.; Rudin, M.; Chen, Z.; Crimi, A.; Voigt, F.F.; Nilsson, K.P.R.; Helmchen, F.; Nitsch, R. Whole brain optoacoustic tomography reveals strain-specific regional beta-amyloid densities in Alzheimer's disease amyloidosis models. *bioRxiv* **2020**. [CrossRef]
20. Saito, T.; Matsuba, Y.; Mihira, N.; Takano, J.; Nilsson, P.; Itohara, S.; Iwata, N.; Saido, T.C. Single App knock-in mouse models of Alzheimer's disease. *Nat. Neurosci.* **2014**, *17*, 661–663. [CrossRef] [PubMed]
21. Serneels, L.; T'Syen, D.; Perez-Benito, L.; Theys, T.; Holt, M.G.; De Strooper, B. Modeling the β-secretase cleavage site and humanizing amyloid-beta precursor protein in rat and mouse to study Alzheimer's disease. *Mol. Neurodegener.* **2020**, *15*, 60. [CrossRef]
22. Sato, K.; Watamura, N.; Fujioka, R.; Mihira, N.; Sekiguchi, M.; Nagata, K.; Ohshima, T.; Saito, T.; Saido, T.C.; Sasaguri, H. A 3(rd) generation mouse model of Alzheimer's disease shows early and increased cored plaque pathology composed of wild-type human amyloid β peptide. *J. Biol. Chem.* **2021**, *297*, 101004. [CrossRef] [PubMed]
23. Baglietto-Vargas, D.; Forner, S.; Cai, L.; Martini, A.C.; Trujillo-Estrada, L.; Swarup, V.; Nguyen, M.M.T.; Do Huynh, K.; Javonillo, D.I.; Tran, K.M.; et al. Generation of a humanized Aβ expressing mouse demonstrating aspects of Alzheimer's disease-like pathology. *Nat. Commun.* **2021**, *12*, 2421. [CrossRef] [PubMed]
24. Latimer, C.S.; Shively, C.A.; Keene, C.D.; Jorgensen, M.J.; Andrews, R.N.; Register, T.C.; Montine, T.J.; Wilson, A.M.; Neth, B.J.; Mintz, A.; et al. A nonhuman primate model of early Alzheimer's disease pathologic change: Implications for disease pathogenesis. *Alzheimer Dement.* **2019**, *15*, 93–105. [CrossRef] [PubMed]
25. Whitesell, J.D.; Buckley, A.R.; Knox, J.E.; Kuan, L.; Graddis, N.; Pelos, A.; Mukora, A.; Wakeman, W.; Bohn, P.; Ho, A.; et al. Whole brain imaging reveals distinct spatial patterns of amyloid beta deposition in three mouse models of Alzheimer's disease. *J. Comp. Neurol.* **2019**, *527*, 2122–2145. [CrossRef]

26. Liu, P.; Reichl, J.H.; Rao, E.R.; McNellis, B.M.; Huang, E.S.; Hemmy, L.S.; Forster, C.L.; Kuskowski, M.A.; Borchelt, D.R.; Vassar, R.; et al. Quantitative Comparison of Dense-Core Amyloid Plaque Accumulation in Amyloid-β Protein Precursor Transgenic Mice. *J. Alzheimers Dis.* **2017**, *56*, 743–761. [CrossRef]

27. Sasaguri, H.; Nilsson, P.; Hashimoto, S.; Nagata, K.; Saito, T.; De Strooper, B.; Hardy, J.; Vassar, R.; Winblad, B.; Saido, T.C. APP mouse models for Alzheimer's disease preclinical studies. *EMBO J.* **2017**, *36*, 2473–2487. [CrossRef]

28. Robbins, E.M.; Betensky, R.A.; Domnitz, S.B.; Purcell, S.M.; Garcia-Alloza, M.; Greenberg, C.; Rebeck, G.W.; Hyman, B.T.; Greenberg, S.M.; Frosch, M.P.; et al. Kinetics of cerebral amyloid angiopathy progression in a transgenic mouse model of Alzheimer disease. *J. Neurosci.* **2006**, *26*, 365–371. [CrossRef]

29. Jäkel, L.; Van Nostrand, W.E.; Nicoll, J.A.R.; Werring, D.J.; Verbeek, M.M. Animal models of cerebral amyloid angiopathy. *Clin. Sci.* **2017**, *131*, 2469–2488. [CrossRef]

30. Ni, R.; Chen, Z.; Shi, G.; Villois, A.; Zhou, Q.; Arosio, P.; Nitsch, R.M.; Nilsson, K.P.R.; Klohs, J.; Razansky, D. Transcranial in vivo detection of amyloid-beta at single plaque resolution with large-field multifocal illumination fluorescence microscopy. *bioRxiv* **2020**. [CrossRef]

31. Cheng, Y.; Ono, M.; Kimura, H.; Kagawa, S.; Nishii, R.; Saji, H. A novel ^{18}F-labeled pyridyl benzofuran derivative for imaging of β-amyloid plaques in Alzheimer's brains. *Bioorg. Med. Chem. Lett.* **2010**, *20*, 6141–6144. [CrossRef]

32. Hostetler, E.D.; Sanabria-Bohórquez, S.; Fan, H.; Zeng, Z.; Gammage, L.; Miller, P.; O'Malley, S.; Connolly, B.; Mulhearn, J.; Harrison, S.T.; et al. [18]F]Fluoroazabenzoxazoles as potential amyloid plaque PET tracers: Synthesis and in vivo evaluation in rhesus monkey. *Nucl. Med. Biol.* **2011**, *38*, 1193–1203. [CrossRef] [PubMed]

33. Snellman, A.; Rokka, J.; Lopez-Picon, F.R.; Helin, S.; Re, F.; Loyttyniemi, E.; Pihlaja, R.; Forloni, G.; Salmona, M.; Masserini, M.; et al. Applicability of ^{11}C]PIB micro-PET imaging for in vivo follow-up of anti-amyloid treatment effects in APP23 mouse model. *Neurobiol. Aging* **2017**, *57*, 84–94. [CrossRef] [PubMed]

34. Oh, S.J.; Lee, H.-J.; Kang, K.J.; Han, S.J.; Lee, Y.J.; Lee, K.C.; Lim, S.M.; Chi, D.Y.; Kim, K.M.; Park, J.-A.; et al. Early Detection of Aβ Deposition in the 5xFAD Mouse by Amyloid PET. *Contrast Media Mol. Imaging* **2018**, *2018*, 5272014. [CrossRef]

35. Oh, S.J.; Kim, M.H.; Han, S.J.; Kang, K.J.; Ko, I.O.; Kim, Y.; Park, J.-A.; Choi, J.Y.; Lee, K.C.; Chi, D.Y.; et al. Preliminary PET Study of ^{18}F-FC119S in Normal and Alzheimer's Disease Models. *Mol. Pharm.* **2017**, *14*, 3114–3120. [CrossRef] [PubMed]

36. Yousefi, B.H.; von Reutern, B.; Scherubl, D.; Manook, A.; Schwaiger, M.; Grimmer, T.; Henriksen, G.; Forster, S.; Drzezga, A.; Wester, H.J. FIBT versus florbetaben and PiB: A preclinical comparison study with amyloid-PET in transgenic mice. *EJNMMI Res.* **2015**, *5*, 20. [CrossRef]

37. Snellman, A.; Rokka, J.; Lopez-Picon, F.R.; Eskola, O.; Wilson, I.; Farrar, G.; Scheinin, M.; Solin, O.; Rinne, J.O.; Haaparanta-Solin, M. Pharmacokinetics of ^{18}F]flutemetamol in wild-type rodents and its binding to beta amyloid deposits in a mouse model of Alzheimer's disease. *Eur. J. Nucl. Med. Mol. Imaging* **2012**, *39*, 1784–1795. [CrossRef]

38. Snellman, A.; Rokka, J.; López-Picón, F.R.; Eskola, O.; Salmona, M.; Forloni, G.; Scheinin, M.; Solin, O.; Rinne, J.O.; Haaparanta-Solin, M. In vivo PET imaging of beta-amyloid deposition in mouse models of Alzheimer's disease with a high specific activity PET imaging agent ^{18}F]flutemetamol. *EJNMMI Res.* **2014**, *4*, 37. [CrossRef]

39. Huang, Y.; Cho, H.-J.; Bandara, N.; Sun, L.; Tran, D.; Rogers, B.E.; Mirica, L.M. Metal-chelating benzothiazole multifunctional compounds for the modulation and 64Cu PET imaging of Aβ aggregation. *Chem. Sci.* **2020**, *11*, 7789–7799. [CrossRef]

40. Xu, M.; Guo, J.; Gu, J.; Zhang, L.; Liu, Z.; Ding, L.; Fu, H.; Ma, Y.; Liang, S.; Wang, H. Preclinical and clinical study on ^{18}F]DRKXH1: A novel β-amyloid PET tracer for Alzheimer's disease. *Eur. J. Nucl. Med. Mol. Imaging* **2021**, 1–12. [CrossRef]

41. Liang, S.H.; Holland, J.P.; Stephenson, N.A.; Kassenbrock, A.; Rotstein, B.H.; Daignault, C.P.; Lewis, R.; Collier, L.; Hooker, J.M.; Vasdev, N. PET neuroimaging studies of ^{18}F]CABS13 in a double transgenic mouse model of Alzheimer's disease and nonhuman primates. *ACS Chem. Neurosci.* **2015**, *6*, 535–541. [CrossRef]

42. Juréus, A.; Swahn, B.M.; Sandell, J.; Jeppsson, F.; Johnson, A.E.; Johnström, P.; Neelissen, J.A.; Sunnemark, D.; Farde, L.; Svensson, S.P. Characterization of AZD4694, a novel fluorinated Abeta plaque neuroimaging PET radioligand. *J. Neurochem.* **2010**, *114*, 784–794. [CrossRef]

43. Parent, M.J.; Zimmer, E.R.; Shin, M.; Kang, M.S.; Fonov, V.S.; Mathieu, A.; Aliaga, A.; Kostikov, A.; Do Carmo, S.; Dea, D.; et al. Multimodal Imaging in Rat Model Recapitulates Alzheimer's Disease Biomarkers Abnormalities. *J. Neurosci.* **2017**, *37*, 12263–12271. [CrossRef] [PubMed]

44. Cho, H.J.; Huynh, T.T.; Rogers, B.E.; Mirica, L.M. Design of a multivalent bifunctional chelator for diagnostic (64)Cu PET imaging in Alzheimer's disease. *Proc. Natl. Acad. Sci. USA* **2020**, *117*, 30928–30933. [CrossRef] [PubMed]

45. Ni, R.; Villois, A.; Dean-Ben, X.L.; Chen, Z.; Vaas, M.; Stavrakis, S.; Shi, G.; deMello, A.; Ran, C.; Razansky, D.; et al. In-vitro and in-vivo characterization of CRANAD-2 for multi-spectral optoacoustic tomography and fluorescence imaging of amyloid-beta deposits in Alzheimer mice. *Photoacoustics* **2021**, *23*, 100285. [CrossRef] [PubMed]

46. Ni, R.; Gillberg, P.-G.; Bogdanovic, N.; Viitanen, M.; Myllykangas, L.; Nennesmo, I.; Långström, B.; Nordberg, A. Amyloid tracers binding sites in autosomal dominant and sporadic Alzheimer's disease. *Alzheimer Dement.* **2017**, *13*, 419–430. [CrossRef]

47. Ni, R.; Röjdner, J.; Voytenko, L.; Dyrks, T.; Thiele, A.; Marutle, A.; Nordberg, A. In vitro Characterization of the Regional Binding Distribution of Amyloid PET Tracer Florbetaben and the Glia Tracers Deprenyl and PK11195 in Autopsy Alzheimer's Brain Tissue. *J. Alzheimers Dis.* **2021**, *80*, 1723–1737. [CrossRef]

48. Snellman, A.; López-Picón, F.R.; Rokka, J.; Salmona, M.; Forloni, G.; Scheinin, M.; Solin, O.; Rinne, J.O.; Haaparanta-Solin, M. Longitudinal amyloid imaging in mouse brain with ^{11}C-PIB: Comparison of APP23, Tg2576, and APPswe-PS1dE9 mouse models of Alzheimer disease. *J. Nucl. Med.* **2013**, *54*, 1434–1441. [CrossRef]

49. Brendel, M.; Jaworska, A.; Grießinger, E.; Rötzer, C.; Burgold, S.; Gildehaus, F.J.; Carlsen, J.; Cumming, P.; Baumann, K.; Haass, C.; et al. Cross-sectional comparison of small animal [18F]-florbetaben amyloid-PET between transgenic AD mouse models. *PLoS ONE* **2015**, *10*, e0116678. [CrossRef]

50. Son, H.J.; Jeong, Y.J.; Yoon, H.J.; Lee, S.Y.; Choi, G.-E.; Park, J.-A.; Kim, M.H.; Lee, K.C.; Lee, Y.J.; Kim, M.K.; et al. Assessment of brain beta-amyloid deposition in transgenic mouse models of Alzheimer's disease with PET imaging agents ^{18}F-flutemetamol and ^{18}F-florbetaben. *BMC Neurosci.* **2018**, *19*, 45. [CrossRef]

51. Catafau, A.M.; Bullich, S. Amyloid PET imaging: Applications beyond Alzheimer's disease. *Clin. Transl. Imaging* **2015**, *3*, 39–55. [CrossRef]

52. Han, B.H.; Zhou, M.-l.; Vellimana, A.K.; Milner, E.; Kim, D.H.; Greenberg, J.K.; Chu, W.; Mach, R.H.; Zipfel, G.J. Resorufin analogs preferentially bind cerebrovascular amyloid: Potential use as imaging ligands for cerebral amyloid angiopathy. *Mol. Neurodegener.* **2011**, *6*, 86. [CrossRef] [PubMed]

53. Abrahamson, E.E.; Stehouwer, J.S.; Vazquez, A.L.; Huang, G.-F.; Mason, N.S.; Lopresti, B.J.; Klunk, W.E.; Mathis, C.A.; Ikonomovic, M.D. Development of a PET radioligand selective for cerebral amyloid angiopathy. *Nucl. Med. Biol.* **2021**, *92*, 85–96. [CrossRef]

54. Biechele, G.; Sebastian Monasor, L.; Wind, K.; Blume, T.; Parhizkar, S.; Arzberger, T.; Sacher, C.; Beyer, L.; Eckenweber, F.; Gildehaus, F.J.; et al. Glitter in the Darkness? Non-fibrillar β-amyloid Plaque Components Significantly Impact the β-amyloid PET Signal in Mouse Models of Alzheimer's Disease. *J. Nucl. Med.* **2021**, *62*. [CrossRef] [PubMed]

55. Meier, S.R.; Sehlin, D.; Roshanbin, S.; Lim Falk, V.; Saito, T.; Saido, T.C.; Neumann, U.; Rokka, J.; Eriksson, J.; Syvanen, S. ^{11}C-PIB and [124]I-antibody PET provide differing estimates of brain amyloid-beta after therapeutic intervention. *J. Nucl. Med.* **2021**, *62*. [CrossRef]

56. Brendel, M.; Jaworska, A.; Herms, J.; Trambauer, J.; Rötzer, C.; Gildehaus, F.J.; Carlsen, J.; Cumming, P.; Bylund, J.; Luebbers, T.; et al. Amyloid-PET predicts inhibition of de novo plaque formation upon chronic γ-secretase modulator treatment. *Mol. Psychiatry* **2015**, *20*, 1179–1187. [CrossRef] [PubMed]

57. Brendel, M.; Jaworska, A.; Overhoff, F.; Blume, T.; Probst, F.; Gildehaus, F.J.; Bartenstein, P.; Haass, C.; Bohrmann, B.; Herms, J.; et al. Efficacy of chronic BACE1 inhibition in PS2APP mice depends on the regional Aβ deposition rate and plaque burden at treatment initiation. *Theranostics* **2018**, *8*, 4957–4968. [CrossRef]

58. Deleye, S.; Waldron, A.M.; Verhaeghe, J.; Bottelbergs, A.; Wyffels, L.; Van Broeck, B.; Langlois, X.; Schmidt, M.; Stroobants, S.; Staelens, S. Evaluation of Small-Animal PET Outcome Measures to Detect Disease Modification Induced by BACE Inhibition in a Transgenic Mouse Model of Alzheimer Disease. *J. Nucl. Med.* **2017**, *58*, 1977–1983. [CrossRef]

59. Xu, Y.; Wang, C.; Wey, H.-Y.; Liang, Y.; Chen, Z.; Choi, S.H.; Ran, C.; Rynearson, K.D.; Bernales, D.R.; Koegel, R.E.; et al. Molecular imaging of Alzheimer's disease–related gamma-secretase in mice and nonhuman primates. *J. Exp. Med.* **2020**, *217*, e20182266. [CrossRef]

60. Toyama, H.; Ye, D.; Ichise, M.; Liow, J.S.; Cai, L.; Jacobowitz, D.; Musachio, J.L.; Hong, J.; Crescenzo, M.; Tipre, D.; et al. PET imaging of brain with the beta-amyloid probe, [11C]6-OH-BTA-1, in a transgenic mouse model of Alzheimer's disease. *Eur. J. Nucl. Med. Mol. Imaging* **2005**, *32*, 593–600. [CrossRef]

61. Rojas, S.; Herance, J.R.; Gispert, J.D.; Abad, S.; Torrent, E.; Jiménez, X.; Pareto, D.; Perpiña, U.; Sarroca, S.; Rodríguez, E.; et al. In vivo evaluation of amyloid deposition and brain glucose metabolism of 5XFAD mice using positron emission tomography. *Neurobiol. Aging* **2013**, *34*, 1790–1798. [CrossRef]

62. Klunk, W.E.; Lopresti, B.J.; Ikonomovic, M.D.; Lefterov, I.M.; Koldamova, R.P.; Abrahamson, E.E.; Debnath, M.L.; Holt, D.P.; Huang, G.F.; Shao, L.; et al. Binding of the positron emission tomography tracer Pittsburgh compound-B reflects the amount of amyloid-beta in Alzheimer's disease brain but not in transgenic mouse brain. *J. Neurosci.* **2005**, *25*, 10598–10606. [CrossRef] [PubMed]

63. Manook, A.; Yousefi, B.H.; Willuweit, A.; Platzer, S.; Reder, S.; Voss, A.; Huisman, M.; Settles, M.; Neff, F.; Velden, J.; et al. Small-animal PET imaging of amyloid-beta plaques with [11C]PiB and its multi-modal validation in an APP/PS1 mouse model of Alzheimer's disease. *PLoS ONE* **2012**, *7*, e31310. [CrossRef] [PubMed]

64. Maier, F.C.; Wehrl, H.F.; Schmid, A.M.; Mannheim, J.G.; Wiehr, S.; Lerdkrai, C.; Calaminus, C.; Stahlschmidt, A.; Ye, L.; Burnet, M.; et al. Longitudinal PET-MRI reveals β-amyloid deposition and rCBF dynamics and connects vascular amyloidosis to quantitative loss of perfusion. *Nat. Med.* **2014**, *20*, 1485–1492. [CrossRef] [PubMed]

65. von Reutern, B.; Grünecker, B.; Yousefi, B.H.; Henriksen, G.; Czisch, M.; Drzezga, A. Voxel-based analysis of amyloid-burden measured with [11C]PiB PET in a double transgenic mouse model of Alzheimer's disease. *Mol. Imaging Biol.* **2013**, *15*, 576–584. [CrossRef] [PubMed]

66. Waldron, A.M.; Wintmolders, C.; Bottelbergs, A.; Kelley, J.B.; Schmidt, M.E.; Stroobants, S.; Langlois, X.; Staelens, S. In vivo molecular neuroimaging of glucose utilization and its association with fibrillar amyloid-β load in aged APPPS1-21 mice. *Alzheimers Res. Ther.* **2015**, *7*, 76. [CrossRef]

67. Chiquita, S.; Ribeiro, M.; Castelhano, J.; Oliveira, F.; Sereno, J.; Batista, M.; Abrunhosa, A.; Rodrigues-Neves, A.C.; Carecho, R.; Baptista, F.; et al. A longitudinal multimodal in vivo molecular imaging study of the 3xTg-AD mouse model shows progressive early hippocampal and taurine loss. *Hum. Mol. Genet.* **2019**, *28*, 2174–2188. [CrossRef] [PubMed]

68. Maeda, J.; Ji, B.; Irie, T.; Tomiyama, T.; Maruyama, M.; Okauchi, T.; Staufenbiel, M.; Iwata, N.; Ono, M.; Saido, T.C.; et al. Longitudinal, quantitative assessment of amyloid, neuroinflammation, and anti-amyloid treatment in a living mouse model of Alzheimer's disease enabled by positron emission tomography. *J. Neurosci.* **2007**, *27*, 10957–10968. [CrossRef]

69. Nishiyama, S.; Ohba, H.; Kanazawa, M.; Kakiuchi, T.; Tsukada, H. Comparing α7 nicotinic acetylcholine receptor binding, amyloid-β deposition, and mitochondria complex-I function in living brain: A PET study in aged monkeys. *Synapse* **2015**, *69*, 475–483. [CrossRef]

70. Tsukada, H.; Nishiyama, S.; Ohba, H.; Kanazawa, M.; Kakiuchi, T.; Harada, N. Comparing amyloid-β deposition, neuroinflammation, glucose metabolism, and mitochondrial complex I activity in brain: A PET study in aged monkeys. *Eur. J. Nucl. Med. Mol. Imaging* **2014**, *41*, 2127–2136. [CrossRef]

71. Frost, G.R.; Longo, V.; Li, T.; Jonas, L.A.; Judenhofer, M.; Cherry, S.; Koutcher, J.; Lekaye, C.; Zanzonico, P.; Li, Y.-M. Hybrid PET/MRI enables high-spatial resolution, quantitative imaging of amyloid plaques in an Alzheimer's disease mouse model. *Sci. Rep.* **2020**, *10*, 10379. [CrossRef]

72. Waldron, A.M.; Wyffels, L.; Verhaeghe, J.; Richardson, J.C.; Schmidt, M.; Stroobants, S.; Langlois, X.; Staelens, S. Longitudinal Characterization of [^{18}F]-FDG and [^{18}F]-AV45 Uptake in the Double Transgenic TASTPM Mouse Model. *J. Alzheimers Dis.* **2017**, *55*, 1537–1548. [CrossRef] [PubMed]

73. Poisnel, G.; Dhilly, M.; Moustié, O.; Delamare, J.; Abbas, A.; Guilloteau, D.; Barré, L. PET imaging with [^{18}F]AV-45 in an APP/PS1-21 murine model of amyloid plaque deposition. *Neurobiol. Aging* **2012**, *33*, 2561–2571. [CrossRef]

74. Sacher, C.; Blume, T.; Beyer, L.; Biechele, G.; Sauerbeck, J.; Eckenweber, F.; Deussing, M.; Focke, C.; Parhizkar, S.; Lindner, S.; et al. Asymmetry of fibrillar plaque burden in amyloid mouse models. *J. Nucl. Med.* **2020**, *61*, 1825–1831. [CrossRef]

75. Rominger, A.; Brendel, M.; Burgold, S.; Keppler, K.; Baumann, K.; Xiong, G.; Mille, E.; Gildehaus, F.J.; Carlsen, J.; Schlichtiger, J.; et al. Longitudinal assessment of cerebral β-amyloid deposition in mice overexpressing Swedish mutant β-amyloid precursor protein using ^{18}F-florbetaben PET. *J. Nucl. Med.* **2013**, *54*, 1127–1134. [CrossRef] [PubMed]

76. Sacher, C.; Blume, T.; Beyer, L.; Peters, F.; Eckenweber, F.; Sgobio, C.; Deussing, M.; Albert, N.L.; Unterrainer, M.; Lindner, S.; et al. Longitudinal PET Monitoring of Amyloidosis and Microglial Activation in a Second-Generation Amyloid-β Mouse Model. *J. Nucl. Med.* **2019**, *60*, 1787–1793. [CrossRef] [PubMed]

77. Biechele, G.; Wind, K.; Blume, T.; Sacher, C.; Beyer, L.; Eckenweber, F.; Franzmeier, N.; Ewers, M.; Zott, B.; Lindner, S.; et al. Microglial activation in the right amygdala-entorhinal-hippocampal complex is associated with preserved spatial learning in App(NL-G-F) mice. *Neuroimage* **2021**, *230*, 117707. [CrossRef] [PubMed]

78. Biechele, G.; Franzmeier, N.; Blume, T.; Ewers, M.; Luque, J.M.; Eckenweber, F.; Sacher, C.; Beyer, L.; Ruch-Rubinstein, F.; Lindner, S.; et al. Glial activation is moderated by sex in response to amyloidosis but not to tau pathology in mouse models of neurodegenerative diseases. *J. Neuroinflamm.* **2020**, *17*, 374. [CrossRef] [PubMed]

79. Blume, T.; Focke, C.; Peters, F.; Deussing, M.; Albert, N.L.; Lindner, S.; Gildehaus, F.-J.; von Ungern-Sternberg, B.; Ozmen, L.; Baumann, K.; et al. Microglial response to increasing amyloid load saturates with aging: A longitudinal dual tracer in vivo μPET-study. *J. Neuroinflamm.* **2018**, *15*, 307. [CrossRef] [PubMed]

80. Chaney, A.M.; Lopez-Picon, F.R.; Serrière, S.; Wang, R.; Bochicchio, D.; Webb, S.D.; Vandesquille, M.; Harte, M.K.; Georgiadou, C.; Lawrence, C.; et al. Prodromal neuroinflammatory, cholinergic and metabolite dysfunction detected by PET and MRS in the TgF344-AD transgenic rat model of AD: A collaborative multi-modal study. *Theranostics* **2021**, *11*, 6644–6667. [CrossRef]

81. Franke, T.N.; Irwin, C.; Bayer, T.A.; Brenner, W.; Beindorff, N.; Bouter, C.; Bouter, Y. In vivo Imaging With ^{18}F-FDG- and ^{18}F-Florbetaben-PET/MRI Detects Pathological Changes in the Brain of the Commonly Used 5XFAD Mouse Model of Alzheimer's Disease. *Front. Med.* **2020**, *7*, 529. [CrossRef]

82. Rodriguez-Vieitez, E.; Ni, R.; Gulyas, B.; Toth, M.; Haggkvist, J.; Halldin, C.; Voytenko, L.; Marutle, A.; Nordberg, A. Astrocytosis precedes amyloid plaque deposition in Alzheimer APPswe transgenic mouse brain: A correlative positron emission tomography and in vitro imaging study. *Eur. J. Nucl. Med. Mol. Imaging* **2015**, *42*, 1119–1132. [CrossRef]

83. Johnson, A.E.; Jeppsson, F.; Sandell, J.; Wensbo, D.; Neelissen, J.A.; Juréus, A.; Ström, P.; Norman, H.; Farde, L.; Svensson, S.P. AZD2184: A radioligand for sensitive detection of beta-amyloid deposits. *J. Neurochem.* **2009**, *108*, 1177–1186. [CrossRef]

84. Kudo, Y.; Okamura, N.; Furumoto, S.; Tashiro, M.; Furukawa, K.; Maruyama, M.; Itoh, M.; Iwata, R.; Yanai, K.; Arai, H. 2-(2-[2-Dimethylaminothiazol-5-yl]ethenyl)-6- (2-[fluoro]ethoxy)benzoxazole: A novel PET agent for in vivo detection of dense amyloid plaques in Alzheimer's disease patients. *J. Nucl. Med.* **2007**, *48*, 553–561. [CrossRef]

85. Furumoto, S.; Okamura, N.; Furukawa, K.; Tashiro, M.; Ishikawa, Y.; Sugi, K.; Tomita, N.; Waragai, M.; Harada, R.; Tago, T.; et al. A ^{18}F-Labeled BF-227 Derivative as a Potential Radioligand for Imaging Dense Amyloid Plaques by Positron Emission Tomography. *Mol. Imaging Biol.* **2013**, *15*, 497–506. [CrossRef] [PubMed]

86. Sundaram, G.S.M.; Dhavale, D.D.; Prior, J.L.; Yan, P.; Cirrito, J.; Rath, N.P.; Laforest, R.; Cairns, N.J.; Lee, J.-M.; Kotzbauer, P.T.; et al. Fluselenamyl: A Novel Benzoselenazole Derivative for PET Detection of Amyloid Plaques (Aβ) in Alzheimer's Disease. *Sci. Rep.* **2016**, *6*, 35636. [CrossRef] [PubMed]

87. Sehlin, D.; Fang, X.T.; Cato, L.; Antoni, G.; Lannfelt, L.; Syvanen, S. Antibody-based PET imaging of amyloid beta in mouse models of Alzheimer's disease. *Nat. Commun.* **2016**, *7*, 10759. [CrossRef]

88. Liu, Y.; Yang, Y.; Sun, M.; Cui, M.; Fu, Y.; Lin, Y.; Li, Z.; Nie, L. Highly specific noninvasive photoacoustic and positron emission tomography of brain plaque with functionalized croconium dye labeled by a radiotracer. *Chem. Sci.* **2017**, *8*, 2710–2716. [CrossRef] [PubMed]

89. An, Y.; Varma, V.R.; Varma, S.; Casanova, R.; Dammer, E.; Pletnikova, O.; Chia, C.W.; Egan, J.M.; Ferrucci, L.; Troncoso, J.; et al. Evidence for brain glucose dysregulation in Alzheimer's disease. *Alzheimer Dement.* **2018**, *14*, 318–329. [CrossRef] [PubMed]

90. Foster, N.L.; Heidebrink, J.L.; Clark, C.M.; Jagust, W.J.; Arnold, S.E.; Barbas, N.R.; DeCarli, C.S.; Turner, R.S.; Koeppe, R.A.; Higdon, R.; et al. FDG-PET improves accuracy in distinguishing frontotemporal dementia and Alzheimer's disease. *Brain* **2007**, *130*, 2616–2635. [CrossRef]

91. Bouter, C.; Henniges, P.; Franke, T.N.; Irwin, C.; Sahlmann, C.O.; Sichler, M.E.; Beindorff, N.; Bayer, T.A.; Bouter, Y. ^{18}F-FDG-PET Detects Drastic Changes in Brain Metabolism in the Tg4-42 Model of Alzheimer's Disease. *Front. Aging Neurosci.* **2018**, *10*, 425. [CrossRef]

92. Kuntner, C.; Kesner, A.L.; Bauer, M.; Kremslehner, R.; Wanek, T.; Mandler, M.; Karch, R.; Stanek, J.; Wolf, T.; Müller, M.; et al. Limitations of small animal PET imaging with [18F]FDDNP and FDG for quantitative studies in a transgenic mouse model of Alzheimer's disease. *Mol. Imaging Biol.* **2009**, *11*, 236–240. [CrossRef] [PubMed]

93. Belfiore, R.; Rodin, A.; Ferreira, E.; Velazquez, R.; Branca, C.; Caccamo, A.; Oddo, S. Temporal and regional progression of Alzheimer's disease-like pathology in 3xTg-AD mice. *Aging Cell* **2019**, *18*, e12873. [CrossRef] [PubMed]

94. Adlimoghaddam, A.; Snow, W.M.; Stortz, G.; Perez, C.; Djordjevic, J.; Goertzen, A.L.; Ko, J.H.; Albensi, B.C. Regional hypometabolism in the 3xTg mouse model of Alzheimer's disease. *Neurobiol. Dis.* **2019**, *127*, 264–277. [CrossRef]

95. Bouter, C.; Bouter, Y. ^{18}F-FDG-PET in Mouse Models of Alzheimer's Disease. *Front. Med.* **2019**, *6*, 71. [CrossRef] [PubMed]

96. Snellman, A.; Takkinen, J.S.; López-Picón, F.R.; Eskola, O.; Solin, O.; Rinne, J.O.; Haaparanta-Solin, M. Effect of genotype and age on cerebral [18F]FDG uptake varies between transgenic APPSwe-PS1dE9 and Tg2576 mouse models of Alzheimer's disease. *Sci. Rep.* **2019**, *9*, 5700. [CrossRef] [PubMed]

97. Xiang, X.; Wind, K.; Wiedemann, T.; Blume, T.; Shi, Y.; Briel, N.; Beyer, L.; Biechele, G.; Eckenweber, F.; Zatcepin, A.; et al. Microglial activation states drive glucose uptake and FDG-PET alterations in neurodegenerative diseases. *Sci. Transl. Med.* **2021**, *13*, eabe5640. [CrossRef] [PubMed]

98. Nicholson, R.M.; Kusne, Y.; Nowak, L.A.; LaFerla, F.M.; Reiman, E.M.; Valla, J. Regional cerebral glucose uptake in the 3xTG model of Alzheimer's disease highlights common regional vulnerability across AD mouse models. *Brain Res.* **2010**, *1347*, 179–185. [CrossRef] [PubMed]

99. Sancheti, H.; Akopian, G.; Yin, F.; Brinton, R.D.; Walsh, J.P.; Cadenas, E. Age-dependent modulation of synaptic plasticity and insulin mimetic effect of lipoic acid on a mouse model of Alzheimer's disease. *PLoS ONE* **2013**, *8*, e69830. [CrossRef]

100. Luo, F.; Rustay, N.R.; Ebert, U.; Hradil, V.P.; Cole, T.B.; Llano, D.A.; Mudd, S.R.; Zhang, Y.; Fox, G.B.; Day, M. Characterization of 7- and 19-month-old Tg2576 mice using multimodal in vivo imaging: Limitations as a translatable model of Alzheimer's disease. *Neurobiol. Aging* **2012**, *33*, 933–944. [CrossRef]

101. Lourenço, C.F.; Ledo, A.; Barbosa, R.M.; Laranjinha, J. Neurovascular uncoupling in the triple transgenic model of Alzheimer's disease: Impaired cerebral blood flow response to neuronal-derived nitric oxide signaling. *Exp. Neurol.* **2017**, *291*, 36–43. [CrossRef]

102. Liu, Y.; Xu, Y.; Li, M.; Pan, D.; Li, Y.; Wang, Y.; Wang, L.; Wu, Q.; Yang, M. Multi-target PET evaluation in APP/PS1/tau mouse model of Alzheimer's disease. *Neurosci. Lett.* **2020**, *728*, 134938. [CrossRef] [PubMed]

103. Xu, A.; Zeng, Q.; Tang, Y.; Wang, X.; Yuan, X.; Zhou, Y.; Li, Z. Electroacupuncture Protects Cognition by Regulating Tau Phosphorylation and Glucose Metabolism via the AKT/GSK3β Signaling Pathway in Alzheimer's Disease Model Mice. *Front. Neurosci.* **2020**, *14*, 585476. [CrossRef] [PubMed]

104. Poisnel, G.; Hérard, A.-S.; El Tannir El Tayara, N.; Bourrin, E.; Volk, A.; Kober, F.; Delatour, B.; Delzescaux, T.; Debeir, T.; Rooney, T.; et al. Increased regional cerebral glucose uptake in an APP/PS1 model of Alzheimer's disease. *Neurobiol. Aging* **2012**, *33*, 1995–2005. [CrossRef] [PubMed]

105. Takkinen, J.S.; López-Picón, F.R.; Al Majidi, R.; Eskola, O.; Krzyczmonik, A.; Keller, T.; Löyttyniemi, E.; Solin, O.; Rinne, J.O.; Haaparanta-Solin, M. Brain energy metabolism and neuroinflammation in ageing APP/PS1-21 mice using longitudinal ^{18}F-FDG and ^{18}F-DPA-714 PET imaging. *J. Cereb. Blood Flow Metab.* **2017**, *37*, 2870–2882. [CrossRef]

106. Stojakovic, A.; Trushin, S.; Sheu, A.; Khalili, L.; Chang, S.-Y.; Li, X.; Christensen, T.; Salisbury, J.L.; Geroux, R.E.; Gateno, B.; et al. Partial inhibition of mitochondrial complex I ameliorates Alzheimer's disease pathology and cognition in APP/PS1 female mice. *Commun. Biol.* **2021**, *4*, 61. [CrossRef]

107. Wagner, J.M.; Sichler, M.E.; Schleicher, E.M.; Franke, T.N.; Irwin, C.; Löw, M.J.; Beindorff, N.; Bouter, C.; Bayer, T.A.; Bouter, Y. Analysis of Motor Function in the Tg4-42 Mouse Model of Alzheimer's Disease. *Front. Behav. Neurosci.* **2019**, *13*, 107. [CrossRef]

108. Macdonald, I.R.; DeBay, D.R.; Reid, G.A.; O'Leary, T.P.; Jollymore, C.T.; Mawko, G.; Burrell, S.; Martin, E.; Bowen, C.V.; Brown, R.E.; et al. Early detection of cerebral glucose uptake changes in the 5XFAD mouse. *Curr. Alzheimer Res.* **2014**, *11*, 450–460. [CrossRef]

109. Choi, H.; Choi, Y.; Lee, E.J.; Kim, H.; Lee, Y.; Kwon, S.; Hwang, D.W.; Lee, D.S. Hippocampal glucose uptake as a surrogate of metabolic change of microglia in Alzheimer's disease. *J. Neuroinflamm.* **2021**, *18*, 190. [CrossRef]

110. Teng, E.; Kepe, V.; Frautschy, S.A.; Liu, J.; Satyamurthy, N.; Yang, F.; Chen, P.P.; Cole, G.B.; Jones, M.R.; Huang, S.C.; et al. [F-18]FDDNP microPET imaging correlates with brain Aβ burden in a transgenic rat model of Alzheimer disease: Effects of aging, in vivo blockade, and anti-Aβ antibody treatment. *Neurobiol. Dis.* **2011**, *43*, 565–575. [CrossRef]

111. Winkeler, A.; Waerzeggers, Y.; Klose, A.; Monfared, P.; Thomas, A.V.; Schubert, M.; Heneka, M.T.; Jacobs, A.H. Imaging noradrenergic influence on amyloid pathology in mouse models of Alzheimer's disease. *Eur. J. Nucl. Med. Mol. Imaging* **2008**, *35*, S107–S113. [CrossRef]

112. Deleye, S.; Waldron, A.M.; Richardson, J.C.; Schmidt, M.; Langlois, X.; Stroobants, S.; Staelens, S. The Effects of Physiological and Methodological Determinants on ^{18}F-FDG Mouse Brain Imaging Exemplified in a Double Transgenic Alzheimer Model. *Mol. Imaging* **2016**, *15*. [CrossRef]

113. Toyonaga, T.; Smith, L.M.; Finnema, S.J.; Gallezot, J.D.; Naganawa, M.; Bini, J.; Mulnix, T.; Cai, Z.; Ropchan, J.; Huang, Y.; et al. In Vivo Synaptic Density Imaging with ^{11}C-UCB-J Detects Treatment Effects of Saracatinib in a Mouse Model of Alzheimer Disease. *J. Nucl. Med.* **2019**, *60*, 1780–1786. [CrossRef]

114. Xiong, M.; Roshanbin, S.; Rokka, J.; Schlein, E.; Ingelsson, M.; Sehlin, D.; Eriksson, J.; Syvänen, S. In vivo imaging of synaptic density with [^{11}C]UCB-J PET in two mouse models of neurodegenerative disease. *NeuroImage* **2021**, *239*, 118302. [CrossRef] [PubMed]

115. Sadasivam, P.; Fang, X.T.; Toyonaga, T.; Lee, S.; Xu, Y.; Zheng, M.Q.; Spurrier, J.; Huang, Y.; Strittmatter, S.M.; Carson, R.E.; et al. Quantification of SV2A Binding in Rodent Brain Using [^{18}F]SynVesT-1 and PET Imaging. *Mol. Imaging Biol.* **2021**, *23*, 372–381. [CrossRef] [PubMed]

116. Lee, M.; Lee, H.J.; Jeong, Y.J.; Oh, S.J.; Kang, K.J.; Han, S.J.; Nam, K.R.; Lee, Y.J.; Lee, K.C.; Ryu, Y.H.; et al. Age dependency of mGluR5 availability in 5xFAD mice measured by PET. *Neurobiol. Aging* **2019**, *84*, 208–216. [CrossRef] [PubMed]

117. Son, Y.; Jeong, Y.J.; Shin, N.-R.; Oh, S.J.; Nam, K.R.; Choi, H.-D.; Choi, J.Y.; Lee, H.-J. Inhibition of Colony-Stimulating Factor 1 Receptor by PLX3397 Prevents Amyloid Beta Pathology and Rescues Dopaminergic Signaling in Aging 5xFAD Mice. *Int. J. Mol. Sci.* **2020**, *21*, 5553. [CrossRef] [PubMed]

118. Varlow, C.; Murrell, E.; Holland, J.P.; Kassenbrock, A.; Shannon, W.; Liang, S.H.; Vasdev, N.; Stephenson, N.A. Revisiting the Radiosynthesis of [^{18}F]FPEB and Preliminary PET Imaging in a Mouse Model of Alzheimer's Disease. *Molecules* **2020**, *25*, 982. [CrossRef] [PubMed]

119. Fang, X.T.; Eriksson, J.; Antoni, G.; Yngve, U.; Cato, L.; Lannfelt, L.; Sehlin, D.; Syvänen, S. Brain mGluR5 in mice with amyloid beta pathology studied with in vivo [^{11}C]ABP688 PET imaging and ex vivo immunoblotting. *Neuropharmacology* **2017**, *113*, 293–300. [CrossRef] [PubMed]

120. Heneka, M.T.; Ramanathan, M.; Jacobs, A.H.; Dumitrescu-Ozimek, L.; Bilkei-Gorzo, A.; Debeir, T.; Sastre, M.; Galldiks, N.; Zimmer, A.; Hoehn, M.; et al. Locus ceruleus degeneration promotes Alzheimer pathogenesis in amyloid precursor protein 23 transgenic mice. *J. Neurosci.* **2006**, *26*, 1343–1354. [CrossRef]

121. Rejc, L.; Gómez-Vallejo, V.; Joya, A.; Moreno, O.; Egimendia, A.; Castellnou, P.; Ríos-Anglada, X.; Cossío, U.; Baz, Z.; Passannante, R.; et al. Longitudinal evaluation of a novel BChE PET tracer as an early in vivo biomarker in the brain of a mouse model for Alzheimer disease. *Theranostics* **2021**, *11*, 6542–6559. [CrossRef] [PubMed]

122. Chen, Y.A.; Lu, C.H.; Ke, C.C.; Chiu, S.J.; Chang, C.W.; Yang, B.H.; Gelovani, J.G.; Liu, R.S. Evaluation of Class IIa Histone Deacetylases Expression and In Vivo Epigenetic Imaging in a Transgenic Mouse Model of Alzheimer's Disease. *Int. J. Mol. Sci.* **2021**, *22*, 8633. [CrossRef] [PubMed]

123. Satoru, Y.; Yurika, I.; Shunsuke, I.; Takeharu, K.; Hiroyuki, O.; Shingo, N.; Masakatsu, K.; Hideo, T.; Kohji, S.; Yasuomi, O. In Vivo elevation of mitochondrial activity and amyloid deposition, but inversely correlated, in early-stage senescence-accelerated mice: A positron emission tomography study. *Res. Sq.* **2021**. [CrossRef]

124. Terada, T.; Therriault, J.; Kang, M.S.P.; Savard, M.; Pascoal, T.A.; Lussier, F.; Tissot, C.; Wang, Y.-T.; Benedet, A.; Matsudaira, T.; et al. Mitochondrial complex I abnormalities is associated with tau and clinical symptoms in mild Alzheimer's disease. *Mol. Neurodegener.* **2021**, *16*, 28. [CrossRef] [PubMed]

125. Torres, J.B.; Andreozzi, E.M.; Dunn, J.T.; Siddique, M.; Szanda, I.; Howlett, D.R.; Sunassee, K.; Blower, P.J. PET Imaging of Copper Trafficking in a Mouse Model of Alzheimer Disease. *J. Nucl. Med.* **2016**, *57*, 109–114. [CrossRef] [PubMed]

126. Sai, K.S.; Damuka, N.; Mintz, A.; Whitlow, C.T.; Craft, S.; Macauley-Rambach, S. [^{11}C]MPC-6827, a microtubule-based PET imaging tracer: A potential early imaging biomarker for AD and other ADRDs. *Alzheimer's Dement.* **2020**, *16*, e037790. [CrossRef]

127. Bernard-Gauthier, V.; Mossine, A.V.; Knight, A.; Patnaik, D.; Zhao, W.N.; Cheng, C.; Krishnan, H.S.; Xuan, L.L.; Chindavong, P.S.; Reis, S.A.; et al. Structural Basis for Achieving GSK-3β Inhibition with High Potency, Selectivity, and Brain Exposure for Positron Emission Tomography Imaging and Drug Discovery. *J. Med. Chem.* **2019**, *62*, 9600–9617. [CrossRef]

128. Giglio, J.; Fernandez, S.; Martinez, A.; Zeni, M.; Reyes, L.; Rey, A.; Cerecetto, H. Glycogen Synthase Kinase-3 Maleimide Inhibitors As Potential PET-Tracers for Imaging Alzheimer's Disease: ^{11}C-Synthesis and In Vivo Proof of Concept. *J. Med. Chem.* **2021**. [CrossRef]

129. Luzi, F.; Savickas, V.; Taddei, C.; Hader, S.; Singh, N.; Gee, A.D.; Bongarzone, S. Radiolabeling of [^{11}C]FPS-ZM1, a receptor for advanced glycation end products-targeting positron emission tomography radiotracer, using a [^{11}C]CO(2)-to-[^{11}C]CO chemical conversion. *Future Med. Chem.* **2020**, *12*, 511–521. [CrossRef]

130. Zoufal, V.; Mairinger, S.; Krohn, M.; Wanek, T.; Filip, T.; Sauberer, M.; Stanek, J.; Kuntner, C.; Pahnke, J.; Langer, O. Measurement of cerebral ABCC1 transport activity in wild-type and APP/PS1-21 mice with positron emission tomography. *J. Cereb. Blood Flow Metab.* **2020**, *40*, 954–965. [CrossRef]

131. Wanek, T.; Zoufal, V.; Brackhan, M.; Krohn, M.; Mairinger, S.; Filip, T.; Sauberer, M.; Stanek, J.; Pekar, T.; Pahnke, J.; et al. Brain Distribution of Dual ABCB1/ABCG2 Substrates Is Unaltered in a Beta-Amyloidosis Mouse Model. *Int. J. Mol. Sci.* **2020**, *21*, 8245. [CrossRef]

132. Zoufal, V.; Mairinger, S.; Brackhan, M.; Krohn, M.; Filip, T.; Sauberer, M.; Stanek, J.; Wanek, T.; Tournier, N.; Bauer, M.; et al. Imaging P-Glycoprotein Induction at the Blood-Brain Barrier of a β-Amyloidosis Mouse Model with ^{11}C-Metoclopramide PET. *J. Nucl. Med.* **2020**, *61*, 1050–1057. [CrossRef] [PubMed]

133. Zoufal, V.; Wanek, T.; Krohn, M.; Mairinger, S.; Filip, T.; Sauberer, M.; Stanek, J.; Pekar, T.; Bauer, M.; Pahnke, J.; et al. Age dependency of cerebral P-glycoprotein function in wild-type and APPPS1 mice measured with PET. *J. Cereb. Blood Flow Metab.* **2020**, *40*, 150–162. [CrossRef] [PubMed]

134. Brendel, M.; Probst, F.; Jaworska, A.; Overhoff, F.; Korzhova, V.; Albert, N.L.; Beck, R.; Lindner, S.; Gildehaus, F.J.; Baumann, K.; et al. Glial Activation and Glucose Metabolism in a Transgenic Amyloid Mouse Model: A Triple-Tracer PET Study. *J. Nucl. Med.* **2016**, *57*, 954–960. [CrossRef] [PubMed]

135. DeKosky, S.T.; Scheff, S.W. Synapse loss in frontal cortex biopsies in Alzheimer's disease: Correlation with cognitive severity. *Ann. Neurol.* **1990**, *27*, 457–464. [CrossRef] [PubMed]

136. Lynch, B.A.; Lambeng, N.; Nocka, K.; Kensel-Hammes, P.; Bajjalieh, S.M.; Matagne, A.; Fuks, B. The synaptic vesicle protein SV2A is the binding site for the antiepileptic drug levetiracetam. *Proc. Natl. Acad. Sci. USA* **2004**, *101*, 9861–9866. [CrossRef]

137. Heurling, K.; Ashton, N.J.; Leuzy, A.; Zimmer, E.R.; Blennow, K.; Zetterberg, H.; Eriksson, J.; Lubberink, M.; Schöll, M. Synaptic vesicle protein 2A as a potential biomarker in synaptopathies. *Mol. Cell. Neurosci.* **2019**, *97*, 34–42. [CrossRef]

138. Kong, Y.; Huang, L.; Li, W.; Liu, X.; Zhou, Y.; Liu, C.; Zhang, S.; Xie, F.; Zhang, Z.; Jiang, D.; et al. The Synaptic Vesicle Protein 2A Interacts With Key Pathogenic Factors in Alzheimer's Disease: Implications for Treatment. *Front. Cell Dev. Biol.* **2021**, *9*, 609908. [CrossRef]

139. Chen, M.K.; Mecca, A.P.; Naganawa, M.; Finnema, S.J.; Toyonaga, T.; Lin, S.F.; Najafzadeh, S.; Ropchan, J.; Lu, Y.; McDonald, J.W.; et al. Assessing Synaptic Density in Alzheimer Disease With Synaptic Vesicle Glycoprotein 2A Positron Emission Tomographic Imaging. *JAMA Neurol.* **2018**, *75*, 1215–1224. [CrossRef]

140. Finnema, S.J.; Nabulsi, N.B.; Eid, T.; Detyniecki, K.; Lin, S.F.; Chen, M.K.; Dhaher, R.; Matuskey, D.; Baum, E.; Holden, D.; et al. Imaging synaptic density in the living human brain. *Sci. Transl. Med.* **2016**, *8*, 348ra396. [CrossRef]

141. Nowack, A.; Malarkey, E.B.; Yao, J.; Bleckert, A.; Hill, J.; Bajjalieh, S.M. Levetiracetam reverses synaptic deficits produced by overexpression of SV2A. *PLoS ONE* **2011**, *6*, e29560. [CrossRef]

142. Bahri, M.A.; Plenevaux, A.; Aerts, J.; Bastin, C.; Becker, G.; Mercier, J.; Valade, A.; Buchanan, T.; Mestdagh, N.; Ledoux, D.; et al. Measuring brain synaptic vesicle protein 2A with positron emission tomography and [^{18}F]UCB-H. *Alzheimer's Dement.* **2017**, *3*, 481–486. [CrossRef] [PubMed]

143. Naganawa, M.; Li, S.; Nabulsi, N.; Henry, S.; Zheng, M.Q.; Pracitto, R.; Cai, Z.; Gao, H.; Kapinos, M.; Labaree, D.; et al. First-in-Human Evaluation of ^{18}F-SynVesT-1, a Radioligand for PET Imaging of Synaptic Vesicle Glycoprotein 2A. *J. Nucl. Med.* **2021**, *62*, 561–567. [CrossRef] [PubMed]

144. Li, S.; Cai, Z.; Wu, X.; Holden, D.; Pracitto, R.; Kapinos, M.; Gao, H.; Labaree, D.; Nabulsi, N.; Carson, R.E.; et al. Synthesis and in Vivo Evaluation of a Novel PET Radiotracer for Imaging of Synaptic Vesicle Glycoprotein 2A (SV2A) in Nonhuman Primates. *ACS Chem. Neurosci.* **2019**, *10*, 1544–1554. [CrossRef] [PubMed]

145. Constantinescu, C.C.; Tresse, C.; Zheng, M.; Gouasmat, A.; Carroll, V.M.; Mistico, L.; Alagille, D.; Sandiego, C.M.; Papin, C.; Marek, K.; et al. Development and In Vivo Preclinical Imaging of Fluorine-18-Labeled Synaptic Vesicle Protein 2A (SV2A) PET Tracers. *Mol. Imaging Biol.* **2019**, *21*, 509–518. [CrossRef]

146. O'Dell, R.S.; Mecca, A.P.; Chen, M.-K.; Naganawa, M.; Toyonaga, T.; Lu, Y.; Godek, T.A.; Harris, J.E.; Bartlett, H.H.; Banks, E.R.; et al. Association of Aβ deposition and regional synaptic density in early Alzheimer's disease: A PET imaging study with [^{11}C]UCB-J. *Alzheimer Res. Ther.* **2021**, *13*, 11. [CrossRef]

147. Bertoglio, D.; Verhaeghe, J.; Miranda, A.; Kertesz, I.; Cybulska, K.; Korat, Š.; Wyffels, L.; Stroobants, S.; Mrzljak, L.; Liu, L.; et al. Validation and noninvasive kinetic modeling of [^{11}C]UCB-J PET imaging in mice. *J. Cereb. Blood Flow Metab.* **2019**, *40*, 0271678X1986408. [CrossRef]

148. Cai, Z.; Li, S.; Zhang, W.; Pracitto, R.; Wu, X.; Baum, E.; Finnema, S.J.; Holden, D.; Toyonaga, T.; Lin, S.-f.; et al. Synthesis and Preclinical Evaluation of an ^{18}F-Labeled Synaptic Vesicle Glycoprotein 2A PET Imaging Probe: [^{18}F]SynVesT-2. *ACS Chem. Neurosci.* **2020**, *11*, 592–603. [CrossRef]

149. Iacobucci, G.J.; Popescu, G.K. NMDA receptors: Linking physiological output to biophysical operation. *Nat. Rev. Neurosci.* **2017**, *18*, 236–249. [CrossRef]

150. Nedergaard, M.; Takano, T.; Hansen, A.J. Beyond the role of glutamate as a neurotransmitter. *Nat. Rev. Neurosci.* **2002**, *3*, 748–755. [CrossRef]

151. Palop, J.J.; Mucke, L. Network abnormalities and interneuron dysfunction in Alzheimer disease. *Nat. Rev. Neurosci.* **2016**, *17*, 777–792. [CrossRef]

152. Wang, R.; Reddy, P.H. Role of Glutamate and NMDA Receptors in Alzheimer's Disease. *J. Alzheimers Dis.* **2017**, *57*, 1041–1048. [CrossRef] [PubMed]

153. Tanzi, R.E. The synaptic Abeta hypothesis of Alzheimer disease. *Nat. Neurosci.* **2005**, *8*, 977–979. [CrossRef] [PubMed]

154. Snyder, E.M.; Nong, Y.; Almeida, C.G.; Paul, S.; Moran, T.; Choi, E.Y.; Nairn, A.C.; Salter, M.W.; Lombroso, P.J.; Gouras, G.K.; et al. Regulation of NMDA receptor trafficking by amyloid-beta. *Nat. Neurosci.* **2005**, *8*, 1051–1058. [CrossRef] [PubMed]
155. Shankar, G.M.; Bloodgood, B.L.; Townsend, M.; Walsh, D.M.; Selkoe, D.J.; Sabatini, B.L. Natural oligomers of the Alzheimer amyloid-beta protein induce reversible synapse loss by modulating an NMDA-type glutamate receptor-dependent signaling pathway. *J. Neurosci.* **2007**, *27*, 2866–2875. [CrossRef] [PubMed]
156. Um, J.W.; Kaufman, A.C.; Kostylev, M.; Heiss, J.K.; Stagi, M.; Takahashi, H.; Kerrisk, M.E.; Vortmeyer, A.; Wisniewski, T. Metabotropic Glutamate Receptor 5 Is a Coreceptor for Alzheimer Aβ Oligomer Bound to Cellular Prion Protein. *Neuron* **2013**, *79*, 887–902. [CrossRef]
157. Reinders, N.R.; Pao, Y.; Renner, M.C.; da Silva-Matos, C.M.; Lodder, T.R.; Malinow, R.; Kessels, H.W. Amyloid-β effects on synapses and memory require AMPA receptor subunit GluA3. *Proc. Natl. Acad. Sci. USA* **2016**, *113*, E6526–E6534. [CrossRef] [PubMed]
158. Tanaka, H.; Sakaguchi, D.; Hirano, T. Amyloid-β oligomers suppress subunit-specific glutamate receptor increase during LTP. *Alzheimers Dement.* **2019**, *5*, 797–808. [CrossRef] [PubMed]
159. Zott, B.; Simon, M.M.; Hong, W.; Unger, F.; Chen-Engerer, H.J.; Frosch, M.P.; Sakmann, B.; Walsh, D.M.; Konnerth, A. A vicious cycle of β amyloid-dependent neuronal hyperactivation. *Science* **2019**, *365*, 559–565. [CrossRef]
160. Hamilton, A.; Vasefi, M.; Vander Tuin, C.; McQuaid, R.J.; Anisman, H.; Ferguson, S.S. Chronic Pharmacological mGluR5 Inhibition Prevents Cognitive Impairment and Reduces Pathogenesis in an Alzheimer Disease Mouse Model. *Cell Rep.* **2016**, *15*, 1859–1865. [CrossRef]
161. Miyazaki, T.; Nakajima, W.; Hatano, M.; Shibata, Y.; Kuroki, Y.; Arisawa, T.; Serizawa, A.; Sano, A.; Kogami, S.; Yamanoue, T.; et al. Visualization of AMPA receptors in living human brain with positron emission tomography. *Nat. Med.* **2020**, *26*, 281–288. [CrossRef]
162. Takahata, K.; Kimura, Y.; Seki, C.; Tokunaga, M.; Ichise, M.; Kawamura, K.; Ono, M.; Kitamura, S.; Kubota, M.; Moriguchi, S.; et al. A human PET study of [^{11}C]HMS011, a potential radioligand for AMPA receptors. *EJNMMI Res.* **2017**, *7*, 63. [CrossRef] [PubMed]
163. Vibholm, A.K.; Landau, A.M.; Møller, A.; Jacobsen, J.; Vang, K.; Munk, O.L.; Orlowski, D.; Sørensen, J.C.; Brooks, D.J. NMDA receptor ion channel activation detected in vivo with [^{18}F]GE-179 PET after electrical stimulation of rat hippocampus. *J. Cereb. Blood Flow Metab.* **2021**, *41*, 1301–1312. [CrossRef] [PubMed]
164. van der Aart, J.; Golla, S.S.V.; van der Pluijm, M.; Schwarte, L.A.; Schuit, R.C.; Klein, P.J.; Metaxas, A.; Windhorst, A.D.; Boellaard, R.; Lammertsma, A.A.; et al. First in human evaluation of [^{18}F]PK-209, a PET ligand for the ion channel binding site of NMDA receptors. *EJNMMI Res.* **2018**, *8*, 69. [CrossRef] [PubMed]
165. Krämer, S.D.; Betzel, T.; Mu, L.; Haider, A.; Herde, A.M.; Boninsegni, A.K.; Keller, C.; Szermerski, M.; Schibli, R.; Wünsch, B.; et al. Evaluation of ^{11}C-Me-NB1 as a Potential PET Radioligand for Measuring GluN2B-Containing NMDA Receptors, Drug Occupancy, and Receptor Cross Talk. *J. Nucl. Med.* **2018**, *59*, 698–703. [CrossRef]
166. Abd-Elrahman, K.S.; Albaker, A.; de Souza, J.M.; Ribeiro, F.M.; Schlossmacher, M.G.; Tiberi, M.; Hamilton, A.; Ferguson, S.S.G. Aβ oligomers induce pathophysiological mGluR5 signaling in Alzheimer's disease model mice in a sex-selective manner. *Sci. Signal* **2020**, *13*. [CrossRef] [PubMed]
167. Wong, D.F.; Waterhouse, R.; Kuwabara, H.; Kim, J.; Brašić, J.R.; Chamroonrat, W.; Stabins, M.; Holt, D.P.; Dannals, R.F.; Hamill, T.G.; et al. ^{18}F-FPEB, a PET radiopharmaceutical for quantifying metabotropic glutamate 5 receptors: A first-in-human study of radiochemical safety, biokinetics, and radiation dosimetry. *J. Nucl. Med.* **2013**, *54*, 388–396. [CrossRef] [PubMed]
168. Ametamey, S.M.; Kessler, L.J.; Honer, M.; Wyss, M.T.; Buck, A.; Hintermann, S.; Auberson, Y.P.; Gasparini, F.; Schubiger, P.A. Radiosynthesis and preclinical evaluation of ^{11}C-ABP688 as a probe for imaging the metabotropic glutamate receptor subtype 5. *J. Nucl. Med.* **2006**, *47*, 698–705. [PubMed]
169. Warnock, G.; Sommerauer, M.; Mu, L.; Pla Gonzalez, G.; Geistlich, S.; Treyer, V.; Schibli, R.; Buck, A.; Krämer, S.D.; Ametamey, S.M. A first-in-man PET study of [^{18}F]PSS232, a fluorinated ABP688 derivative for imaging metabotropic glutamate receptor subtype 5. *Eur. J. Nucl. Med. Mol. Imaging* **2018**, *45*, 1041–1051. [CrossRef]
170. Mecca, A.P.; McDonald, J.W.; Michalak, H.R.; Godek, T.A.; Harris, J.E.; Pugh, E.A.; Kemp, E.C.; Chen, M.K.; Salardini, A.; Nabulsi, N.B.; et al. PET imaging of mGluR5 in Alzheimer's disease. *Alzheimers Res. Ther.* **2020**, *12*, 15. [CrossRef]
171. Treyer, V.; Gietl, A.F.; Suliman, H.; Gruber, E.; Meyer, R.; Buchmann, A.; Johayem, A.; Unschuld, P.G.; Nitsch, R.M.; Buck, A.; et al. Reduced uptake of [^{11}C]-ABP688, a PET tracer for metabolic glutamate receptor 5 in hippocampus and amygdala in Alzheimer's dementia. *Brain Behav.* **2020**, *10*, e01632. [CrossRef]
172. Nordberg, A. Nicotinic receptor abnormalities of Alzheimer's disease: Therapeutic implications. *Biol. Psychiatry* **2001**, *49*, 200–210. [CrossRef]
173. Hampel, H.; Mesulam, M.M.; Cuello, A.C.; Farlow, M.R.; Giacobini, E.; Grossberg, G.T.; Khachaturian, A.S.; Vergallo, A.; Cavedo, E.; Snyder, P.J.; et al. The cholinergic system in the pathophysiology and treatment of Alzheimer's disease. *Brain* **2018**, *141*, 1917–1933. [CrossRef] [PubMed]
174. Wang, H.; Yu, M.; Ochani, M.; Amella, C.A.; Tanovic, M.; Susarla, S.; Li, J.H.; Yang, H.; Ulloa, L.; Al-Abed, Y.; et al. Nicotinic acetylcholine receptor alpha7 subunit is an essential regulator of inflammation. *Nature* **2003**, *421*, 384–388. [CrossRef] [PubMed]
175. Marutle, A.; Gillberg, P.G.; Bergfors, A.; Yu, W.F.; Ni, R.; Nennesmo, I.; Voytenko, L.; Nordberg, A. H-3-Deprenyl and H-3-PIB autoradiography show different laminar distributions of astroglia and fibrillar beta-amyloid in Alzheimer brain. *J. Neuroinflamm.* **2013**, *10*, S491–S496. [CrossRef]

176. Ikonomovic, M.D.; Wecker, L.; Abrahamson, E.E.; Wuu, J.; Counts, S.E.; Ginsberg, S.D.; Mufson, E.J.; Dekosky, S.T. Cortical alpha7 nicotinic acetylcholine receptor and beta-amyloid levels in early Alzheimer disease. *Arch. Neurol.* **2009**, *66*, 646–651. [CrossRef]

177. Yi, J.H.; Whitcomb, D.J.; Park, S.J.; Martinez-Perez, C.; Barbati, S.A.; Mitchell, S.J.; Cho, K. M1 muscarinic acetylcholine receptor dysfunction in moderate Alzheimer's disease pathology. *Brain Commun.* **2020**, *2*, fcaa058. [CrossRef]

178. Ni, R.; Marutle, A.; Nordberg, A. Modulation of α7 nicotinic acetylcholine receptor and fibrillar amyloid-β interactions in Alzheimer's disease brain. *J. Alzheimers Dis.* **2013**, *33*, 841–851. [CrossRef]

179. Wang, H.-Y.; Stucky, A.; Liu, J.; Shen, C.; Trocme-Thibierge, C.; Morain, P. Dissociating beta-amyloid from alpha 7 nicotinic acetylcholine receptor by a novel therapeutic agent, S 24795, normalizes alpha 7 nicotinic acetylcholine and NMDA receptor function in Alzheimer's disease brain. *J. Neurosci.* **2009**, *29*, 10961–10973. [CrossRef]

180. Medeiros, R.; Castello, N.A.; Cheng, D.; Kitazawa, M.; Baglietto-Vargas, D.; Green, K.N.; Esbenshade, T.A.; Bitner, R.S.; Decker, M.W.; LaFerla, F.M. α7 Nicotinic receptor agonist enhances cognition in aged 3xTg-AD mice with robust plaques and tangles. *Am. J. Pathol.* **2014**, *184*, 520–529. [CrossRef]

181. George, A.A.; Vieira, J.M.; Xavier-Jackson, C.; Gee, M.T.; Cirrito, J.R.; Bimonte-Nelson, H.A.; Picciotto, M.R.; Lukas, R.J.; Whiteaker, P. Implications of Oligomeric Amyloid-Beta (oAβ(42)) Signaling through α7β2-Nicotinic Acetylcholine Receptors (nAChRs) on Basal Forebrain Cholinergic Neuronal Intrinsic Excitability and Cognitive Decline. *J. Neurosci.* **2021**, *41*, 555–575. [CrossRef]

182. Ettrup, A.; Mikkelsen, J.D.; Lehel, S.; Madsen, J.; Nielsen, E.; Palner, M.; Timmermann, D.B.; Peters, D.; Knudsen, G.M. ^{11}C-NS14492 as a novel PET radioligand for imaging cerebral alpha7 nicotinic acetylcholine receptors: In vivo evaluation and drug occupancy measurements. *J. Nucl. Med.* **2011**, *52*, 1449–1456. [CrossRef] [PubMed]

183. Gao, Y.; Kellar, K.J.; Yasuda, R.P.; Tran, T.; Xiao, Y.; Dannals, R.F.; Horti, A.G. Derivatives of dibenzothiophene for positron emission tomography imaging of α7-nicotinic acetylcholine receptors. *J. Med. Chem.* **2013**, *56*, 7574–7589. [CrossRef] [PubMed]

184. Yamamoto, S.; Nishiyama, S.; Kawamata, M.; Ohba, H.; Wakuda, T.; Takei, N.; Tsukada, H.; Domino, E.F. Muscarinic Receptor Occupancy and Cognitive Impairment: A PET Study with [11]C(+)3-MPB and Scopolamine in Conscious Monkeys. *Neuropsychopharmacology* **2011**, *36*, 1455–1465. [CrossRef] [PubMed]

185. Rowe, C.C.; Krishnadas, N.; Ackermann, U.; Doré, V.; Goh, R.Y.W.; Guzman, R.; Chong, L.; Bozinovski, S.; Mulligan, R.; Kanaan, R.; et al. PET Imaging of brain muscarinic receptors with ^{18}F-Fluorobenzyl-Dexetimide: A first in human study. *Psychiatry Res. Neuroimaging* **2021**, *316*, 111354. [CrossRef]

186. Nabulsi, N.B.; Holden, D.; Zheng, M.Q.; Bois, F.; Lin, S.F.; Najafzadeh, S.; Gao, H.; Ropchan, J.; Lara-Jaime, T.; Labaree, D.; et al. Evaluation of ^{11}C-LSN3172176 as a Novel PET Tracer for Imaging M(1) Muscarinic Acetylcholine Receptors in Nonhuman Primates. *J. Nucl. Med.* **2019**, *60*, 1147–1153. [CrossRef]

187. Tong, L.; Li, W.; Lo, M.M.-C.; Gao, X.; Wai, J.M.-C.; Rudd, M.; Tellers, D.; Joshi, A.; Zeng, Z.; Miller, P.; et al. Discovery of ^{11}C]MK-6884: A Positron Emission Tomography (PET) Imaging Agent for the Study of M4Muscarinic Receptor Positive Allosteric Modulators (PAMs) in Neurodegenerative Diseases. *J. Med. Chem.* **2020**, *63*, 2411–2425. [CrossRef] [PubMed]

188. Kadir, A.; Almkvist, O.; Wall, A.; Långström, B.; Nordberg, A. PET imaging of cortical ^{11}C-nicotine binding correlates with the cognitive function of attention in Alzheimer's disease. *Psychopharmacology* **2006**, *188*, 509–520. [CrossRef]

189. Montagne, A.; Nikolakopoulou, A.M.; Huuskonen, M.T.; Sagare, A.P.; Lawson, E.J.; Lazic, D.; Rege, S.V.; Grond, A.; Zuniga, E.; Barnes, S.R.; et al. APOE4 accelerates advanced-stage vascular and neurodegenerative disorder in old Alzheimer's mice via cyclophilin A independently of amyloid-β. *Nat. Aging* **2021**, *1*, 506–520. [CrossRef]

190. Montagne, A.; Nation, D.A.; Sagare, A.P.; Barisano, G.; Sweeney, M.D.; Chakhoyan, A.; Pachicano, M.; Joe, E.; Nelson, A.R.; D'Orazio, L.M.; et al. APOE4 leads to blood-brain barrier dysfunction predicting cognitive decline. *Nature* **2020**, *581*, 71–76. [CrossRef]

191. Bien-Ly, N.; Boswell, C.A.; Jeet, S.; Beach, T.G.; Hoyte, K.; Luk, W.; Shihadeh, V.; Ulufatu, S.; Foreman, O.; Lu, Y.; et al. Lack of Widespread BBB Disruption in Alzheimer's Disease Models: Focus on Therapeutic Antibodies. *Neuron* **2015**, *88*, 289–297. [CrossRef]

192. Merlini, M.; Meyer, E.P.; Ulmann-Schuler, A.; Nitsch, R.M. Vascular β-amyloid and early astrocyte alterations impair cerebrovascular function and cerebral metabolism in transgenic arcAβ mice. *Acta Neuropathol.* **2011**, *122*, 293–311. [CrossRef] [PubMed]

193. Erdő, F.; Denes, L.; de Lange, E. Age-associated physiological and pathological changes at the blood-brain barrier: A review. *J. Cereb. Blood Flow Metab.* **2017**, *37*, 4–24. [CrossRef] [PubMed]

194. Cirrito, J.R.; Deane, R.; Fagan, A.M.; Spinner, M.L.; Parsadanian, M.; Finn, M.B.; Jiang, H.; Prior, J.L.; Sagare, A.; Bales, K.R.; et al. P-glycoprotein deficiency at the blood-brain barrier increases amyloid-beta deposition in an Alzheimer disease mouse model. *J. Clin. Investig.* **2005**, *115*, 3285–3290. [CrossRef] [PubMed]

195. Mossel, P.; Garcia Varela, L.; Arif, W.M.; van der Weijden, C.W.J.; Boersma, H.H.; Willemsen, A.T.M.; Boellaard, R.; Elsinga, P.H.; Borra, R.J.H.; Colabufo, N.A.; et al. Evaluation of P-glycoprotein function at the blood-brain barrier using ^{18}F]MC225-PET. *Eur. J. Nucl. Med. Mol. Imaging* **2021**, 1–2. [CrossRef]

196. Raaphorst, R.M.; Luurtsema, G.; Schuit, R.C.; Kooijman, E.J.M.; Elsinga, P.H.; Lammertsma, A.A.; Windhorst, A.D. Synthesis and Evaluation of New Fluorine-18 Labeled Verapamil Analogs To Investigate the Function of P-Glycoprotein in the Blood-Brain Barrier. *ACS Chem. Neurosci.* **2017**, *8*, 1925–1936. [CrossRef]

197. García-Varela, L.; Arif, W.M.; Vállez García, D.; Kakiuchi, T.; Ohba, H.; Harada, N.; Tago, T.; Elsinga, P.H.; Tsukada, H.; Colabufo, N.A.; et al. Pharmacokinetic Modeling of [^{18}F]MC225 for Quantification of the P-Glycoprotein Function at the Blood–Brain Barrier in Non-Human Primates with PET. *Mol. Pharm.* **2020**, *17*, 3477–3486. [CrossRef]

198. Savolainen, H.; Windhorst, A.D.; Elsinga, P.H.; Cantore, M.; Colabufo, N.A.; Willemsen, A.T.; Luurtsema, G. Evaluation of [^{18}F]MC225 as a PET radiotracer for measuring P-glycoprotein function at the blood-brain barrier in rats: Kinetics, metabolism, and selectivity. *J. Cereb. Blood Flow Metab.* **2017**, *37*, 1286–1298. [CrossRef]

199. Schmidt, A.M.; Yan, S.D.; Yan, S.F.; Stern, D.M. The biology of the receptor for advanced glycation end products and its ligands. *Biochim. Biophys. Acta* **2000**, *1498*, 99–111. [CrossRef]

200. Yan, S.D.; Chen, X.; Fu, J.; Chen, M.; Zhu, H.; Roher, A.; Slattery, T.; Zhao, L.; Nagashima, M.; Morser, J.; et al. RAGE and amyloid-beta peptide neurotoxicity in Alzheimer's disease. *Nature* **1996**, *382*, 685–691. [CrossRef]

201. Kong, Y.; Hua, F.; Guan, Y.; Zhao, B. RAGE-specific probe ^{18}F -FPS-ZM1 may be a promising biomarker for early detection of Diabetes with Alzheimer's disease. *J. Nucl. Med.* **2016**, *57*, 1049.

202. Cary, B.P.; Brooks, A.F.; Fawaz, M.V.; Drake, L.R.; Desmond, T.J.; Sherman, P.; Quesada, C.A.; Scott, P.J. Synthesis and Evaluation of [^{18}F]RAGER: A First Generation Small-Molecule PET Radioligand Targeting the Receptor for Advanced Glycation Endproducts. *ACS Chem. Neurosci.* **2016**, *7*, 391–398. [CrossRef]

203. Drake, L.R.; Brooks, A.F.; Stauff, J.; Sherman, P.S.; Arteaga, J.; Koeppe, R.A.; Reed, A.; Montavon, T.J.; Skaddan, M.B.; Scott, P.J.H. Strategies for PET imaging of the receptor for advanced glycation endproducts (RAGE). *J. Pharm. Anal.* **2020**, *10*, 452–465. [CrossRef]

204. Konopka, C.J.; Wozniak, M.; Hedhli, J.; Ploska, A.; Schwartz-Duval, A.; Siekierzycka, A.; Pan, D.; Munirathinam, G.; Dobrucki, I.T.; Kalinowski, L.; et al. Multimodal imaging of the receptor for advanced glycation end-products with molecularly targeted nanoparticles. *Theranostics* **2018**, *8*, 5012–5024. [CrossRef]

205. Kreisl, W.C.; Kim, M.J.; Coughlin, J.M.; Henter, I.D.; Owen, D.R.; Innis, R.B. PET imaging of neuroinflammation in neurological disorders. *Lancet Neurol.* **2020**, *19*, 940–950. [CrossRef]

206. Leng, F.; Edison, P. Neuroinflammation and microglial activation in Alzheimer disease: Where do we go from here? *Nat. Rev. Neurol.* **2021**, *17*, 157–172. [CrossRef] [PubMed]

207. Janssen, B.; Mach, R.H. Development of brain PET imaging agents: Strategies for imaging neuroinflammation in Alzheimer's disease. *Prog. Mol. Biol. Transl. Sci.* **2019**, *165*, 371–399. [CrossRef]

208. Van Camp, N.; Lavisse, S.; Roost, P.; Gubinelli, F.; Hillmer, A.; Boutin, H. TSPO imaging in animal models of brain diseases. *Eur. J. Nucl. Med. Mol. Imaging* **2021**, 1–33. [CrossRef] [PubMed]

209. Bellaver, B.; Ferrari-Souza, J.P.; Uglione da Ros, L.; Carter, S.F.; Rodriguez-Vieitez, E.; Nordberg, A.; Pellerin, L.; Rosa-Neto, P.; Leffa, D.T.; Zimmer, E.R. Astrocyte Biomarkers in Alzheimer Disease: A Systematic Review and Meta-analysis. *Neurology* **2021**, *96*, e2944–e2955. [CrossRef]

210. Zhou, R.; Ji, B.; Kong, Y.; Qin, L.; Ren, W.; Guan, Y.; Ni, R. PET Imaging of Neuroinflammation in Alzheimer's Disease. *Front. Immunol.* **2021**, *12*, 3750. [CrossRef]

211. Pascoal, T.A.; Benedet, A.L.; Ashton, N.J.; Kang, M.S.; Therriault, J.; Chamoun, M.; Savard, M.; Lussier, F.Z.; Tissot, C.; Karikari, T.K.; et al. Microglial activation and tau propagate jointly across Braak stages. *Nat. Med.* **2021**, *27*, 1592–1599. [CrossRef]

212. Ising, C.; Venegas, C.; Zhang, S.; Scheiblich, H.; Schmidt, S.V.; Vieira-Saecker, A.; Schwartz, S.; Albasset, S.; McManus, R.M.; Tejera, D.; et al. NLRP3 inflammasome activation drives tau pathology. *Nature* **2019**, *575*, 669–673. [CrossRef] [PubMed]

213. Heneka, M.T.; Carson, M.J.; El Khoury, J.; Landreth, G.E.; Brosseron, F.; Feinstein, D.L.; Jacobs, A.H.; Wyss-Coray, T.; Vitorica, J.; Ransohoff, R.M.; et al. Neuroinflammation in Alzheimer's disease. *Lancet Neurol.* **2015**, *14*, 388–405. [CrossRef]

214. Huang, Y.; Happonen, K.E.; Burrola, P.G.; O'Connor, C.; Hah, N.; Huang, L.; Nimmerjahn, A.; Lemke, G. Microglia use TAM receptors to detect and engulf amyloid β plaques. *Nat. Immunol.* **2021**, *22*, 586–594. [CrossRef] [PubMed]

215. Deczkowska, A.; Keren-Shaul, H.; Weiner, A.; Colonna, M.; Schwartz, M.; Amit, I. Disease-Associated Microglia: A Universal Immune Sensor of Neurodegeneration. *Cell* **2018**, *173*, 1073–1081. [CrossRef] [PubMed]

216. Keren-Shaul, H.; Spinrad, A.; Weiner, A.; Matcovitch-Natan, O.; Dvir-Szternfeld, R.; Ulland, T.K.; David, E.; Baruch, K.; Lara-Astaiso, D.; Toth, B.; et al. A Unique Microglia Type Associated with Restricting Development of Alzheimer's Disease. *Cell* **2017**, *169*, 1276–1290.e17. [CrossRef]

217. Song, W.M.; Colonna, M. The identity and function of microglia in neurodegeneration. *Nat. Immunol.* **2018**, *19*, 1048–1058. [CrossRef]

218. Venneti, S.; Lopresti, B.J.; Wang, G.; Hamilton, R.L.; Mathis, C.A.; Klunk, W.E.; Apte, U.M.; Wiley, C.A. PK11195 labels activated microglia in Alzheimer's disease and in vivo in a mouse model using PET. *Neurobiol. Aging* **2009**, *30*, 1217–1226. [CrossRef]

219. Mirzaei, N.; Tang, S.P.; Ashworth, S.; Coello, C.; Plisson, C.; Passchier, J.; Selvaraj, V.; Tyacke, R.J.; Nutt, D.J.; Sastre, M. In vivo imaging of microglial activation by positron emission tomography with [^{11}C]PBR28 in the 5XFAD model of Alzheimer's disease. *Glia* **2016**, *64*, 993–1006. [CrossRef]

220. Ikawa, M.; Lohith, T.G.; Shrestha, S.; Telu, S.; Zoghbi, S.S.; Castellano, S.; Taliani, S.; Da Settimo, F.; Fujita, M.; Pike, V.W.; et al. ^{11}C-ER176, a Radioligand for 18-kDa Translocator Protein, Has Adequate Sensitivity to Robustly Image All Three Affinity Genotypes in Human Brain. *J. Nucl. Med.* **2017**, *58*, 320–325. [CrossRef]

221. Wright, A.L.; Zinn, R.; Hohensinn, B.; Konen, L.M.; Beynon, S.B.; Tan, R.P.; Clark, I.A.; Abdipranoto, A.; Vissel, B. Neuroinflammation and Neuronal Loss Precede Aβ Plaque Deposition in the hAPP-J20 Mouse Model of Alzheimer's Disease. *PLoS ONE* **2013**, *8*, e59586. [CrossRef]
222. López-Picón, F.R.; Snellman, A.; Eskola, O.; Helin, S.; Solin, O.; Haaparanta-Solin, M.; Rinne, J.O. Neuroinflammation Appears Early on PET Imaging and Then Plateaus in a Mouse Model of Alzheimer Disease. *J. Nucl. Med.* **2018**, *59*, 509. [CrossRef]
223. Brendel, M.; Kleinberger, G.; Probst, F.; Jaworska, A.; Overhoff, F.; Blume, T.; Albert, N.L.; Carlsen, J.; Lindner, S.; Gildehaus, F.J.; et al. Increase of TREM2 during Aging of an Alzheimer's Disease Mouse Model Is Paralleled by Microglial Activation and Amyloidosis. *Front. Aging Neurosci.* **2017**, *9*, 8. [CrossRef]
224. Focke, C.; Blume, T.; Zott, B.; Shi, Y.; Deussing, M.; Peters, F.; Schmidt, C.; Kleinberger, G.; Lindner, S.; Gildehaus, F.J.; et al. Early and Longitudinal Microglial Activation but Not Amyloid Accumulation Predicts Cognitive Outcome in PS2APP Mice. *J. Nucl. Med.* **2019**, *60*, 548–554. [CrossRef] [PubMed]
225. Chaney, A.; Cropper, H.C.; Johnson, E.M.; Lechtenberg, K.J.; Peterson, T.C.; Stevens, M.Y.; Buckwalter, M.S.; James, M.L. ^{11}C-DPA-713 Versus ^{18}F-GE-180: A Preclinical Comparison of Translocator Protein 18 kDa PET Tracers to Visualize Acute and Chronic Neuroinflammation in a Mouse Model of Ischemic Stroke. *J. Nucl. Med.* **2019**, *60*, 122–128. [CrossRef]
226. Ji, B.; Ono, M.; Yamasaki, T.; Fujinaga, M.; Zhang, M.R.; Seki, C.; Aoki, I.; Kito, S.; Sawada, M.; Suhara, T.; et al. Detection of Alzheimer's disease-related neuroinflammation by a PET ligand selective for glial versus vascular translocator protein. *J. Cereb. Blood Flow Metab.* **2021**, 271678x21992457. [CrossRef] [PubMed]
227. Beaino, W.; Janssen, B.; Vugts, D.J.; de Vries, H.E.; Windhorst, A.D. Toward PET imaging of the dynamic phenotypes of microglia. *Clin. Exp. Immunol.* **2021**, *206*, 282–300. [CrossRef] [PubMed]
228. Ni, R.; Müller Herde, A.; Haider, A.; Keller, C.; Louloudis, G.; Vaas, M.; Schibli, R.; Ametamey, S.M.; Klohs, J.; Mu, L. In vivo Imaging of Cannabinoid Type 2 Receptors: Functional and Structural Alterations in Mouse Model of Cerebral Ischemia by PET and MRI. *bioRxiv* **2021**. [CrossRef]
229. Hagens, M.H.J.; Golla, S.S.V.; Janssen, B.; Vugts, D.J.; Beaino, W.; Windhorst, A.D.; O'Brien-Brown, J.; Kassiou, M.; Schuit, R.C.; Schwarte, L.A.; et al. The P2X7 receptor tracer [11]CSMW139 as an in vivo marker of neuroinflammation in multiple sclerosis: A first-in man study. *Eur. J. Nucl. Med. Mol. Imaging* **2020**, *47*, 379–389. [CrossRef]
230. Janssen, B.; Vugts, D.J.; Wilkinson, S.M.; Ory, D.; Chalon, S.; Hoozemans, J.J.M.; Schuit, R.C.; Beaino, W.; Kooijman, E.J.M.; van den Hoek, J.; et al. Identification of the allosteric P2X7 receptor antagonist [11]CSMW139 as a PET tracer of microglial activation. *Sci. Rep.* **2018**, *8*, 6580. [CrossRef]
231. Maeda, J.; Minamihisamatsu, T.; Shimojo, M.; Zhou, X.; Ono, M.; Matsuba, Y.; Ji, B.; Ishii, H.; Ogawa, M.; Akatsu, H.; et al. Distinct microglial response against Alzheimer's amyloid and tau pathologies characterized by P2Y12 receptor. *Brain Commun.* **2021**, *3*, fcab011. [CrossRef]
232. Horti, A.G.; Naik, R.; Foss, C.A.; Minn, I.; Misheneva, V.; Du, Y.; Wang, Y.; Mathews, W.B.; Wu, Y.; Hall, A.; et al. PET imaging of microglia by targeting macrophage colony-stimulating factor 1 receptor (CSF1R). *Proc. Natl. Acad. Sci. USA* **2019**, *116*, 1686–1691. [CrossRef] [PubMed]
233. Zhou, X.; Ji, B.; Seki, C.; Nagai, Y.; Minamimoto, T.; Fujinaga, M.; Zhang, M.R.; Saito, T.; Saido, T.C.; Suhara, T.; et al. PET imaging of colony-stimulating factor 1 receptor: A head-to-head comparison of a novel radioligand, ^{11}C-GW2580, and ^{11}C-CPPC, in mouse models of acute and chronic neuroinflammation and a rhesus monkey. *J. Cereb. Blood Flow Metab.* **2021**, 271678x211004146. [CrossRef] [PubMed]
234. Shukuri, M.; Mawatari, A.; Ohno, M.; Suzuki, M.; Doi, H.; Watanabe, Y.; Onoe, H. Detection of Cyclooxygenase-1 in Activated Microglia During Amyloid Plaque Progression: PET Studies in Alzheimer's Disease Model Mice. *J. Nucl. Med.* **2016**, *57*, 291–296. [CrossRef] [PubMed]
235. Meier, S.R.; Sehlin, D.; Hultqvist, G.; Syvänen, S. Pinpointing Brain TREM2 Levels in Two Mouse Models of Alzheimer's Disease. *Mol. Imaging Biol.* **2021**, 1–11. [CrossRef]
236. Thomsen, M.B.; Jacobsen, J.; Lillethorup, T.P.; Schacht, A.C.; Simonsen, M.; Romero-Ramos, M.; Brooks, D.J.; Landau, A.M. In vivo imaging of synaptic SV2A protein density in healthy and striatal-lesioned rats with [11]CUCB-J PET. *J. Cereb. Blood Flow Metab.* **2021**, *41*, 819–830. [CrossRef] [PubMed]
237. Parbo, P.; Ismail, R.; Hansen, K.V.; Amidi, A.; Mårup, F.H.; Gottrup, H.; Brændgaard, H.; Eriksson, B.O.; Eskildsen, S.F.; Lund, T.E.; et al. Brain inflammation accompanies amyloid in the majority of mild cognitive impairment cases due to Alzheimer's disease. *Brain* **2017**, *140*, 2002–2011. [CrossRef] [PubMed]
238. Hou, C.; Hsieh, C.-J.; Li, S.; Lee, H.; Graham, T.J.; Xu, K.; Weng, C.-C.; Doot, R.K.; Chu, W.; Chakraborty, S.K.; et al. Development of a Positron Emission Tomography Radiotracer for Imaging Elevated Levels of Superoxide in Neuroinflammation. *ACS Chem. Neurosci.* **2018**, *9*, 578–586. [CrossRef]
239. Schützmann, M.P.; Hasecke, F.; Bachmann, S.; Zielinski, M.; Hänsch, S.; Schröder, G.F.; Zempel, H.; Hoyer, W. Endo-lysosomal Aβ concentration and pH trigger formation of Aβ oligomers that potently induce Tau missorting. *Nat. Commun.* **2021**, *12*, 4634. [CrossRef]
240. Kumar, J.S.D.; Solingapuram Sai, K.K.; Prabhakaran, J.; Oufkir, H.R.; Ramanathan, G.; Whitlow, C.T.; Dileep, H.; Mintz, A.; Mann, J.J. Radiosynthesis and in Vivo Evaluation of [11]CMPC-6827, the First Brain Penetrant Microtubule PET Ligand. *J. Med. Chem.* **2018**, *61*, 2118–2123. [CrossRef]

241. Solingapuram Sai, K.K.; Prabhakaran, J.; Ramanathan, G.; Rideout, S.; Whitlow, C.; Mintz, A.; Mann, J.J.; Kumar, J.S.D. Radiosynthesis and Evaluation of [^{11}C]HD-800, a High Affinity Brain Penetrant PET Tracer for Imaging Microtubules. *ACS Med. Chem. Lett.* **2018**, *9*, 452–456. [CrossRef]

242. Baum, E.; Cai, Z.; Bois, F.; Holden, D.; Lin, S.F.; Lara-Jaime, T.; Kapinos, M.; Chen, Y.; Deuther-Conrad, W.; Fischer, S.; et al. PET Imaging Evaluation of Four σ(1) Radiotracers in Nonhuman Primates. *J. Nucl. Med.* **2017**, *58*, 982–988. [CrossRef] [PubMed]

243. Lepelletier, F.-X.; Vandesquille, M.; Asselin, M.-C.; Prenant, C.; Robinson, A.C.; Mann, D.M.A.; Green, M.; Barnett, E.; Banister, S.D.; Mottinelli, M.; et al. Evaluation of ^{18}F-IAM6067 as a sigma-1 receptor PET tracer for neurodegeneration in vivo in rodents and in human tissue. *Theranostics* **2020**, *10*, 7938–7955. [CrossRef] [PubMed]

244. Lan, Y.; Bai, P.; Chen, Z.; Neelamegam, R.; Placzek, M.S.; Wang, H.; Fiedler, S.A.; Yang, J.; Yuan, G.; Qu, X.; et al. Novel radioligands for imaging sigma-1 receptor in brain using positron emission tomography (PET). *Acta Pharm. Sin. B* **2019**, *9*, 1204–1215. [CrossRef] [PubMed]

245. Knight, A.C.; Varlow, C.; Tong, J.; Vasdev, N. In Vitro and In Vivo Evaluation of GSK-3 Radioligands in Alzheimer's Disease: Preliminary Evidence of Sex Differences. *ACS Pharmacol. Transl. Sci.* **2021**, *4*, 1287–1294. [CrossRef]

246. Escartin, C.; Galea, E.; Lakatos, A.; O'Callaghan, J.P.; Petzold, G.C.; Serrano-Pozo, A.; Steinhäuser, C.; Volterra, A.; Carmignoto, G.; Agarwal, A.; et al. Reactive astrocyte nomenclature, definitions, and future directions. *Nat. Neurosci.* **2021**, *24*, 312–325. [CrossRef]

247. Joshi, A.U.; Minhas, P.S.; Liddelow, S.A.; Haileselassie, B.; Andreasson, K.I.; Dorn, G.W., 2nd; Mochly-Rosen, D. Fragmented mitochondria released from microglia trigger A1 astrocytic response and propagate inflammatory neurodegeneration. *Nat. Neurosci.* **2019**, *22*, 1635–1648. [CrossRef]

248. Castellani, G.; Schwartz, M. Immunological Features of Non-neuronal Brain Cells: Implications for Alzheimer's Disease Immunotherapy. *Trends Immunol.* **2020**, *41*, 794–804. [CrossRef]

249. McAlpine, C.S.; Park, J.; Griciuc, A.; Kim, E.; Choi, S.H.; Iwamoto, Y.; Kiss, M.G.; Christie, K.A.; Vinegoni, C.; Poller, W.C.; et al. Astrocytic interleukin-3 programs microglia and limits Alzheimer's disease. *Nature* **2021**, *595*, 701–706. [CrossRef]

250. Damisah, E.C.; Hill, R.A.; Rai, A.; Chen, F.; Rothlin, C.V.; Ghosh, S.; Grutzendler, J. Astrocytes and microglia play orchestrated roles and respect phagocytic territories during neuronal corpse removal in vivo. *Sci. Adv.* **2020**, *6*, eaba3239. [CrossRef]

251. Habib, N.; McCabe, C.; Medina, S.; Varshavsky, M.; Kitsberg, D.; Dvir-Szternfeld, R.; Green, G.; Dionne, D.; Nguyen, L.; Marshall, J.L.; et al. Disease-associated astrocytes in Alzheimer's disease and aging. *Nat. Neurosci.* **2020**, *23*, 701–706. [CrossRef]

252. Olsen, M.; Aguilar, X.; Sehlin, D.; Fang, X.T.; Antoni, G.; Erlandsson, A.; Syvänen, S. Astroglial Responses to Amyloid-Beta Progression in a Mouse Model of Alzheimer's Disease. *Mol. Imaging Biol.* **2018**, *20*, 605–614. [CrossRef]

253. Harada, R.; Hayakawa, Y.; Ezura, M.; Lerdsirisuk, P.; Du, Y.; Ishikawa, Y.; Iwata, R.; Shidahara, M.; Ishiki, A.; Kikuchi, A.; et al. ^{18}F-SMBT-1: A Selective and Reversible PET Tracer for Monoamine Oxidase-B Imaging. *J. Nucl. Med.* **2021**, *62*, 253–258. [CrossRef] [PubMed]

254. Alzghool, O.M.; Rokka, J.; López-Picón, F.R.; Snellman, A.; Helin, J.S.; Okamura, N.; Solin, O.; Rinne, J.O.; Haaparanta-Solin, M. (S)-[^{18}F]THK5117 brain uptake is associated with Aβ plaques and MAO-B enzyme in a mouse model of Alzheimer's disease. *Neuropharmacology* **2021**, *196*, 108676. [CrossRef] [PubMed]

255. Dukić-Stefanović, S.; Hang Lai, T.; Toussaint, M.; Clauß, O.; Jevtić, I.I.; Penjišević, J.Z.; Andrić, D.; Ludwig, F.A.; Gündel, D.; Deuther-Conrad, W.; et al. In vitro and in vivo evaluation of fluorinated indanone derivatives as potential positron emission tomography agents for the imaging of monoamine oxidase B in the brain. *Bioorg. Med. Chem. Lett.* **2021**, *48*, 128254. [CrossRef] [PubMed]

256. Kumar, A.; Koistinen, N.A.; Malarte, M.-L.; Nennesmo, I.; Ingelsson, M.; Ghetti, B.; Lemoine, L.; Nordberg, A. Astroglial tracer BU99008 detects multiple binding sites in Alzheimer's disease brain. *Mol. Psychiatry* **2021**, 1–15. [CrossRef] [PubMed]

257. Livingston, N.R.; Calsolaro, V.; Hinz, R.; Nowell, J.; Raza, S.; Gentleman, S.; Tyacke, R.J.; Myers, J.; Venkataraman, A.V.; Perneczky, R.; et al. Relationship between astrocyte reactivity, using novel ^{11}C-BU99008 PET, and glucose metabolism, grey matter volume and amyloid load in cognitively impaired individuals. *medRxiv* **2021**. [CrossRef]

258. Calsolaro, V.; Matthews, P.M.; Donat, C.K.; Livingston, N.R.; Femminella, G.D.; Guedes, S.S.; Myers, J.; Fan, Z.; Tyacke, R.J.; Venkataraman, A.V.; et al. Astrocyte reactivity with late-onset cognitive impairment assessed in vivo using ^{11}C-BU99008 PET and its relationship with amyloid load. *Mol. Psychiatry* **2021**, 1–8. [CrossRef]

259. Ni, R.; Ji, B.; Ono, M.; Sahara, N.; Zhang, M.R.; Aoki, I.; Nordberg, A.; Suhara, T.; Higuchi, M. Comparative In Vitro and In Vivo Quantifications of Pathologic Tau Deposits and Their Association with Neurodegeneration in Tauopathy Mouse Models. *J. Nucl. Med.* **2018**, *59*, 960–966. [CrossRef]

260. Ishikawa, A.; Tokunaga, M.; Maeda, J.; Minamihisamatsu, T.; Shimojo, M.; Takuwa, H.; Ono, M.; Ni, R.; Hirano, S.; Kuwabara, S.; et al. In Vivo Visualization of Tau Accumulation, Microglial Activation, and Brain Atrophy in a Mouse Model of Tauopathy rTg4510. *J. Alzheimers Dis.* **2018**, *61*, 1037–1052. [CrossRef]

261. Ni, R.; Rudin, M.; Klohs, J. Cortical hypoperfusion and reduced cerebral metabolic rate of oxygen in the arcAβ mouse model of Alzheimer's disease. *Photoacoustics* **2018**, *10*, 38–47. [CrossRef]

262. Colom-Cadena, M.; Spires-Jones, T.; Zetterberg, H.; Blennow, K.; Caggiano, A.; DeKosky, S.T.; Fillit, H.; Harrison, J.E.; Schneider, L.S.; Scheltens, P.; et al. The clinical promise of biomarkers of synapse damage or loss in Alzheimer's disease. *Alzheimers Res. Ther.* **2020**, *12*, 21. [CrossRef]

263. Neuner, S.M.; Heuer, S.E.; Huentelman, M.J.; O'Connell, K.M.S.; Kaczorowski, C.C. Harnessing Genetic Complexity to Enhance Translatability of Alzheimer's Disease Mouse Models: A Path toward Precision Medicine. *Neuron* **2019**, *101*, 399–411.e5. [CrossRef]

264. Hodge, R.D.; Bakken, T.E.; Miller, J.A.; Smith, K.A.; Barkan, E.R.; Graybuck, L.T.; Close, J.L.; Long, B.; Johansen, N.; Penn, O.; et al. Conserved cell types with divergent features in human versus mouse cortex. *Nature* **2019**, *573*, 61–68. [CrossRef]
265. Rosen, R.F.; Tomidokoro, Y.; Farberg, A.S.; Dooyema, J.; Ciliax, B.; Preuss, T.M.; Neubert, T.A.; Ghiso, J.A.; LeVine, H., 3rd; Walker, L.C. Comparative pathobiology of β-amyloid and the unique susceptibility of humans to Alzheimer's disease. *Neurobiol. Aging* **2016**, *44*, 185–196. [CrossRef] [PubMed]
266. Vitek, M.P.; Araujo, J.A.; Fossel, M.; Greenberg, B.D.; Howell, G.R.; Rizzo, S.J.S.; Seyfried, N.T.; Tenner, A.J.; Territo, P.R.; Windisch, M.; et al. Translational animal models for Alzheimer's disease: An Alzheimer's Association Business Consortium Think Tank. *Alzheimer's Dement.* **2020**, *6*, e12114. [CrossRef] [PubMed]
267. Oblak, A.L.; Forner, S.; Territo, P.R.; Sasner, M.; Carter, G.W.; Howell, G.R.; Sukoff-Rizzo, S.J.; Logsdon, B.A.; Mangravite, L.M.; Mortazavi, A.; et al. Model organism development and evaluation for late-onset Alzheimer's disease: MODEL-AD. *Alzheimers Dement.* **2020**, *6*, e12110. [CrossRef] [PubMed]
268. Preuss, C.; Pandey, R.; Piazza, E.; Fine, A.; Uyar, A.; Perumal, T.; Garceau, D.; Kotredes, K.P.; Williams, H.; Mangravite, L.M.; et al. A novel systems biology approach to evaluate mouse models of late-onset Alzheimer's disease. *Mol. Neurodegener.* **2020**, *15*, 67. [CrossRef]

 *pharmaceuticals*

Article

Production of GMP-Compliant Clinical Amounts of Copper-61 Radiopharmaceuticals from Liquid Targets

Alexandra I. Fonseca [1], Vítor H. Alves [1,2], Sérgio J. C. do Carmo [1,3], Magda Silva [1], Ivanna Hrynchak [1], Francisco Alves [3,4], Amílcar Falcão [1,5] and Antero J. Abrunhosa [1,3,*]

[1] ICNAS Produção Unipessoal, Lda., Ed. ICNAS, Polo das Ciências da Saúde, University of Coimbra, 3000-548 Coimbra, Portugal; alexandrafonseca@icnas.uc.pt (A.I.F.); vitoralves@uc.pt (V.H.A.); sergiocarmo@uc.pt (S.J.C.d.C.); magdasilva@icnas.uc.pt (M.S.); ivanna.ua@icnas.uc.pt (I.H.); amilcar.falcao@uc.pt (A.F.)

[2] Fluidomica, Lda., 3060-197 Cantanhede, Portugal

[3] CIBIT/ICNAS, Coimbra Institute for Biomedical Imaging and Translational Research/Institute for Nuclear Sciences Applied to Health—University of Coimbra, 3000-548 Coimbra, Portugal; franciscoalves@uc.pt

[4] Instituto Politécnico de Coimbra, ESTeSC—Coimbra Health School, 3040-854 Coimbra, Portugal

[5] Faculty of Pharmacy, University of Coimbra, 3000-548 Coimbra, Portugal

[*] Correspondence: antero@pet.uc.pt

Abstract: PET imaging has gained significant momentum in the last few years, especially in the area of oncology, with an increasing focus on metal radioisotopes owing to their versatile chemistry and favourable physical properties. Copper-61 ($t_{1/2}$ = 3.33 h, 61% β^+, E_{max} = 1.216 MeV) provides unique advantages versus the current clinical standard (i.e., gallium-68) even though, until now, no clinical amounts of ^{61}Cu-based radiopharmaceuticals, other than thiosemicarbazone-based molecules, have been produced. This study aimed to establish a routine production, using a standard medical cyclotron, for a series of widely used somatostatin analogues, currently labelled with gallium-68, that could benefit from the improved characteristics of copper-61. We describe two possible routes to produce the radiopharmaceutical precursor, either from natural zinc or enriched zinc-64 liquid targets and further synthesis of [^{61}Cu]Cu-DOTA-NOC, [^{61}Cu]Cu-DOTA-TOC and [^{61}Cu]Cu-DOTA-TATE with a fully automated GMP-compliant process. The production from enriched targets leads to twice the amount of activity (3.28 ± 0.41 GBq vs. 1.84 ± 0.24 GBq at EOB) and higher radionuclidic purity (99.97% vs. 98.49% at EOB). Our results demonstrate, for the first time, that clinical doses of ^{61}Cu-based radiopharmaceuticals can easily be obtained in centres with a typical biomedical cyclotron optimised to produce ^{18}F-based radiopharmaceuticals.

Keywords: radiometals; copper-61; liquid targets; post-processing; [^{61}Cu]Cu-DOTA-NOC; [^{61}Cu]Cu-DOTA-TOC; [^{61}Cu]Cu-DOTA-TATE

Citation: Fonseca, A.I.; Alves, V.H.; do Carmo, S.J.C.; Silva, M.; Hrynchak, I.; Alves, F.; Falcão, A.; Abrunhosa, A.J. Production of GMP-Compliant Clinical Amounts of Copper-61 Radiopharmaceuticals from Liquid Targets. *Pharmaceuticals* **2022**, *15*, 723. https://doi.org/10.3390/ph15060723

Academic Editor: Giorgio Treglia

Received: 22 April 2022
Accepted: 4 June 2022
Published: 7 June 2022

Publisher's Note: MDPI stays neutral with regard to jurisdictional claims in published maps and institutional affiliations.

Copyright: © 2022 by the authors. Licensee MDPI, Basel, Switzerland. This article is an open access article distributed under the terms and conditions of the Creative Commons Attribution (CC BY) license (https://creativecommons.org/licenses/by/4.0/).

1. Introduction

The use of advanced imaging technologies, especially nuclear medicine (i.e., PET and SPECT), can enhance diagnosis, staging, treatment planning and evaluation of treatment response in cancer care. Over the last two decades, an emerging quantity of small biomolecules (e.g., peptides, antibodies, antibodies fragments or nanoparticles) have been labelled with beta- or alpha-emitting metal radionuclides (e.g., gallium-68, copper-64, luthetium-177, actinium-225 and astatine-211) for imaging and therapeutic applications [1–5]. The wide variety of physical decay properties and half-lives, the simple and fast one-step radiolabelling chemistry [6,7]—easily adaptable to any type of vector for any target delivery—and the easy translation of metal-based radiopharmaceuticals into a theranostic approach [8] have primarily contributed to their interest and, currently, are major contributors to its success.

^{68}Ge/^{68}Ga generators play a substantial role in this growing phenomenon by allowing worldwide access to gallium-68 ($t_{1/2}$ = 68 min, 89% β^+, E_{max} = 1.899 MeV)—even in small hospital radiopharmacies (not requiring an onsite cyclotron)—which simplifies the translation of ^{68}Ga-conjugated peptides from the bench to routine clinical use [9]. Notwithstanding this, because of the worldwide shortage of gallium-68 generators, they are gradually losing out to more cost-effective production methods (i.e., accelerator-produced radiometals by the irradiation of natural or enriched targets). These methods are able to produce higher amounts of activity, without waiting time between productions (unlike the typical 3–4 h interval between elutions of ^{68}Ge/^{68}Ga generators), aiming at fulfilling the ever-increasing clinical needs of gallium-68 [10,11]. The recently approved monograph of gallium-68 chloride solution produced from zinc-68 irradiation (Eur. Ph. 3109) [12] is a clear sign of the need for accelerator-produced methods in radiochemistry centres worldwide, which have access to this technology. Furthermore, since ^{68}Ge/^{68}Ga generators are no longer a discriminatory advantage for using gallium-68 over other metal radionuclides on a routine basis, new promising radionuclides with better physicochemical properties are arising (e.g., scandium-43, scandium-44, copper-61, copper-64 and zirconium-89) with significant advantages over gallium-68: (1) easier distribution to centres that do not have onsite cyclotrons, (2) lower maximum positron emission energies that meet the requirements for a new generation of tomographs with higher resolution and (3) nuclides having a close therapeutic match, which is determinant for personalised medicine as we enter the theranostic era [13,14].

Copper-61 ($t_{1/2}$ = 3.33 h, 61% β^+, E_{max} = 1.216 MeV) [15] is a positron-emitting radionuclide presenting decay characteristics comparable to gallium-68 but with the advantage of presenting lower maximum positron energy (E_{max} = 1.216 MeV vs. E_{max} = 1.899 MeV) and a substantially more practical half-life (3.33 h vs. 68 min). In the past few years, several groups have attempted to find the best production and purification methods for cyclotron-produced copper-61. Liquid and solid target irradiations have both been explored. In 2012, the production of copper-61 from natural cobalt solid targets, following the ^{nat}Co(α,xn)^{61}Cu nuclear reaction, resulted in high-purity copper-61 [16]. Later, Asad et al. [17,18] and Thieme et al. [19] detailed the production of copper-61 from natural zinc and zinc-64, also in solid targets. In 2017, our group described the production of copper-61 from the irradiation of liquid targets at low proton energies [20] and later its automated purification [21]. More recently, the possibility of producing copper-61 from solid natural nickel targets, following ^{nat}Ni(d,x)^{61}Cu nuclear reaction, was also outlined, along with its fully automated purification process [22]. Despite increasing efforts being made in the development of the above-mentioned methods towards being capable of producing high-purity copper-61, to date, only a handful of molecules have been labelled with this radioisotope—mostly thiosemicarbazone-based molecules (i.e., ATSM, PTSM, APTS and TATS) [23–27], which are well known for presenting high affinity for copper.

Considering the above, the aim of the current work is to demonstrate that the production of chelator-based copper-61 labelled radiopharmaceuticals can easily be performed and can be made Good Manufacturing Practise (GMP)-compliant for routine clinical use. For that purpose, we present the production, synthesis and quality control of [^{61}Cu]Cu-DOTA-TATE, [^{61}Cu]Cu-DOTA-TOC and [^{61}Cu]Cu-DOTA-NOC, the copper-61 equivalents of the somatostatin (SST) analogues extensively used with gallium-68, in current clinical practice [28].

Initial work on targeting and staging neuroendocrine tumours (NETs) through the labelling of SST analogues begun with [^{123}I]I-Tyr3-octreotide [29]. The first Food and Drug Administration (FDA)-approved radiopharmaceutical was Octreoscan®, in 1994 ([^{111}In]In-DTPA-Octreotide) [30,31]. Today, ^{68}Ga-labelled radiopharmaceuticals such as [^{68}Ga]Ga-DOTA-NOC, [^{68}Ga]Ga-DOTA-TOC and [^{68}Ga]Ga-DOTA-TATE are in current clinical practice to diagnose, with PET, solid tumours which over-express SST receptors (SSTRs) [32,33]. The ready-to-label "cold kit" with DOTA-TATE was approved by the FDA in 2016 (Netspot™), and the equivalent with DOTA-TOC (Somakit-TOC) was approved by the EMA in 2017 [34].

Most previous works regarding the labelling of SST analogues with copper focused on copper-64 ($t_{1/2}$ = 12.7 h, 18% β^+, E_{max} = 0.653 MeV). A first-in-human study with [^{64}Cu]Cu-DOTA-TATE revealed several advantages, i.e., higher lesion detection, better image quality and lower radiation doses, when compared with [^{111}In]In-DTPA-octreotide when used for SPECT imaging [35]. More recent clinical studies, particularly head-to-head comparisons of [^{64}Cu]Cu-DOTA-TATE with both [^{111}In]In-DOTA-TATE [36] and [^{68}Ga]Ga-DOTA-TOC [37] revealed an overall better performance of the ^{64}Cu-conjugated in terms of sensitivity, resolution and rate of lesion detection. Additionally, other first-in-human studies with [^{64}Cu]Cu-DOTA-TOC also showed high lesion detection rate, safety of use and high effectiveness for predicting treatment planning [38]. Even more recently, Loft et.al. confirmed the extended imaging window of [^{64}Cu]Cu-DOTA-TATE from 1 h to 3 h [39], without a decrease in performance. There is still work in progress aiming at clarifying the dosimetric parameters and predicting both the overall survival (OS) and progression-free survival (PFS) ability of these radiopharmaceuticals labelled with copper-64 [40–42]. These results lead us to conclude that the substitution of gallium for copper on these SST analogues has most likely a positive impact on their performance as PET radiopharmaceuticals. Moreover, the favourable physical properties of copper-61 when compared with copper-64 (shorter half-life, higher β^+-emission) makes it an ideal nuclide for this purpose.

In this context, the simple, cost-effective production and separation methods herein described could pave the way for the widespread clinical use of copper-61 radiopharmaceuticals, providing an even better alternative to the scarce and expensive-to-obtain gallium-68.

2. Results and Discussion

2.1. [^{61}Cu]CuCl$_2$ Production

Copper-61 was produced using the target system previously described in [20] and then in [43]. Several cyclotron irradiations were performed with both natural and enriched zinc. Table 1 summarises the number of runs, the irradiation conditions and the activity produced for each target. The same solution of zinc-64, with an initial concentration of 200 mg/mL, was irradiated a maximum of four times.

Table 1. Irradiation conditions applied to each target and total activity produced (GBq) at EOB.

Target	n	[HNO$_3$] (M)	[Zn] (mg/mL)	I (μAh)	Irrad. Time (min)	Act. Produced (GBq)
^{nat}Zn(p,α)^{61}Cu	20	0.01	200	70.1 ± 0.3	180	1.84 ± 0.24
^{64}Zn(p,α)^{61}Cu	32	0.01	200[1]	67.4 ± 2.9	180	3.28 ± 0.41

Initial concentration before recycling. EOB: End Of Bombardment.

A direct comparison between the 180 min long irradiation of natural zinc and first-time irradiated enriched zinc-64 showed that, under the same irradiation conditions, i.e., time, concentration, current and pressure, the use of zinc-64 allowed the production of twice the activity of copper-61 than natural zinc: 3.65 ± 0.18 GBq (N = 8) and 1.84 ± 0.24 GBq (N = 20), respectively. These correspond to low yields when compared to solid targets, as stated in [44]; however, the latter also come with high cost and tremendous operational complexity. These studies confirmed the expected higher activities of copper-61 from the enriched target, considering the 49.2% abundance of the zinc-64 isotope in the natural zinc. Although higher activities of copper-61 are produced using enriched target material, zinc-64 is approximately 200 times more expensive than the natural target (550–669 €/g zinc-64 vs. 2.92 €/g natural zinc). Given this tremendous difference, a cost–benefit analysis is required.

2.2. Recovery and Recycling of ^{64}Zn

One of the advantages of using liquid targets is that the recycling of enriched material is simplified. This is especially important when considering the high cost of zinc-64.

Notwithstanding, few authors have actually described this. In this study, the zinc-64 target was recovered from the CU resin waste container several days after been irradiated. It was then evaporated and re-dissolved into the initial form of 10 mM HNO_3. Moreover, the recycling process was simple, since no solvent other than HNO_3 was introduced during the purification process, and the zinc-64 solution could be re-used directly after filtration. The percentage of zinc-64 recovered and re-irradiated was collectively determined to be higher than 90% each time it was recycled. We found a slight variation in the activity produced (corrected at EOB), depending on how many times the zinc-64 solution was recycled and subsequently irradiated (Table S1). The second irradiation of the same batch of zinc-64 did not show a significant decrease in the amounts of produced nor purified copper-61. On the other hand, with more than two irradiations, there was a statistically significant decrease in the amount of copper-61 produced and, consequently, in purified copper (Table S1). This decrease in the activity of copper-61 is explained by the loss of zinc-64 during the several steps of the process: recycling, purification, evaporation and final filtration of the solution. Regarding isotopic enrichment of the recycled solutions (Table S1), the recovery process did not lead to a significant decrease in zinc-64 enrichment.

The purification process was performed as described earlier [45] without further modifications.

2.3. [^{61}Cu]Cu-DOTA-NOC, [^{61}Cu]Cu-DOTA-TATE and [^{61}Cu]Cu-DOTA-TOC Production Activity Distribution

[^{61}Cu]Cu-DOTA-NOC, [^{61}Cu]Cu-DOTA-TOC and [^{61}Cu]Cu-DOTA-TATE were produced using the Synthera® Extension automated module (IBA, Louvain-la-Neuve, Belgium). This fully automated process complies with GMPs to produce radiopharmaceuticals (EudraLex, Volume 4, Annex 3) (i.e., the use of disposable cassettes and tubing systems, ensuring high quality and reproducibility of the final radiopharmaceutical product and the narrowing of the risk of radioactive cross-contamination). A total of 50 µg of DOTA-NOC (N = 10), DOTA-TATE (N = 3) and DOTA-TOC (N = 3) were labelled with purified [^{61}Cu]CuCl$_2$ at a 85–100 °C reaction temperature and 10 min reaction time. Specifications are summarised in Table 2.

Table 2. Summary of activities and Yields (i.e., Labelling Yield and RCY) achieved in the radiopharmaceutical synthesis of [^{61}Cu]Cu-DOTA-NOC, [^{61}Cu]Cu-DOTA-TATE and [^{61}Cu]Cu-DOTA-TOC produced from either natural or enriched zinc.

Radiopharmaceutical	Target	Process Duration (min)	Activity @EOS (GBq)	Labelling Yield (%)	RCY (%)
[^{61}Cu]Cu-DOTA-NOC N = 5	Natural Zinc	32 ± 4	0.99 ± 0.16	98.48 ± 0.89	94.73 ± 3.03
[^{61}Cu]Cu-DOTA-NOC N = 5	Zinc-64	38 ± 2	1.95 ± 0.21	97.72 ± 2.01	94.03 ± 1.84
[^{61}Cu]Cu-DOTA-TATE N = 3	Zinc-64	37 ± 6	2.06 ± 0.08	98.61 ± 0.84	95.91 ± 1.50
[^{61}Cu]Cu-DOTA-TOC N = 3	Zinc-64	38 ± 4	1.77 ± 0.12	97.87 ± 1.10	94.67 ± 1.19

RCY: radiochemical yield. EOS: End Of Synthesis.

As expected, depending on the production route of [^{61}Cu]CuCl$_2$ used, the greatest difference found was in the amount of [^{61}Cu]Cu-DOTA-NOC activity at the End Of Synthesis (EOS): 0.99 ± 0.16 GBq or 1.95 ± 0.21 GBq from natural or enriched targets, respectively. To evaluate the efficacy and reproducibility of this synthesis method, we determined the radiochemical and labelling yields of the process. Radiochemical Yield (RCY) refers to the final activity in the product of [^{61}Cu]Cu-DOTA-NOC/TOC/TATE, expressed as the percentage (%) of starting activity of [^{61}Cu]CuCl$_2$ obtained after purification [46]. The quantity of both was decay-corrected to the same time point. All radioactivity lost during transfer, labelling reaction, solid phase purification (SPE) and dispensing were accounted for in the RCY. Whereas the labelling yield indicated the direct yield of the labelling reaction.

Nonetheless, labelling yield referred only to the extent of the labelling reaction, comparing the amount of $[^{61}Cu]CuCl_2$ that reacted into $[^{61}Cu]Cu$-DOTA-NOC/TOC/TATE, and did not consider any other process losses. The data showed that neither labelling yields nor RCY were affected by the amount of activity, which confirms that activities of copper-61 up to 2.7 GBq (at the End Of Purification (EOP)) do not have negative effects on the synthesis process of these radiopharmaceuticals.

We also compared the distribution patterns regarding the different cassette components for all radiopharmaceuticals (Figure 1). Activity distribution revealed similar results for all peptides. It is important to note the low residual activity in the different components and the small SD of its values, which reflects both high reproducibility and efficacy of the automated synthesis process.

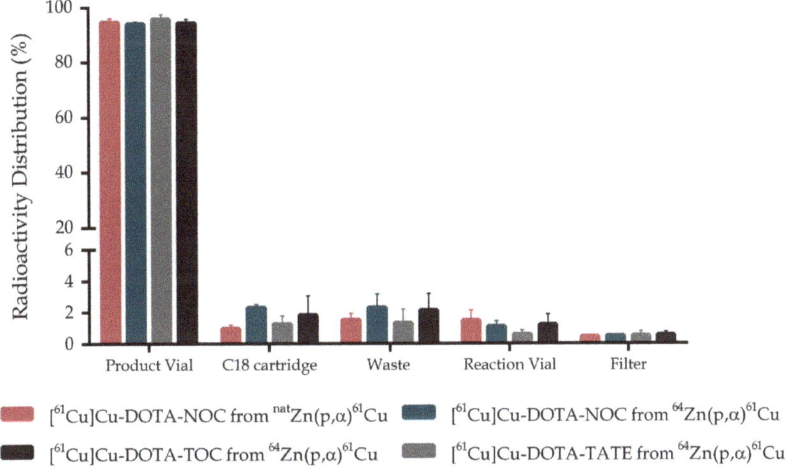

Figure 1. Activity distribution of the different cassette components after synthesis on Synthera® Extension module: Final Product Vial, C18 SPE cartridge, Waste and Reaction vial. Data comprises the different radiopharmaceuticals produced (mean ± SD, N ≥ 3).

2.4. Quality Control

Table 3 outlines the final product specifications obtained, including radiochemical and radionuclidic purity, radionuclidic identity and pH. Radiochemical purity was evaluated by radio-HPLC, using the methods described in the next section (Table 4 in Materials and Methods). Radionuclidic purity was evaluated using a High Purity Germanium (HPGe) detector, several hours after EOS. A single value of radionuclidic purity is presented for $[^{61}Cu]Cu$-DOTA-NOC, $[^{61}Cu]Cu$-DOTA-TOC and $[^{61}Cu]Cu$-DOTA-TATE produced from the enriched target, as radionuclidic purity is only dependent on the method of copper-61 production, regardless of the subsequent synthesis process.

As expected, the choice of target material has an impact on radionuclidic purity. When copper-61 is produced from natural zinc, 1.5% of copper-64 (at EOB) is produced simultaneously with copper-61 [20], whereas when an enriched target is used, almost no copper impurities are expected to be produced; however, small amounts of copper-64 are present as a side product, resulting from the (p,α) nuclear reactions on residuals zinc-67 and zinc-68 present in the enriched target material. This amounts to about 0.03% of copper-64 co-produced when irradiating zinc-64. Figure 2 indicates the impact of this percentage of copper-64 on the shelf life of the product when each target is used.

Table 3. Final product specifications for [^{61}Cu]Cu-DOTA-NOC, [^{61}Cu]Cu-DOTA-TATE and [^{61}Cu]Cu-DOTA-TOC (mean ± SD, N ≥ 3).

Production Route	^{nat}Zn(p,α)^{61}Cu	^{64}Zn(p,α)^{61}Cu		
TEST	**[^{61}Cu]Cu-DOTA-NOC**	**[^{61}Cu]Cu-DOTA-NOC**	**[^{61}Cu]Cu-DOTA-TATE**	**[^{61}Cu]Cu-DOTA-TOC**
MA (MBq/nmol)	28.93 ± 4.58	56.82 ± 6.25	52.31 ± 9.83	50.27 ± 3.40
Activity at EOS (GBq)	0.99 ± 0.16	1.95 ± 0.21	2.06 ± 0.08	1.77 ± 0.12
RCP (%)	99.48 ± 0.51	98.71 ± 0.57	99.90 ± 0.03	99.77 ± 0.16
RNP (%)	98.49 ± 0.07		99.97 ± 0.03	
Radionuclidic identity (h)	3.33 ± 0.04	3.33 ± 0.04	3.33 ± 0.04	3.33 ± 0.04
pH	3–5	3–5	3–5	3–5
Visual Inspection	Clear, Colourless	Clear, Colourless	Clear, Colourless	Clear, Colourless
Volume (mL)	5–10	5–10	5–10	5–10

MA: Molar activity. RCP: Radiochemical Purity. RNP: Radionuclidic Purity.

Table 4. HPLC methods for RCP determination.

	Time (min)	Mobile Phase A (Per Cent *v/v*)	Mobile Phase B (Per Cent *v/v*)
Solvents		Water/0.1% TFA	ACN/0.1% TFA
Method A	0–11	74 → 60	26 → 40
	11–12	60 → 40	40 → 60
	12–14	40	60
Method B	0–8	78	22
	8–9	78 → 40	22 → 60
	9–14	40	60

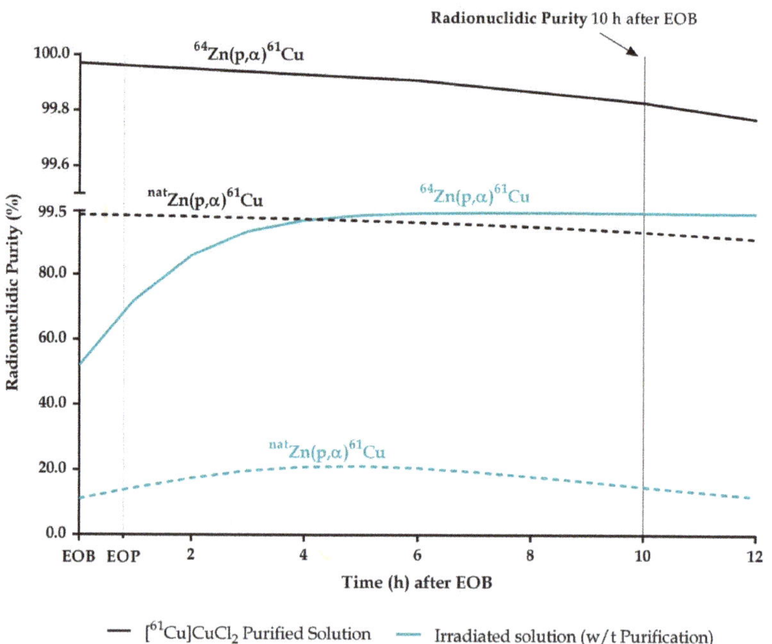

Figure 2. Radionuclidic purity (%) of copper-61 produced from liquid targets before purification (blue lines obtain from calculations [20]) and after purification (black lines obtain from HPGe measurements) either from natural Zinc (dashed lines) or enriched Zinc-64 (solid lines) experimental values determined by HPGe.

Currently, no European Pharmacopeia (Ph. Eur.) monograph exists for copper-61 [47]. Taking into consideration the limits set for radionuclidic impurities in the Gallium (^{68}Ga) Chloride (accelerator produced) monograph of 2% (mon. 3109), the production of copper-61 from natural Zinc would require it to be used immediately after purification. On the other hand, when produced from zinc-64, the radionuclidic purity of copper-61 is higher than 99% and remains at this level for many hours after production (Figure 2).

2.5. In Vitro Stability

The stability of [^{61}Cu]Cu-DOTA-NOC, [^{61}Cu]Cu-DOTA-TOC and [^{61}Cu]Cu-DOTA-TATE in aqueous solution (NaCl 0.9% or PBS) and in mouse serum was evaluated up to 12 h after the EOS. Figure 3 shows that all radiopharmaceuticals were stable under the conditions tested. Radiochemical purity results indicated that these compounds are highly stable (over 95%) at 37 °C up to 12 h after the EOS, in the final formulation (NaCl 0.9%), PBS and serum.

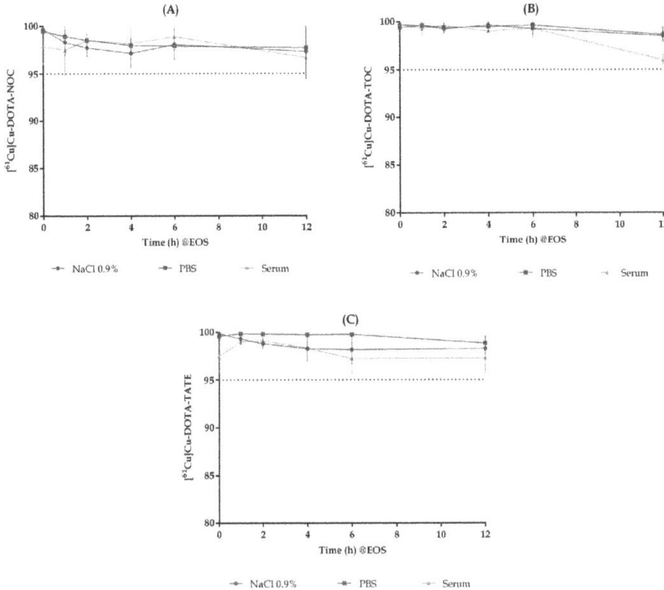

Figure 3. Stability of [^{61}Cu]Cu-DOTA-NOC (**A**), [^{61}Cu]Cu-DOTA-TATE (**B**) and [^{61}Cu]Cu-DOTA-TOC (**C**) in NaCl 0.9%, PBS and mouse serum. Radiochemical purity results were obtained by radioHPLC at: T0, T0 + 1 h, T0 + 2 h, T0 + 4 h, T0 + 6 h and T0 + 12 h, where T0 represents the EOS.

3. Materials and Methods

All chemicals and solvents used for purification of [^{61}Cu]CuCl$_2$ and synthesis of ^{61}Cu-conjugated peptides were trace metal grade, and HPLC solvents were HPLC grade. The remaining solvents and reagents (i.e., hydrochloric acid > 30% and nitric acid > 69% (Honeywell Fluka, Charlotte, NC, USA), bi-distilled water (BBraun, Melsungen, Germany), ethanol (Rotem, Israel), sodium acetate anhydrous (Honeyweell Fluka, Germany), *L*-ascorbic acid (Sigma-Aldrich, St. Louis, MO, USA) and DTPA (Alfa Aesar, Kandel, Germany)) were also trace metal basis, to prevent metal cross-contamination.

Zinc (99.998%) was acquired from Alfa Aesar, whereas the enriched zinc metal form (^{64}Zn—99.89%) was obtained from CMR (Moscow, Russia). Purification and labelling disposable kits were purchased from Fluidomica (Cantanhede, Portugal) and purification resins (i.e., CU-B25-A resin and SAX 1 × 8 200–400 mesh, Cl^{-} form resin) from Triskem (Bruz, Belgium). Peptides DOTA-NOC acetate, DOTA-TATE acetate and DOTA-TOC acetate, fractioned and kept at −20 °C in an aqueous solution, were manufactured by

ABX (Radeberg, Germany). The usage of polyethylene and polypropylene materials was favoured over that of glass materials.

3.1. Irradiation and Purification of $[^{61}Cu]CuCl_2$

Irradiation of zinc liquid targets, both natural and enriched, and further copper-61 purification was conducted following the previously published and described methodology [48,49]. Briefly, copper-61 was obtained from the irradiation of both highly pure zinc nitrate hexahydrate and enriched zinc-64 solutions using an IBA Cyclone 18/9 (IBA, Louvain-la-Neuve, Belgium) [20]. Zinc nitrate was directly dissolved in 10 mM nitric acid, yielding a concentration of 0.2 g/mL, whereas zinc-64 (metal form) had to be initially dissolved with highly concentrated nitric acid, left overnight, evaporated to dryness, and finally re-dissolved in 10 mM nitric acid, yielding likewise a concentration of 0.2 g/mL. Zinc-64 solutions were recycled and re-irradiated up to 4 times. Since only water and HNO_3 were added to the original zinc-64 solution during the purification process, the recycling process was made possible simply by evaporating the excess of water. These solutions were irradiated at 65–75 µA for 180 min. Copper-61 automatic purification was conducted using a Synthera® Extension module (IBA, Louvain-la-Neuve, Belgium) without any manual intervention, and it was completed in less than 40 min from the EOB [45].

3.2. EtOH as Radiolytic Scavenger

The considerably high percentage of radiolysis in the final product vial (FPV), caused by the presence of free radicals in solution (e.g., superoxide or hydroxyl radicals) [50], impaired the establishment of the most favourable labelling conditions. Although not explicitly measured, for higher activity concentrations, the radiolysis percentage was anticipated to increase. Several compounds are known to act as radiolytic stabilizers and protect against radiolysis. Antioxidant compounds, such as ascorbic acid (AA) and gentisic acid (GA), are commonly known to protect against radiolysis and are mainly described in the literature for radiolabelling biomolecules with β^--emitting radionuclides (e.g., yttrium-90 and luthetium-177) [51,52]. Notwithstanding, these compounds might have a negative impact on copper-based radiopharmaceuticals, given the redox properties of copper. More recently, EtOH also gained relevance in ^{68}Ga-based radiopharmaceuticals and proved to be of great value, as confirmed by Eppard et al. [53–55]. To evaluate the applicability of EtOH as a radiolytic scavenger for ^{61}Cu-based radiopharmaceuticals, a single test with and without EtOH was performed before establishing optimal labelling conditions. Figure 4 shows the results attained when using EtOH up to 300 µL (maximum 5 vol% ethanol). Compared with the labelling of $[^{61}Cu]$Cu-DOTA-NOC without EtOH (Figure 4A), it is evident that there is a significant decrease in the rate of radiolysis using the ethanol-based method (Figure 4B–D). This method showed that 5 vol% EtOH leads to a decrease in radiolysis from more than 15% in $[^{61}Cu]$Cu-DOTA-NOC to less than 5% for $[^{61}Cu]$Cu-DOTA-NOC and less than 1% for both $[^{61}Cu]$Cu-DOTA-TOC and $[^{61}Cu]$Cu-DOTA-TATE. We found that $[^{61}Cu]$Cu-DOTA-NOC is the most sensitive peptide to radiolysis, even in the presence of EtOH. Based on these findings, the use of EtOH was implemented in all further labelling formulations to act as a radiolytic stabilizing agent during the labelling reaction.

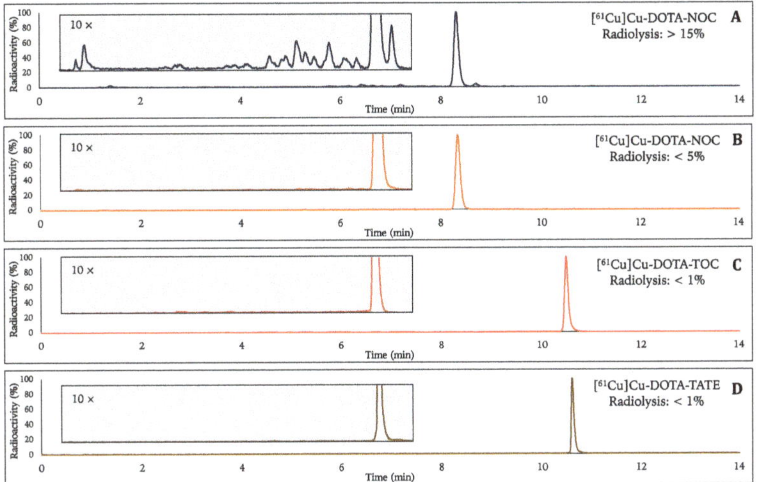

Figure 4. Representative chromatograms of [61Cu]Cu-DOTA-NOC, [61Cu]Cu-DOTA-TOC and [61Cu]Cu-DOTA-TATE without (**A**) and with (**B**–**D**) using EtOH (maximum 5 vol% EtOH) during the labelling reaction. Two different HPLC methods were used, as described in Table 4. For a more practical comparative analysis between chromatograms, raw data was normalised as percentage of total radioactivity. Percentage of radiolysis is the ratio of radiolysis counts to total counts.

3.3. Synthesis of [61Cu]Cu-Conjugated Peptides on the IBA Synthera® Extension Module

Fully automated post-processing synthesis was performed using a Synthera® Extension module (Figure 5) and completed within a maximum of 25 min from the EOP, without any manual intervention. After the purification process, [61Cu]CuCl$_2$ was automatically transferred to the reaction vial (B). Then, the peptide (DOTA-NOC acetate, DOTA-TOC acetate or DOTA-TATE acetate), dissolved in 2.5 M sodium acetate buffer, was transferred to the same reaction vial, where the reaction occurred. After the labelling reaction, the mixture was cooled down with water, and the product was then purified using a C18 cartridge (Sep-Pak Plus Short C18, Waters, Milford, Massachusetts, USA). After a rinse step, the 61Cu-conjugated peptide was eluted from the C18 cartridge with a mixture of (50/50%) water/ethanol.

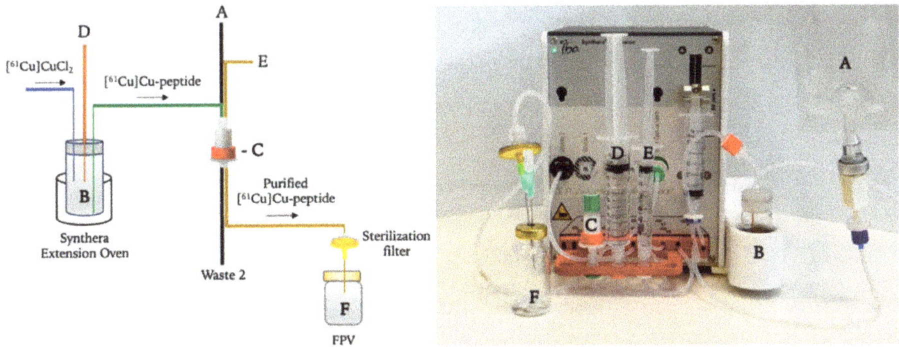

Figure 5. Schematic flow and Synthera® Extension module device and disposable cassette: (A) Water, (B) Reaction vial and oven, (C) SPE C18 cartridge, (D) Buffer and peptide, (E) Ethanol/Water (50%/50%) solution and (F) product vial.

The general automated synthesis/radiolabelling steps are as follows:

1. The C18 cartridge (C) is preconditioned with ethanol (10 mL) followed by water (10 mL) prior to use;
2. Purified [^{61}Cu]CuCl$_2$ (3 mL, 0.5 M HCl) is transferred to the reaction vial (B);
3. Peptide (50 µg), previously diluted in 2.5 M sodium acetate (3 mL) and EtOH (200–300 µL) (D) to prevent radiolysis, is transferred to the reaction vial (B) and mixed with [^{61}Cu]CuCl$_2$ for 10 s;
4. Radiolabelling reaction is conducted for 10 min, with variable temperature (85–100 °C) and pH fixed between 4 and 5;
5. Reaction mixture is cooled down with water (12 mL) (A) and passed through a C18 cartridge at 3 mL/min flow to the waste container (Waste 2);
6. C18 cartridge is then rinsed with water (10 mL) (A), which rinses the column at a 3 mL/min flow;
7. [^{61}Cu]Cu-labelled peptide is finally eluted from the C18 column with a solution of water/EtOH (50/50%) (E) to the final product vial (F) with a 3 mL/min flow.

After labelling and purification, the FPV was transferred to the Quality Control (QC) laboratory, and all the components were measured, after which the radiochemical yield was determined.

3.4. Quality Control

3.4.1. Radionuclidic Purity (HPGe)

The RNP of copper-61 at EOB was determined through γ-spectroscopy of the final solution using a High Purity Germanium detector (HPGe), several hours after the EOP. The HPGe was calibrated with ^{154}Eu and ^{133}Ba radioactive sources and placed in a low-background shielding. γ-spectra were acquired using point-source-like samples with a dead-time below 4%. GammaVision (ORTEC Inc., Easley, SC, USA) software was used to determine photopeak areas.

3.4.2. Radiochemical Purity (Radio-HPLC)

RCP was measured by HPLC (Agilent 1200 series HPLC system, Agilent Technologies, Santa Clara, CA, USA) equipped with a GABIStar NaI(Tl) radiometric detector (Raytest Isotopenmessgeraete GmbH, Straubenhardt, Germany) (20 µL sample volume). Two different methods (Table 4) were used, one for [^{61}Cu]Cu-DOTA-NOC (Method A) and a second to evaluate both [^{61}Cu]Cu-DOTA-TATE and [^{61}Cu]Cu-DOTA-TOC (Method B). An ACE 3 C18 150 × 3 mm HPLC column (ACE, Reading, UK) was used in both methods, and the flow was fixed at 0.6 mL/min.

3.4.3. Stability Experiments

The stability of ^{61}Cu-conjugated peptides was evaluated under various conditions: in the final formulation (10% EtOH/0.9% NaCl), in the presence of PBS and in mouse serum. All stability measurements were quantified by HPLC, as incubation solutions could affect the accuracy of Thin Layer Chromatography (TLC). The HPLC methods used to evaluate stability were previously described (in Table 4), with exception of [^{61}Cu]Cu-DOTA-NOC. In this case, stability was evaluated using a faster method, with the following gradient: 0–5 min Mobile Phase A (100% to 0%).

3.4.4. Stability in Aqueous Solvents

The published protocol was followed with minor changes [56]. Briefly, 50 µL of the final purified solution (Water/EtOH: 50%/50%) containing the radiolabelled ^{61}Cu-conjugated compound under study was added to 450 µL of each medium (0.9% NaCl or PBS), and the mixtures were incubated at 37 °C (T0). At different time points (T0, T0 + 1 h, T0 + 2 h, T0 + 4 h, T0 + 6 h and T0 + 12 h), aliquots were taken and measured using the HPLC methods formerly characterised (Table 4).

3.4.5. Stability in Mice Serum

For stability in mice serum, 500 μL of serum was incubated with 50 μL of ^{61}Cu-conjugated peptides dissolved in the final formulation, at 37 °C. At different time points (T0, T0 + 1 h, T0 + 2 h, T0 + 4 h, T0 + 6 h and T0 + 12 h), 50 μL aliquots were taken, and 150 μL of ethanol was added to precipitate the plasma proteins. The mixture was centrifuged at 3000 rpm for 10 min and the supernatant was collected and diluted in NaCl 0.9% for HPLC analysis.

4. Conclusions

In this study, we demonstrated that clinical amounts of ^{61}Cu-based radiopharmaceuticals can be produced, under GMP, in a medical cyclotron, using liquid targets. Production yields are higher using enriched target in comparison to irradiating natural zinc. The high radionuclidic and radiochemical purity of the produced ^{61}Cu-labelled radiopharmaceuticals ([^{61}Cu]Cu-DOTA-NOC, [^{61}Cu]Cu-DOTA-TOC and [^{61}Cu]Cu-DOTA-TATE), opens the possibility for them to be used as an alternative to the current clinically used versions with gallium-68. This work serves as background for future preclinical in vitro and in vivo studies aiming at bringing copper-61 radiopharmaceuticals to the clinical setting in the near future.

Supplementary Materials: The following supporting information can be downloaded at: https://www.mdpi.com/article/10.3390/ph15060723/s1, Table S1: Comparison of copper-61 activity produced and purified, corrected at EOB and EOP, respectively, when using non-recycled, once recycled, twice recycled, and three times recycled zinc-64 solution (mean ± SD, N = 8). Isotopic enrichment of the irradiated zinc-64 recycled solution determined by ICP-MS analysis.

Author Contributions: Conceptualization, A.I.F., V.H.A. and F.A.; methodology, A.I.F., V.H.A. and S.J.C.d.C.; resources, M.S., I.H. and S.J.C.d.C.; investigation, A.I.F.; writing—original draft preparation, A.I.F.; writing—review and editing, A.J.A., F.A. and V.H.A.; supervision—A.J.A., F.A. and A.F.; project administration, A.J.A. and F.A. All authors have read and agreed to the published version of the manuscript.

Funding: This work was founded by the Portuguese Foundation for Science and Technology (FCT) through PhD grants (grant number PD/BDE/150681/2020 and PD/BDE/150331/2019) and by ICNAS-P.

Institutional Review Board Statement: Not applicable.

Informed Consent Statement: Not applicable.

Data Availability Statement: Data is contained within the article and supplementary material.

Conflicts of Interest: The authors declare no conflict of interest.

References

1. Eder, M.; Hennrich, U. [^{68}Ga]Ga-PSMA-11: The first FDA-Approved ^{68}Ga-Radiopharmaceutical for pET Imaging of Prostate Cancer. *Pharmaceuticals* **2021**, *14*, 713. [CrossRef]
2. Suzuki, H.; Kise, S.; Kaizuka, Y.; Watanabe, R.; Sugawa, T.; Furukawa, T.; Fujii, H.; Uehara, T. Copper-64-Labeled Antibody Fragments for Immuno-PET/Radioimmunotherapy with Low Renal Radioactivity Levels and Amplified Tumor-Kidney Ratios. *ACS Omega* **2021**, *6*, 21556–21562. [CrossRef] [PubMed]
3. Huynh, T.T.; Sreekumar, S.; Mpoy, C.; Rogers, B.E. First-In-Human Results on the Biodistribution, Pharmacokinetics, and Dosimetry of [^{177}Lu]Lu-DOTA.SA.FAPi and [^{177}Lu]Lu-DOTAGA.(SA.FAPi)2. *Pharmaceuticals* **2021**, *14*, 229. [CrossRef]
4. Merkx, R.I.J.; Rijpkema, M.; Franssen, G.M.; Kip, A.; Smeets, B.; Morgenstern, A.; Bruchertseifer, F.; Yan, E.; Wheatcroft, M.P.; Oosterwijk, E.; et al. Carbonic Anhydrase IX-Targeted α-Radionuclide Therapy with ^{225}Ac Inhibits Tumor Growth in a Renal Cell Carcinoma Model. *Pharmaceuticals* **2022**, *15*, 570. [CrossRef] [PubMed]
5. Zalutsky, M.R.; Reardon, D.A.; Pozzi, O.R. Targeted alpha-particle radiotherapy with ^{211}At-labeled monoclonal antibodies. *Nucl. Med. Biol.* **2007**, *34*, 779–785. [CrossRef]
6. Price, E.W.; Orvig, C. Matching chelators to radiometals for radiopharmaceuticals. *Chem. Soc. Rev.* **2014**, *43*, 260–290. [CrossRef]
7. Krasikova, R.N.; Aliev, R.A.; Kalmykov, S.N. The next generation of positron emission tomography radiopharmaceuticals labeled with non-conventional radionuclides. *Mendeleev Commun.* **2016**, *26*, 85–94. [CrossRef]

8.   Boros, E.; Packard, A.B. Radioactive Transition Metals for Imaging and Therapy. *Chem. Rev.* **2019**, *119*, 870–901. [CrossRef]

9.   Dash, A.; Chakravarty, R. Radionuclide generators: The prospect of availing PET radiotracers to meet current clinical needs and future research demands. *Am. J. Nucl. Med. Mol. Imaging* **2019**, *9*, 30–66.

10.   Thisgaard, H.; Kumlin, J.; Langkjær, N.; Chua, J.; Hook, B.; Jensen, M.; Kassaian, A.; Zeisler, S.; Borjian, S.; Cross, M.; et al. Multi-curie production of gallium-68 on a biomedical cyclotron and automated radiolabelling of PSMA-11 and DOTATATE. *EJNMMI Radiopharm. Chem.* **2021**, *6*, 1. [CrossRef] [PubMed]

11.   International Atomic Energy Agency. *AIEA-TECDOC-1863: Gallium-68 Cyclotron Production*; AIEA: Vienna, Austria, 2019.

12.   Gallium (68Ga) Chloride (Accelerator-Produced) Solution for Radiolabelling. Ph Eur. PA/PH/Exp. 14/T (18) 13 ANP(Monograph 3109). 2017.

13.   International Atomic Energy Agency. *AIEA-TECDOC-1955: Production of Emerging Radionuclides towards Theranostic Applications: Copper-61, Scandium-43 and -44, and Yttrium-86*; AIEA: Vienna, Austria, 2021.

14.   Chaple, I.F.; Lapi, S.E. Production and Use of the First-Row Transition Metal PET Radionuclides ^{43,44}Sc, ^{52}Mn, and ^{45}Ti. *J. Nucl. Med.* **2018**, *59*, 1655–1659. [CrossRef]

15.   McCarthy, D.W.; A Bass, L.; Cutler, P.; E Shefer, R.; E Klinkowstein, R.; Herrero, P.; Lewis, J.; Cutler, C.S.; Anderson, C.; Welch, M.J. High purity production and potential applications of copper-60 and copper-61. *Nucl. Med. Biol.* **1999**, *26*, 351–358. [CrossRef]

16.   Das, S.S.; Chattopadhyay, S.; Barua, L.; Das, M.K. Production of ^{61}Cu using natural cobalt target and its separation using ascorbic acid and common anion exchange resin. *Appl. Radiat. Isot.* **2012**, *70*, 365–368. [CrossRef]

17.   Asad, A.H.; Smith, S.V.; Chan, S.; Jeffery, C.M.; Morandeau, L.; Price, R.I. Cyclotron production of ^{61}Cu using natural Zn & enriched ^{64}Zn targets. *AIP Conf. Proc.* **2012**, *1509*, 91–95. [CrossRef]

18.   Asad, A.H.; Smith, S.V.; Morandeau, L.M.; Chan, S.; Jeffery, C.M.; Price, R.I. Production of ^{61}Cu by the ^{nat}Zn(p,α) reaction: Improved separation and specific activity determination by titration with three chelators. *J. Radioanal. Nucl. Chem.* **2015**, *305*, 899–906. [CrossRef]

19.   Thieme, S.; Walther, M.; Preusche, S.; Rajander, J.; Pietzsch, H.-J.; Lill, J.-O.; Kaden, M.; Solin, O.; Steinbach, J. High specific activity ^{61}Cu via ^{64}Zn(p,α)^{61}Cu reaction at low proton energies. *Appl. Radiat. Isot.* **2013**, *72*, 169–176. [CrossRef]

20.   Do Carmo, S.J.C.; Alves, V.H.; Alves, F.; Abrunhosa, A.J. Fast and cost-effective cyclotron production of ^{61}Cu using a ^{nat}Zn liquid target: An opportunity for radiopharmaceutical production and R&D. *Dalton Trans.* **2017**, *46*, 14556–14560. [CrossRef]

21.   Fonseca, A.I.; Alves, V.H.; Carmo, S.J.C.D.; Falcão, A.; Abrunhosa, A.J.; Alves, F. GMP automated purification of copper-61 produced in cyclotron liquid targets: Methodological aspects. *Curr. Radiopharm.* **2021**, *14*, 420–428. [CrossRef] [PubMed]

22.   Svedjehed, J.; Kutyreff, C.J.; Engle, J.W.; Gagnon, K. Automated, cassette-based isolation and formulation of high-purity [^{61}Cu]CuCl$_2$ from solid Ni targets. *EJNMMI Radiopharm. Chem.* **2020**, *5*, 21. [CrossRef]

23.   Jalilian, A.; Rowshanfarzad, P.; Sabet, M.; Shafiee, A. Preparation of [^{61}Cu]-2-acetylpyridine thiosemicarbazone complex as a possible PET tracer for malignancies. *Appl. Radiat. Isot.* **2006**, *64*, 337–341. [CrossRef]

24.   Jalilian, A.; Rowshanfarzad, P.; Sabet, M. Preparation of [^{61}Cu]pyruvaldehyde-bis (N4-methylthiosemicarbazone) complex as a possible PET radiopharmaceutical. *Radiochim. Acta* **2006**, *94*, 113–117. [CrossRef]

25.   Fukumura, T.; Okada, K.; Szelecsényi, F.; Kovács, Z.; Suzuki, K. Practical production of ^{61}Cu using natural Co target and its simple purification with a chelating resin for ^{61}Cu-ATSM. *Radiochim. Acta* **2004**, *92*, 209–214. [CrossRef]

26.   Jalilian, A.; Rostampour, N.; Rowshanfarzad, P.; Shafaii, K.; Kamali-Dehghan, M.; Akhlaghi, M. Preclinical studies of [^{61}Cu]ATSM as a PET radiopharmaceutical for fibrosarcoma imaging. *Acta Pharm.* **2009**, *59*, 45–55. [CrossRef] [PubMed]

27.   Jalilian, A.R.; Nikzad, M.; Zandi, H. Preparation and evaluation of [^{61}Cu]-thiophene-2-aldehyde thiosemicarbazone for PET studies. *Nucl. Med. Rev.* **2008**, *11*, 41–47.

28.   Virgolini, I.; Ambrosini, V.; Bomanji, J.B.; Baum, R.P.; Fanti, S.; Gabriel, M.; Papathanasiou, N.D.; Pepe, G.; Oyen, W.; De Cristoforo, C.; et al. Procedure guidelines for PET/CT tumour imaging with ^{68}Ga-DOTA- conjugated peptides: ^{68}Ga-DOTA-TOC, ^{68}Ga-DOTA-NOC, ^{68}Ga-DOTA-TATE. *Eur. J. Nucl. Med. Mol. Imaging* **2010**, *37*, 2004–2010. [CrossRef]

29.   Krenning, E.; Breeman, W.; Kooij, P.; Lameris, J.; Bakker, W.; Koper, J.; Ausema, L.; Reubi, J.; Lamberts, S. Localisation of Endocrine-Related Tumours With Radioiodinated Analogue of Somatostatin. *Lancet* **1989**, *333*, 242–244. [CrossRef]

30.   Kwekkeboom, D.J.; Krenning, E.P.; Scheidhauer, K.; Lewington, V.; Lebtahi, R.; Grossman, A.; Vitek, P.; Sundin, A.; Plöckinger, U. ENETS Consensus Guidelines for the Standards of Care in Neuroendocrine Tumors: Somatostatin Receptor Imaging with ^{111}In-Pentetreotide. *Neuroendocrinology* **2009**, *90*, 184–189. [CrossRef] [PubMed]

31.   Krenning, E.P.; Bakker, W.H.; Kooij, P.P.M.; Breeman, W.A.P.; Oei, H.Y.; de Jong, M.; Reubi, J.C.; Visser, T.J.; Bruns, C.; Kwekkeboom, D.J.; et al. Somatostatin Receptor Scintigraphy with Indium 111-DTPA-D-Phe-1-Octreotide in Man: Metabolism, Dosimetry and Comparison with Iodine- 123-Tyr-3-Octreotide. *J. Nucl. Med.* **1992**, *33*, 652–658.

32.   Sundin, A.; Arnold, R.; Baudin, E.; Cwikla, J.B.; Eriksson, B.; Fanti, S.; Fazio, N.; Giammarile, F.; Hicks, R.J.; Kjaer, A.; et al. ENETS Consensus Guidelines for the Standards of Care in Neuroendocrine Tumors: Radiological, Nuclear Medicine and Hybrid Imaging. *Neuroendocrinology* **2017**, *105*, 212–244. [CrossRef]

33.   Hicks, R.J.; Kwekkeboom, D.J.; Krenning, E.; Bodei, L.; Grozinsky-Glasberg, S.; Arnold, R.; Borbath, I.; Cwikla, J.B.; Toumpanakis, C.; Kaltsas, G.; et al. ENETS Consensus Guidelines for the Standards of Care in Neuroendocrine Neoplasms: Peptide Receptor Radionuclide Therapy with Radiolabelled Somatostatin Analogues. *Neuroendocrinology* **2017**, *105*, 295–309. [CrossRef]

34. Seemann, J.; Waldron, B.; Parker, D.; Roesch, F. DATATOC: A novel conjugate for kit-type ^{68}Ga labelling of TOC at ambient temperature. *EJNMMI Radiopharm. Chem.* **2016**, *1*, 12. [CrossRef] [PubMed]

35. Pfeifer, A.; Knigge, U.; Mortensen, J.; Oturai, P.; Berthelsen, A.K.; Loft, A.; Binderup, T.; Rasmussen, P.; Elema, D.; Klausen, T.L.; et al. Clinical PET of Neuroendocrine Tumors Using ^{64}Cu-DOTATATE: First-in-Humans Study. *J. Nucl. Med.* **2012**, *53*, 1207–1215. [CrossRef]

36. Pfeifer, A.; Knigge, U.; Binderup, T.; Mortensen, J.; Oturai, P.; Loft, A.; Berthelsen, A.K.; Langer, S.W.; Rasmussen, P.; Elema, D.; et al. ^{64}Cu-DOTATATE PET for Neuroendocrine Tumors: A Prospective Head-to-Head Comparison with ^{111}In-DTPA-Octreotide in 112 Patients. *J. Nucl. Med.* **2015**, *56*, 847–854. [CrossRef]

37. Johnbeck, C.B.; Knigge, U.; Loft, A.; Berthelsen, A.K.; Mortensen, J.; Oturai, P.; Langer, S.W.; Elema, D.R.; Kjaer, A. Head-to-Head Comparison of ^{64}Cu-DOTATATE and ^{68}Ga-DOTATOC PET/CT: A Prospective Study of 59 Patients with Neuroendocrine Tumors. *J. Nucl. Med.* **2017**, *58*, 451–457. [CrossRef]

38. Mirzaei, S.; Revheim, M.-E.; Raynor, W.; Zehetner, W.; Knoll, P.; Zandieh, S.; Alavi, A. ^{64}Cu-DOTATOC PET-CT in Patients with Neuroendocrine Tumors. *Oncol. Ther.* **2020**, *8*, 125–131. [CrossRef] [PubMed]

39. Loft, M.; Carlsen, E.A.; Johnbeck, C.B.; Johannesen, H.H.; Binderup, T.; Pfeifer, A.; Mortensen, J.; Oturai, P.; Loft, A.; Berthelsen, A.K.; et al. ^{64}Cu-DOTATATE PET in patients with neuroendocrine neoplasms: Prospective, head-to-head comparison of imaging at 1 hour and 3 hours after injection. *J. Nucl. Med.* **2021**, *62*, 62–68. [CrossRef]

40. Delpassand, E.S.; Ranganathan, D.; Wagh, N.; Shafie, A.; Gaber, A.; Abbasi, A.; Kjaer, A.; Tworowska, I.; Núñez, R. ^{64}Cu-DOTATATE PET/CT for imaging patients with known or suspected somatostatin receptor-Positive neuroendocrine tumors: Results of the first U.S. prospective, reader-masked clinical trial. *J. Nucl. Med.* **2020**, *61*, 890–896. [CrossRef]

41. Carlsen, E.A.; Johnbeck, C.B.; Binderup, T.; Loft, M.; Pfeifer, A.; Mortensen, J.; Oturai, P.; Loft, A.; Berthelsen, A.K.; Langer, S.W.; et al. ^{64}Cu-DOTATATE PET/CT and prediction of overall and progression-free survival in patients with neuroendocrine neoplasms. *J. Nucl. Med.* **2020**, *61*, 1491–1497. [CrossRef]

42. Hicks, R.J.; Jackson, P.; Kong, G.; Ware, R.E.; Hofman, M.S.; Pattison, D.A.; Akhurst, T.A.; Drummond, E.; Roselt, P.; Callahan, J.; et al. ^{64}Cu-SARTATE PET imaging of patients with neuroendocrine tumors demonstrates high tumor uptake and retention, potentially allowing prospective dosimetry for peptide receptor radionuclide therapy. *J. Nucl. Med.* **2019**, *60*, 777–785. [CrossRef]

43. Do Carmo, S.J.C.; Alves, V.H.; Alves, F.; Abrunhosa, A.J. Oral presentation at the 17th Workshop on Targetry and Target Chemistry (WTTC17). 2018. Available online: https://slideslive.com/38910264/production-of-cu61-in-liquid-targets (accessed on 21 April 2022).

44. Do Carmo, S.J.C.; Scott, P.J.H.; Alves, F. Production of radiometals in liquid targets. *EJNMMI Radiopharm. Chem.* **2020**, *5*, 21. [CrossRef]

45. Fonseca, A.I.; Alves, V.H.; do Carmo, S.J.C.; Alves, F.; Abrunhosa, A.J. Copper-61 from liquid targets: Optimized purification for GMP labelling. *Nucl. Med. Biol.* **2021**, *96*, 96–97. [CrossRef]

46. Coenen, H.H.; Gee, A.D.; Adam, M.; Antoni, G.; Cutler, C.S.; Fujibayashi, Y.; Jeong, J.M.; Mach, R.H.; Mindt, T.L.; Pike, V.W.; et al. Consensus nomenclature rules for radiopharmaceutical chemistry—Setting the record straight. *Nucl. Med. Biol.* **2017**, *55*, i-xi. [CrossRef]

47. Decristoforo, C.; Neels, O.; Patt, M. Emerging Radionuclides in a Regulatory Framework for Medicinal Products – How Do They Fit? *Front. Med.* **2021**, *8*, 769. [CrossRef]

48. Alves, V.H.; Carmo, S.J.C.D.; Alves, F.; Abrunhosa, A.J. Automated Purification of Radiometals Produced by Liquid Targets. *Instruments* **2018**, *2*, 17. [CrossRef]

49. Alves, F.; Alves, V.H.P.; Do Carmo, S.J.C.; Neves, A.C.B.; Silva, M.; Abrunhosa, A.J. Production of copper-64 and gallium-68 with a medical cyclotron using liquid targets. *Mod. Phys. Lett. A* **2017**, *32*, 1740013. [CrossRef]

50. Garrison, W.M. Reaction mechanisms in the radiolysis of peptides, polypeptides, and proteins. *Chem. Rev.* **1987**, *87*, 381–398. [CrossRef]

51. Liu, S.; Ellars, C.E.; Edwards, D.S. Ascorbic acid: Useful as a buffer agent and radiolytic stabilizer for metalloradiopharmaceuticals. *Bioconjug. Chem.* **2003**, *14*, 1052–1056. [CrossRef]

52. Liu, S.; Edwards, D.S. Bifunctional chelators for therapeutic lanthanide radiopharmaceuticals. *Bioconjug. Chem.* **2001**, *12*, 7–34. [CrossRef]

53. Meisenheimer, M.; Kürpig, S.; Essler, M.; Eppard, E. Manual vs. automated 68Ga-radiolabelling—A comparison of optimized processes. *J. Label. Compd. Radiopharm.* **2020**, *63*, 162–173. [CrossRef]

54. Eppard, E.; Pèrez-Malo, M.; Rösch, F. Improved radiolabeling of DOTATOC with trivalent radiometals for clinical application by addition of ethanol. *EJNMMI Radiopharm. Chem.* **2017**, *1*, 6. [CrossRef]

55. Eppard, E.; Wuttke, M.; Nicodemus, P.L.; Rösch, F. Ethanol-based post-processing of generator-derived [68]Ga Toward kit-type preparation of ^{68}Ga-radiopharmaceuticals. *J. Nucl. Med.* **2014**, *55*, 1023–1028. [CrossRef] [PubMed]

56. Domnanich, K.A.; Müller, C.; Farkas, R.; Schmid, R.M.; Ponsard, B.; Schibli, R.; Türler, A.; Van Der Meulen, N.P. [44]Sc for labeling of DOTA- and NODAGA-functionalized peptides: Preclinical in vitro and in vivo investigations. *EJNMMI Radiopharm. Chem.* **2016**, *1*, 8. [CrossRef] [PubMed]

pharmaceuticals

Article

[^{99m}Tc]Tc-iFAP/SPECT Tumor Stroma Imaging: Acquisition and Analysis of Clinical Images in Six Different Cancer Entities

Paola Vallejo-Armenta [1], Guillermina Ferro-Flores [1,*], Clara Santos-Cuevas [1], Francisco Osvaldo García-Pérez [2], Pamela Casanova-Triviño [2], Bayron Sandoval-Bonilla [3], Blanca Ocampo-García [1], Erika Azorín-Vega [1,*] and Myrna Luna-Gutiérrez [1,*]

[1] Department of Radioactive Materials, Instituto Nacional de Investigaciones Nucleares, Ocoyoacac 52750, Mexico; paovallejoarmenta@gmail.com (P.V.-A.); clara.cuevas@inin.gob.mx (C.S.-C.); blanca.ocampo@inin.gob.mx (B.O.-G.)

[2] Department of Nuclear Medicine, Instituto Nacional de Cancerología, Tlalpan, Mexico City 14080, Mexico; fosvaldogarcia@gmail.com (F.O.G.-P.); zafily3@gmail.com (P.C.-T.)

[3] Department of Neurosurgery, Hospital de Especialidades del Centro Médico Nacional Siglo XXI, IMSS, Cuauhtémoc, Mexico City 06720, Mexico; bayronsandoval@gmail.com

* Correspondence: guillermina.ferro@inin.gob.mx (G.F.-F.); erica.azorin@inin.gob.mx (E.A.-V.); myrna.luna@inin.gob.mx (M.L.-G.)

Abstract: Fibroblast activation protein (FAP) is highly expressed on the cancer-associated fibroblasts (CAF) of the tumor stroma. Recently, we reported the preclinical evaluation and clinical biokinetics of a novel ^{99m}Tc-labeled FAP inhibitor radioligand ([^{99m}Tc]Tc-iFAP). This research aimed to evaluate [^{99m}Tc]Tc-iFAP for the tumor stroma imaging of six different cancerous entities and analyze them from the perspective of stromal heterogeneity. [^{99m}Tc]Tc-iFAP was prepared from freeze-dried kits with a radiochemical purity of 98 ± 1%. The study included thirty-two patients diagnosed with glioma ($n = 5$); adrenal cortex neuroendocrine tumor ($n = 1$); and breast ($n = 21$), lung ($n = 2$), colorectal ($n = 1$) and cervical ($n = 3$) cancer. Patients with glioma had been evaluated with a previous cranial MRI scan and the rest of the patients had been involved in a [^{18}F]FDG PET/CT study. All oncological diagnoses were corroborated histopathologically. The patients underwent SPECT/CT brain imaging (glioma) or thoracoabdominal imaging 1 h after [^{99m}Tc]Tc-iFAP administration (i.v., 735 ± 63 MBq). The total lesions ($n = 111$) were divided into three categories: primary tumors (PT), lymph node metastases (LNm), and distant metastases (Dm). [^{99m}Tc]Tc-iFAP brain imaging was positive in four high-grade WHO III–IV gliomas and negative in one treatment-naive low-grade glioma. Both [^{99m}Tc]Tc-iFAP and [^{18}F]FDG detected 26 (100%) PT, although the number of positive LNm and Dm was significantly higher with [^{18}F]FDG [82 (96%)], in comparison to [^{99m}Tc]Tc-iFAP imaging (35 (41%)). Peritoneal carcinomatosis lesions in a patient with recurrent colorectal cancer were only visualized with [^{99m}Tc]Tc-iFAP. In patients with breast cancer, a significant positive correlation was demonstrated among [^{99m}Tc]Tc-iFAP uptake values (Bq/cm^3) of PT and the molecular subtype, being higher for subtypes HER2+ and Luminal B HER2-enriched. Four different CAF subpopulations have previously been described for LNm of breast cancer (from CAF-S1 to CAF-S4). The only subpopulation that expresses FAP is CAF-S1, which is preferentially detected in aggressive subtypes (HER2 and triple-negative), confirming that FAP+ is a marker for poor disease prognosis. The results of this pilot clinical research show that [^{99m}Tc]Tc-iFAP SPECT imaging is a promising tool in the prognostic assessment of some solid tumors, particularly breast cancer.

Keywords: FAP; ^{99m}Tc-FAP inhibitor; ^{99m}Tc-labeled iFAP; tumor microenvironment; SPECT

Citation: Vallejo-Armenta, P.; Ferro-Flores, G.; Santos-Cuevas, C.; García-Pérez, F.O.; Casanova-Triviño, P.; Sandoval-Bonilla, B.; Ocampo-García, B.; Azorín-Vega, E.; Luna-Gutiérrez, M. [^{99m}Tc]Tc-iFAP/SPECT Tumor Stroma Imaging: Acquisition and Analysis of Clinical Images in Six Different Cancer Entities. *Pharmaceuticals* **2022**, *15*, 729. https://doi.org/10.3390/ph15060729

Academic Editor: Xuyi Yue

Received: 12 May 2022
Accepted: 6 June 2022
Published: 9 June 2022

Publisher's Note: MDPI stays neutral with regard to jurisdictional claims in published maps and institutional affiliations.

Copyright: © 2022 by the authors. Licensee MDPI, Basel, Switzerland. This article is an open access article distributed under the terms and conditions of the Creative Commons Attribution (CC BY) license (https://creativecommons.org/licenses/by/4.0/).

1. Introduction

Tumors are pathological complexes composed of tumor cells and the tumor stroma or tumor microenvironment (TME), which consists of cellular and acellular components, such as cancer-associated fibroblasts (CAFs), endothelial cells, adipocytes, mesenchymal stem cells (MSC), macrophages, blood vessels, pericytes, and extracellular matrices (ECM) [1,2]. In fact, CAFs induce a cancer phenotype and are responsible for the production of proteolytic enzymes, growth factors, and extracellular matrix components [2]. CAFs contribute up to 90% of the macroscopic tumor mass, provide mechanical support to tumor cells and control their survival, metastasis, proliferation, and resistance to therapies. CAFs can have different origins, including adipose mesenchymal stem cells, resident tissue fibroblasts and epithelial/endothelial cells, and adipocytes and pericytes that transdifferentiate to mesenchymal cells; therefore, they represent a heterogeneous cell population within the TME [1].

Fibroblast activation protein (FAP) is a membrane-anchored peptidase expressed by CAFs at the stromal level of various tumor entities and contributes to progression and a worse prognosis. FAP degrades denatured collagens and participates in tumor growth via a non-enzymatic mechanism [1–3].

Diagnostic FAP inhibitor radiotracers under clinical evaluation use ^{18}F and ^{68}Ga linked to quinolinoyl-cyanopyrrolidine [3–6] and cyclo-[benzene(trimethanethiol-DOTA)-Met-Pro-Pro-Thr-Glu-Phe-Met] (FAPI-2286) structures [7], which are radiotracers for PET (positron emission tomography), and only one work has reported ^{99m}Tc, also linked to quinolinoyl-cyanopyrrolidine for SPECT (single-photon emission computed tomography) imaging [3].

Internationally, the amount of equipment available for molecular imaging studies is predominantly higher for gamma cameras (SPECT modality), and they represent more than 70% of the total. For SPECT images, the most-employed radionuclide is ^{99m}Tc. Therefore, the need for target-specific radiopharmaceuticals labeled with ^{99m}Tc is increasing within the field of oncology. Our group previously reported [^{99m}Tc]Tc-((R)-1-((6-hydrazinylnicotinoyl)-D-alanyl)pyrrolidin-2-yl) boronic acid ([^{99m}Tc]Tc-iFAP) as a new SPECT radioligand capable of specifically detecting FAP expressed by CAFs located in the cancer stroma and, to our knowledge, the first ligand based on ^{99m}Tc-labeled boron-Pro derivatives [8]. Furthermore, the [^{99m}Tc]Tc-iFAP biokinetic–dosimetric evaluation in healthy volunteers and three cancer patients diagnosed with breast, lung, and cervical cancer showed favorable biokinetics and uptake in primary tumor lesions and lymph node metastases, achieving high-quality and high-contrast molecular images [9].

This research aimed to evaluate [^{99m}Tc]Tc-iFAP for the tumor stroma imaging of six different cancerous entities and analyze them from the perspective of stromal heterogeneity.

2. Results

No adverse events related to the diagnostic use of [^{99m}Tc]Tc-iFAP were observed.

Table 1 shows the general characteristics of patients included in the [^{99m}Tc]Tc-iFAP imaging evaluation. A detailed cancer staging of patients is shown in Table A1 (Appendix A). Patient imaging results were categorized into two groups. Patients with gliomas ($n = 5$), with which SPECT and SPECT/MR images were acquired, were identified as the first group. The second group involves all cases ($n = 27$) of breast, lung, colon, NET, renal cortex, and cervical cancer, in which SPECT/CT and PET/CT images were obtained.

[^{99m}Tc]Tc-iFAP SPECT brain imaging was positive in four high-grade WHO III–IV gliomas (T/Bc range 6.3–13.9) (Table 2) and negative in one treatment-naive low-grade glioma (Figure 1). [^{99m}Tc]Tc-iFAP imaging resolution and contrast were good enough for the high-grade glioma, which could allow the performing of non-invasive diagnoses to differentiate between low- and high-grade gliomas based on their distinct FAP expression [10].

For all cancer cases, a total of 111 lesions were evaluated, which were classified as primary tumors (PT)($n = 26$), lymph node metastases (LNm) ($n = 61$), and distant metastases (Dm) ($n = 24$) (Table 3).

Table 1. General characteristics of the patients included in the [^{99m}Tc]Tc-iFAP imaging study.

Characteristics	Number
No. patient	32
Age (years)	50.8 ± 16.7
Gender (%)	
Female	28 (88%)
Male	4 (12%)
Diagnosis	**Cases (%)**
Breast cancer	21 (66%)
Ductal carcinoma, Luminal A	2
Ductal carcinoma, Luminal B	3
Ductal carcinoma, Luminal B HER2+	5
Ductal carcinoma, HER2+	2
Ductal carcinoma, Triple negative	9
Lung cancer	2 (6%)
NSCLC adenocarcinoma	
Cervical cancer	3 (9%)
Squamous cell carcinoma	
Glioma	5 (16%)
Astrocytoma NOS (WHO II)	1
Anaplastic astrocytoma NOS (WHO III)	3
Glioblastoma NOS (WHO IV)	1
Colorectal cancer	1 (3%)
Adenocarcinoma	1
Adrenal cortical neuroendocrine tumor	1 (3%)
Poorly differentiated, Ki67 30%	1
Clinical setting (%)	
Initial staging	27 (84%)
Restaging	5 (15%)

Table 2. Tumor-to-contralateral tissue background ratio (T/Bc) of [^{99m}Tc]Tc-iFAP in patients with high-grade WHO III–IV gliomas.

Diagnosis	Status Brain SPECT	T/Bc
Astrocytoma NOS (WHO II)	Negative	NA
Anaplastic astrocytoma NOS (WHO III) (n = 2)	Positive	6.3 and 7.8
Anaplastic astrocytoma NOS restaging (WHO III)	Positive	15.4
Glioblastoma NOS (WHO IV)	Positive	13.9

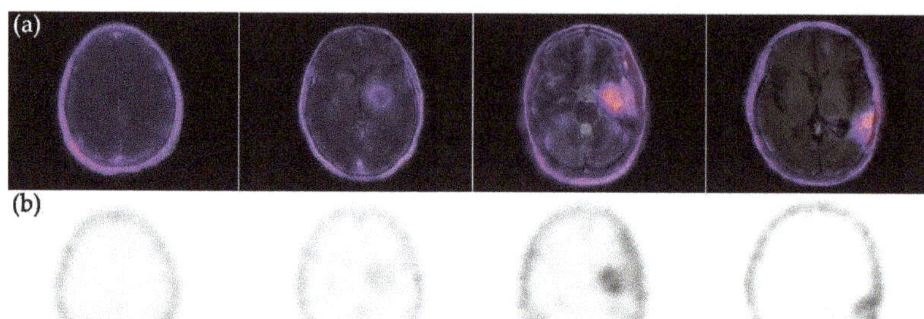

Figure 1. (a) [^{99m}Tc]Tc-iFAP SPECT coregistered to MR images (SPECT/MRI) and (b) [^{99m}Tc]Tc-iFAP SPECT. Note the adequate visualization of [^{99m}Tc]Tc-iFAP uptake in high-grade gliomas (WHO III-IV)-treatment-naive and recurrent (R). However, low-grade glioma (WHO II) did not show uptake.

Table 3. Number of lesions detected with [^{99m}Tc]Tc-iFAP and [^{18}F]FDG in all patients except gliomas.

	Primary Tumor	Lymph Node Metastases	Distant Metastases	Total
All lesions (N)	**26**	**61**	**24**	**111**
[^{99m}Tc]Tc-iFAP	26 (100%)	31 (51%)	4 (17%)	61 (55%)
[^{18}F]FDG	26 (100%)	61 (100%)	21 (88%)	108 (97%)

	Diagnosis	Lymph node metastases	Distant metastases
		$n = 31$ (51%)	$n = 4$ (17%)
[^{99m}Tc]Tc-iFAP	Lung cancer NSCLC	3	0
	Triple-negative BC	10	0
	Luminal A	0	0
	Luminal B HER2+ BC	7	2
	Luminal B BC	4	0
	HER2+ BC	5	0
	Cervical cancer	2	0
	Colorectal cancer	0	3
	Adrenal cortical NT	0	1
		$n = 61$ (100%)	$n = 21$ (88%)
[^{18}F]FDG	Lung cancer NSCLC	3	1
	Triple-negative BC	25	1
	Luminal A BC	5	0
	Luminal B HER2+ BC	8	16
	Luminal B BC	10	1
	HER2+ BC	7	0
	Cervical cancer	2	0
	Colorectal cancer	0	0
	Adrenal cortical NT	1	2

BC: breast cancer; NSCLC: non-small cell lung cancer; NT: neuroendocrine tumor.

All primary tumors were detected with both [^{99m}Tc]Tc-iFAP SPECT/CT and [^{18}F]FDG PET/CT (Figures 2 and 3), which did not occur with LNm and Dm lesions (Figures 4 and 5). That is, [^{99m}Tc]Tc-iFAP SPECT/CT detected PT (100%), LNm (51%), and Dm (17%) in contrast to [^{18}F]FDG PET/CT, which detected PT (100%), LNm (100%), and Dm (88%) (Table 3).

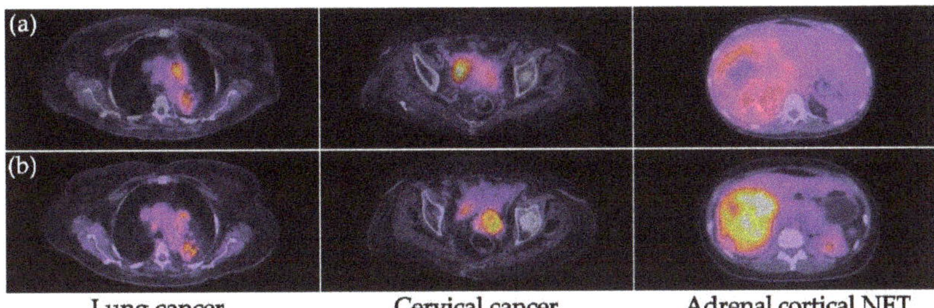

Figure 2. (**a**) [99mTc]Tc-iFAP SPECT/CT and (**b**) [18F]FDG PET/CT images of the primary tumors of three different types of cancers. All primary lesions show concordant uptake between both molecular imaging methods. [99mTc]Tc-iFAP uptake was considerably lower regarding [18F]FDG in patients with cervical cancer and neuroendocrine tumor (NET) of the adrenal cortex.

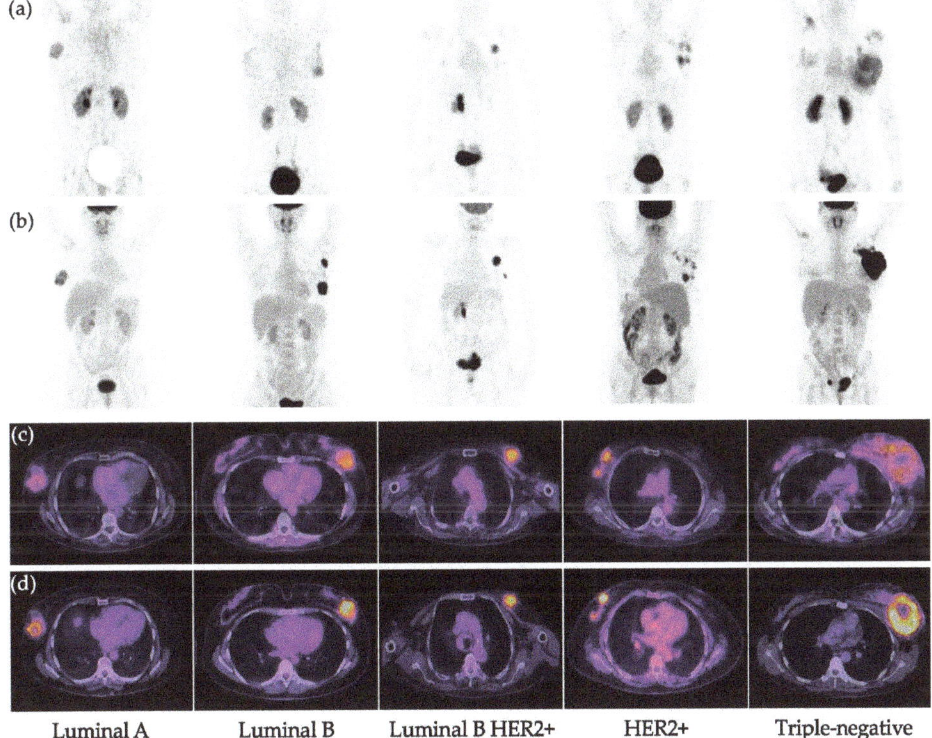

Figure 3. Primary breast cancer tumors. (**a**) [99mTc]Tc-iFAP MIP, (**b**) [18F]FDG MIP, (**c**) [99mTc]Tc-iFAP SPECT/CT, and (**d**) [18F]FDG PET/CT. All primary lesions show concordant uptake between both molecular imaging methods. [99mTc]Tc-iFAP uptake is decreased in pure hormonal molecular subtypes (Luminal A and B) and elevated in subtypes with HER2+ expression (Luminal B HER2+ and pure HER2+). The triple negative subtype shows moderate and heterogeneous uptake. MIP: maximum intensity projection.

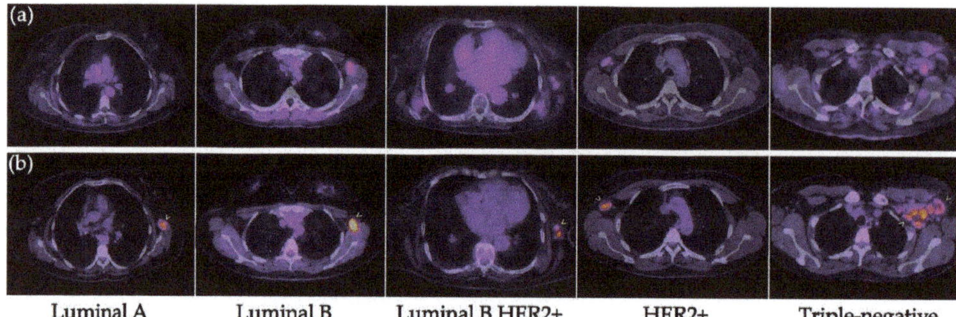

Luminal A   **Luminal B**   **Luminal B HER2+**   **HER2+**   **Triple-negative**

Figure 4. Lymph node metastases in breast cancer. (**a**) [^{99m}Tc]Tc-iFAP SPECT/T and (**b**) [^{18}F]FDG PET/CT. All malignant-appearing axillary lymphadenopathies are hypermetabolic; however, most of them (arrowheads) exhibit reduced or absent [^{99m}Tc]Tc-iFAP uptake in all molecular subtypes of breast cancer (lesion sizes >8 mm).

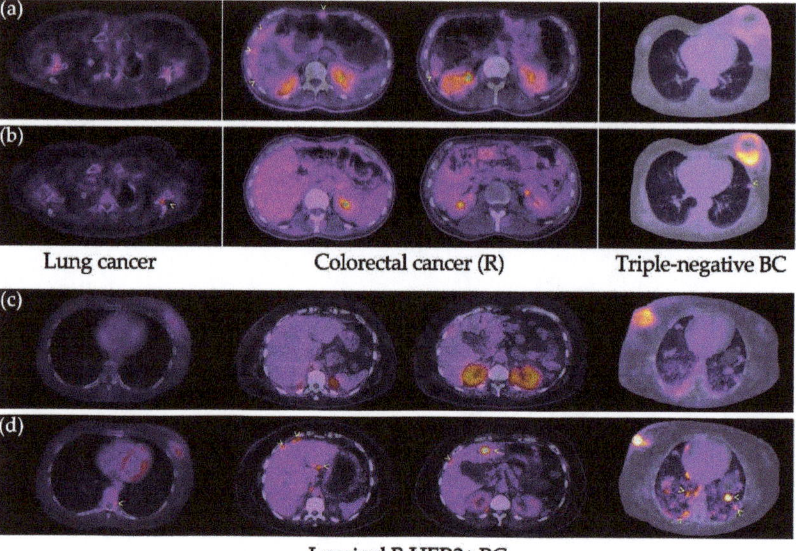

Lung cancer   **Colorectal cancer (R)**   **Triple-negative BC**

Luminal B HER2+ BC

Figure 5. Distant metastases in various types of cancers. (**a,c**) [^{99m}Tc]Tc-iFAP SPECT/CT and (**b,d**) [^{18}F]FDG PET/CT. All distant metastatic lesions are hypermetabolic; however, most of them (arrowhead) exhibit decreased or no uptake of [^{99m}Tc]Tc-iFAP. In the case of the patient with recurrent colon cancer, areas of diffuse [^{99m}Tc]Tc-iFAP uptake were observed in liver subcapsular implants and in the anterior abdominal wall, which were not detected with [^{18}F]FDG. BC: breast cancer. R: recurrence.

The non-detection of LNm and Dm with [^{99m}Tc]Tc-iFAP could be attributed to the lower spatial resolution of the SPECT technique in 61% of the lesions (size < 8 mm), including those not detected in NT and lung cancer, but not in 39% of the lesions with dimensions greater than 8 mm and associated to breast cancer. Additionally, none of the Dm lesions detected by [^{18}F]FDG in patients with triple negative and luminal B HER2+ molecular subtypes at the bone, liver, and lung exhibited [^{99m}Tc]Tc-iFAP uptake.

These results are expected since FAP expression decreases once the cells succeed to invade. FAP is a protein that promotes metastasis; therefore, once the micrometastasis is established in a distant site from the PT, it loses its FAP expression. The signaling mediated by FAP/integrins/PI3K has a negative effect on IGF2 expression (associated with increased glucose uptake). This fact probably explains why as FAP uptake in Dm decreases, FDG uptake increases [11].

As a unique feature of [^{99m}Tc]Tc-iFAP images, a very low background was achieved as previously reported (Figure 3) [9].

[^{99m}Tc]Tc-iFAP uptake was considerably lower regarding [^{18}F]FDG in patients with cervical cancer and neuroendocrine tumor (NET) of the adrenal cortex, which agrees with their relatively low FAP expression in comparison to lung and breast cancer (Figure 2) [12,13].

In general, the Dm lesions detected with [^{18}F]FDG did not show [^{99m}Tc]Tc-iFAP uptake, except for peritoneal carcinomatosis lesions in recurrent colorectal cancer, which only showed [^{99m}Tc]Tc-iFAP uptake, but not [^{18}F]FDG uptake (Figure 5). CAFs are abundant in mesothelial metastases, and, through the mesothelial-mesenchymal transformation mechanism, it is likely that in carcinomatosis there is a greater transdifferentiation of mesenchymal cells towards CAFs FAP+ [14].

When comparing the values obtained from the average tumor-to-background ratios of the different background sites [T/Bm (tumor/mediastinum), T/Bl (tumor/liver) and T/Bp (tumor/psoas muscle)] for all lesions, the highest values were T/Bp for both imaging methods (Figure 6). Although no statistically significant difference was found, the values of the T/Bp ratios were higher with [^{18}F]FDG than with [^{99m}Tc]Tc-iFAP (Figure 6). In the T/B data, the same trend is observed in terms of a higher uptake of [^{99m}Tc]Tc-iFAP in the primary tumors compared to that obtained in the lymph node and distant metastases (Figure 6).

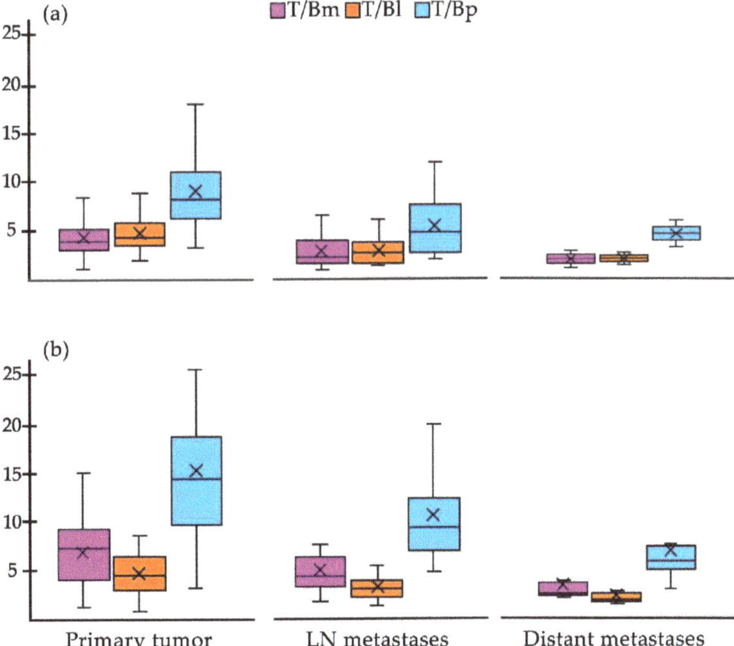

Figure 6. Box plot of the target-to-background ratios of (**a**) [^{99m}Tc]Tc-iFAP and (**b**) [^{18}F]F-FDG in all primary tumors, lymph node metastases, and distant metastases (except gliomas). The T/Bp ratio is higher in all categories of both radiotracers, particularly in the primary tumor. T/Bm (tumor/mediastinum), T/Bl (tumor/liver), and T/Bp (tumor/psoas muscle). × = mode.

In the case of breast cancer, [^{99m}Tc]Tc-iFAP showed a significant positive correlation between the T/Bp value of the primary tumors and the molecular subtype (Pearson correlation coefficient: $r = 0.8085$), where HER2+ and Luminal B HER2+ enriched subtypes showed the highest T/Bp ratios (Figure 7). The [^{99m}Tc]Tc-iFAP uptake in HER2+ could be associated to the Erb2-mediated phosphorylation of Tyr654 of β-catenin, which promotes the activation of Wnt signaling pathways and the consequent promotion of the tumor invasive capacity (FAP expression) through a very common mechanism in breast cancer, the epithelial–mesenchymal transition (EMT) process, induced by the microenvironment, which infers the gain of invasive capacity and the arrest of the cell cycle, while, at the signaling level, it implies the repression of E-cadherin expression through snail/slug [14,15].

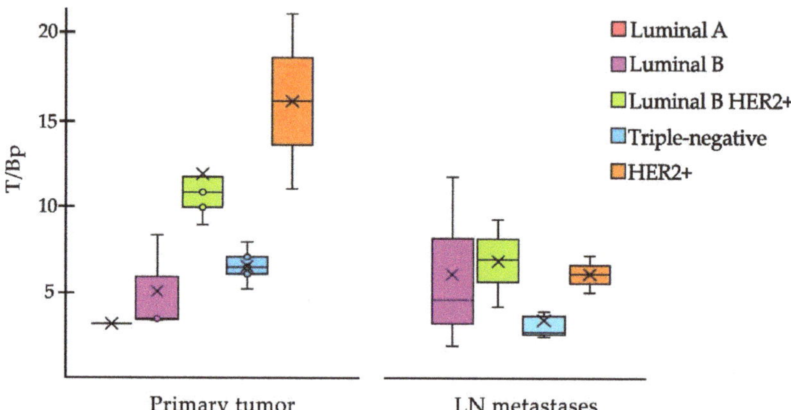

Figure 7. Box plot of the T/Bp ratio of [^{99m}Tc]Tc-iFAP in primary tumors and LN metastases of breast cancer. In primary tumors, the T/Bp (tumor/psoas muscle) ratio is higher in HER2+ and Luminal B HER2+ molecular subtypes. In LN metastases, a decrease in T/Bp is observed and there is no significant difference among the molecular subtypes (Pearson correlation coefficient: $r = 0.4027$). × = mode and ° = outliers.

In LN metastases, a decrease in T/Bp ratio was observed and there was no significant correlation among the molecular subtypes (Pearson correlation coefficient: $r = 0.4027$) (Figure 7).

3. Discussion

The expression of FAP is an indication that the cell is expressing an invasive phenotype associated with an intense process of differentiation, typical of the first stages of carcinogenesis [16]. During this phase, there is an intense activation of signaling pathways aimed at promoting the differentiation of cell precursors towards the activated fibroblast phenotype. As the tumor evolves, the stroma changes genetically and epigenetically to generate the appropriate niche for its stage. Cellular plasticity allows cells to adapt to their microenvironment through reprogramming processes (phenotypic and genotypic modifications) for tumor progression. RNAs produce epigenetic modifications that alter transcription, activating stem cell transformation and EMT processes (including FAP expression), which are essential for invasion to occur [16]. FAP is overexpressed by CAFs from various tumor entities, making it a promising biomarker and target for many medical interventions. CAF subpopulations (from CAF-S1 to CAF-S4) are classified depending on the expression of six markers: integrin b1/CD29, α-SMA (alpha-smooth muscle actin), PDGFR-β (platelet-derived growth factor receptor β), fibroblast activation protein (FAP), CAV1 (caveolin 1), and S100-A4/FSP1 (fibroblast-specific protein 1). The only subpopulation that expresses FAP is CAF-S1 (CAF-S1 FAP+) [17].

In their study, Kratochwil et al. demonstrated the elevated and selective uptake of ^{68}Ga-FAPI-04 in the stroma of multiple tumors, including breast, lung, colorectal, and NET cancer [13]. However, this research demonstrated the tumor stroma imaging with [99mTc]Tc-iFAP as the first SPECT radioligand based on a boron-Pro derivative [8].

The results showed that the detection of primary tumor lesions with [99mTc]Tc-iFAP is consistent when compared with [18F]FDG. However, when detecting LNm and Dm, the superiority of [18F]FDG is clear. This fact can be attributed to the lower spatial resolution of the SPECT technique in 61% of the lesions (size < 8 mm), but not in 39% of the tumors with dimensions greater than 8 mm and associated to breast cancer. Thus, our findings are discussed from the perspective of tumor stroma heterogeneity in lesions with enough size to be detected by SPECT.

As mentioned, the dynamics of differentiation in the tumor microenvironment are attributed to genetic and non-genetic changes in tumor cells, the composition of the extracellular matrix, cell–cell interactions, and cell heterogeneity [16]. Based on this, it is likely that the increased uptake of [99mTc]Tc-iFAP by primary tumors due to the presence of increased amounts of CAF-S1 FAP+ indicates an active EMT process, which is known to happen in the early phases of carcinogenesis, through which the dissemination of cells from the primary mass to distant sites is promoted [15]. EMT involves the regulation of both intercellular adhesions by decreasing E-cadherin and increasing N-cadherin, as well as substrate adhesions through integrin mediated primarily by TGF-B, β-catenin, and the Wnt signaling pathway.

On the other hand, hypoxic and hypoglycemic tumor stroma synergistically promotes the EMT phenotype in carcinomas. Thus, tumors where GLUT1 expression is commonly increased will also have an inability to express an (invasive) EMT phenotype [14]. Accordingly, it is likely that the lack of uptake of [99mTc]Tc-iFAP in LNm and Dm is related to the fact that in this type of lesion there is an increase in the expression of GLUT1 receptors that leads to an increase in glucose metabolism, which produces a rise in the uptake of [18F]FDG and, at the same time, inhibits EMT (including FAP expression).

The neoplasm with the largest number of patients in this study was breast cancer, which showed a significant positive correlation in PT between the T/Bp value and the molecular subtypes, with the highest T/Bp ratios for the HER2+ and Luminal B subtypes HER2+. The T/Bp values in HER2+ breast cancer showed a significant decrease in the LN metastases regarding PT (Figure 7), which may be due to crosstalking (cross-regulation), which occurs between integrins and EGFR receptors, such as HER2 [18,19]. Additionally, it is known that the Wnt signaling pathway promotes the proliferation and invasion of breast cancer cells in a HER2-dependent manner [20]. It was recently confirmed that the expression of HER2 in the cell membrane is heterogeneous and that the accumulation of HER2 occurs in regions where adhesion to the extracellular matrix is dynamic [21]. Therefore, HER2 expression decreases in regions where focal adhesions are concentrated, and the relative local decrease in HER2 expression in LNm, compared to PT, is probably related to the metastatic process.

As a relevant point, it is noted that the presence of CAFs in the tumor stroma of breast cancer is associated with resistance to immunotherapy [22], since the elements secreted by CAFs derived from HER2+ tumors regulate resistance to treatment in a paracrine way. Thus, the decrease in [99mTc]Tc-iFAP uptake by LNm could indicate greater sensitivity to trastuzumab treatment.

Highly-relevant data have been reported on axillary LNm in breast cancer: (1) the stroma represents around 25–30% of the invaded areas (regardless of subtype); (2) the predominant CAF subpopulations are CAF-S1 and CAF-S4 (the latter being the most abundant); (3) CAFs enrichments are different in LNm compared to PT; (4) the secretion of CXCL12β by CAF-S1 and the expression of CXCR4 in cancer cells is involved in the initiation of EMT and in the distant metastatic process, particularly in lung and bone; and (5) the global stromal content in LNm provides a prognostic stratification of breast cancer patients and, therefore, the CAF-S1/CAF-S4 abundance status exhibits a prognostic value, since both present pro-invasive properties with different modes of action; however, CAF-S4 is known to have a greater impact on distant metastatic spread (Dm), particularly on the liver [17,23].

CAF-S1 FAP+ promotes an immunosuppressive environment by secreting CXCL12β, promoting the presence of CD4+CD25+ T cells, increasing T cell survival, and promoting the cell differentiation into CD25+FOXP3+ cells. The ability of regulatory T cells (Tregs) to inhibit the proliferation of effector T cells is also enhanced by CAF-S1. CAF-S4 is highly contractile and induces cancer cell invasion in three dimensions through Notch signaling. CAF-S1 FAP+ is preferentially detected in aggressive subtypes (HER2+ and triple negative), confirming that FAP+ is a poor prognostic marker [17,23].

Summarizing, the relatively low performance of [^{99m}Tc]Tc-iFAP in detecting LNm and Dm may be related to the molecular biology of cancer and the proportion of the enrichment of CAF-S1 FAP+, which is not the most abundant in metastatic lesions (LN or distant) (Figure 8). Even when the FAP expression is associated with a phenotype that tends to transmigration and proliferation, attention must be placed on the fact that its expression is temporary and that it depends largely on the tumor microenvironment dynamics; thus, when the characteristics of the tumor stroma are modified, FAP expression and cancer prognosis can change.

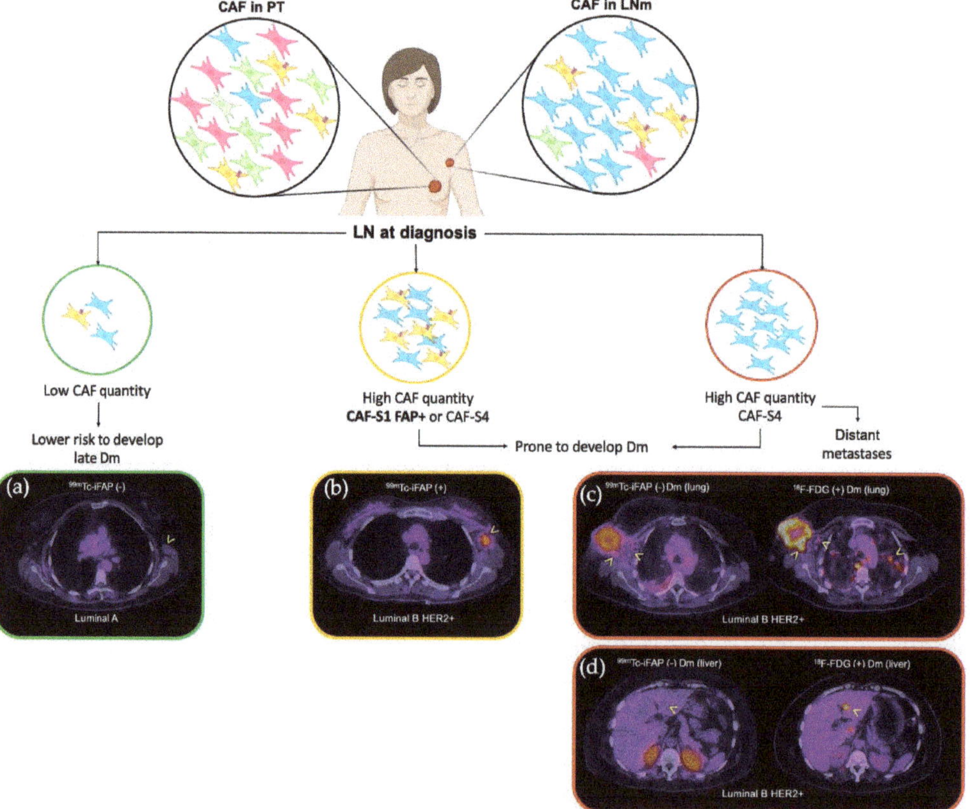

Figure 8. CAF subpopulations as prognostic markers in breast cancer (in diagnosis). Four CAF subpopulations have been reported in the lymph node metastases of breast cancer (CAF-S1 to CAF-S4). The most relevant and predominant are CAF-S1 FAP+ and CAF-S4 FAP-. Pelon et al. [23] established a model of clinical application to the knowledge generated from the different subpopulations, in such

a way that a prognostic impact is proposed according to the predominance of CAFs as follows: if at the time of diagnosis the patient exhibits low content of CAF-S1 FAP+ in LNm, they present a low risk of late Dm ((**a**) [^{99m}Tc]Tc-iFAP SPECT/CT(-), no uptake in left axillary adenopathy of Luminal A breast cancer); on the other hand, if high levels of CAF-S1 FAP+ are demonstrated in LNm, the risk of distant metastasis increases ((**b**) [^{99m}Tc]Tc-iFAP SPECT/CT(+), uptake in left axillary adenopathy of Luminal B HER2+ breast cancer). Finally, in distant metastatic lesions, only CAF-S4 FAP- is expressed [(**c**) [^{99m}Tc]Tc-iFAP SPECT/CT(-) in lung Dm and extremely low uptake in some right axillary lymph nodes that exhibit hypermetabolism with [^{18}F]FDG; likewise, (**d**) multiple lung and liver metastases did not exhibit uptake of [^{99m}Tc]Tc-iFAP]. BC: breast cancer, LNm: lymph node metastasis, Dm: distant metastasis.

Our results differ from the work of Kömek et al. [24], where they showed that PET/CT [^{68}Ga]Ga-FAPI-04 is superior to [^{18}F]FDG in the detection of primary mammary lesions and metastases (ganglionic and visceral) in twenty patients with breast cancer, both in primary and recurrent lesions, although the average size of the evaluated LNm was 10 mm [24]. On the other hand, Backhaus et al. [25] evaluated the use of PET/MRI with the ligand [^{68}Ga]Ga-FAPI-46 PET/CT in 19 women with breast cancer with evidence of high uptake in the primary lesions (mean diameter of 26 mm) and LNm (average diameter of 21 mm). Our results probably vary from the previous research carried out due to the heterogeneity of the sample with respect to the molecular subtypes of breast cancer, the image acquisition time, and the different image acquisition method (SPECT/CT vs. PET/CT vs. PET/MRI). The dynamic behavior of FAP is firmly associated with its functions in the progression phase during cancer evolution (tissue remodeling, extracellular matrix degradation, the promotion of tumor proliferation, and immunomodulation) [15–23], which deserves to be used as a tool for the detection of the heterogeneity of the tumor stroma in the different stages of cancer through molecular imaging with specific radiotracers, such as [^{99m}Tc]Tc-iFAP. Therefore, additional clinical studies must be performed, including the results of the ex vivo FAP expression in tumors (immunohistochemical evaluation) to be correlated with the uptake of FAP inhibitory radiotracers.

Today, CAFs is receiving considerable attention in the field of cancer biology. Targeted CAF therapy can potentially inhibit metastasis and cancer progression by reducing immunosuppression and remodeling the tumor microenvironment. Therapeutic targeting of FAP has been described in different modalities, such as vaccines, oncolytic viruses, and nanoparticles [26]. In preclinical studies, CAF-S1 FAP+ has shown to cause resistance to anti-PD-L1 immunotherapy and reduce antitumor immunity. CAFs from breast, ovarian, lung, pancreas, and colon cancer have shown expression of PD-L1 and/or PD-L2; particularly the CAF-S1 FAP+ subset. Additionally, the CAF-S1 FAP+ subpopulation is an important source of CXCL12 secretion, which plays a crucial role in resistance to anti-PD-1 and anti-CTLA-4 immunotherapies in pancreatic, ovarian, and breast cancer [27,28].

Taking into account the deleterious effect of metastases on the survival of breast cancer patients, our data could heighten the interest in evaluating the abundance of the CAF-S1 FAP+ subpopulation, in vivo, in a non-invasive manner, by means of [^{99m}Tc]Tc-iFAP SPECT in axillary LNm during the initial clinical approach (staging) to determine the prognosis and the benefit of therapies, such as anti-FAP, anti-TGFβ, anti-CXCR4, and/or anti-PD-L1 immunotherapy, in combination with standard therapies (Figure 8). More prospective research is needed to enrich the information obtained so far and we believe that future research can be focused on the function of FAP ligands in different molecular and histological subtypes of breast cancer, as well as their potential in detecting relapse of the disease, in the evaluation of the response to therapy and the prognosis of the patient.

Peritoneal carcinomatosis is a complication of various malignant tumors and is generally associated with a poor prognosis. The superiority of uptake by [^{99m}Tc]Tc-iFAP in peritoneal carcinomatosis, due to recurrent colon cancer observed in the patient included in this study, agrees with the findings previously described, demonstrating a greater sensitivity of [^{68}Ga]Ga-FAPI-04 for the detection of peritoneal carcinomatosis in patients with various types of cancer [29].

The findings observed in patients with glioma coincide with the data previously reported by Röhrich et al., where they showed little or no uptake of ^{68}Ga-FAPI-02 and FAPI-04 in low-grade WHO II gliomas and high uptake in gliomas of high WHO III-IV grade, regardless of HDI status [10]; therefore, its usefulness could lie mainly in the differentiation of tumor recurrence versus post-treatment changes and in surgical and/or radiotherapy planning, for which more prospective studies are needed in this regard.

4. Materials and Methods

4.1. Reagents

An iFAP (boron-Pro ligand) lyophilized kit for ^{99m}Tc labeling was obtained from the National Institute of Nuclear Research (ININ, Ocoyoacac, Mexico) with GMP certification [8]. [^{99m}Tc]TcO$_4$Na was eluted from a generator ([99]Mo/^{99m}Tc GETEC, ININ, Ocoyoacac, Mexico). Other reagents were received from Millipore Co. (Burlington, MA, USA).

4.2. [^{99m}Tc]Tc-iFAP Preparation

After the reconstitution of the iFAP lyophilized kit with a [^{99m}Tc]TcO$_4$Na/0.2 M phosphate buffer (1:1 *v/v*, 2 mL, 740 MBq) solution and incubation in a block heater (92 °C, 15 min), the [^{99m}Tc]Tc-iFAP radioligand was obtained with a radiochemical purity (R.P.) greater than 98% (HPLC, Discovery C18 column, 5 μm particle size, I.D. of 0.46 cm, length of 25 cm; Supelco, Millipore, Burling-ton, MA, USA; coupled to a UV–Vis detector and a radiometric detector), applying the following linear gradient: a flow rate of 1 mL/min, 0.1% TFA/water (A) (from 100 to 50%, over 10 min, maintained for 10 min, 30% over 5 min, and returned to 100% over 5 min) and 0.1% TFA/acetonitrile (B). As previously reported, the lyophilized formulation contains the HYNIC-iFAP (((R)-1-((6-hydrazinylnicotinoyl)-*D*-alanyl)pyrrolidin-2-yl)boronic) ligand with a specific alignment to the corresponding regions of the FAP binding site [8], stannous chloride as a reducing agent, as well as ethylenediaminediacetic acid (EDDA) to complete the coordination sphere of the [Tc(V)]HYNIC core (Figure 9).

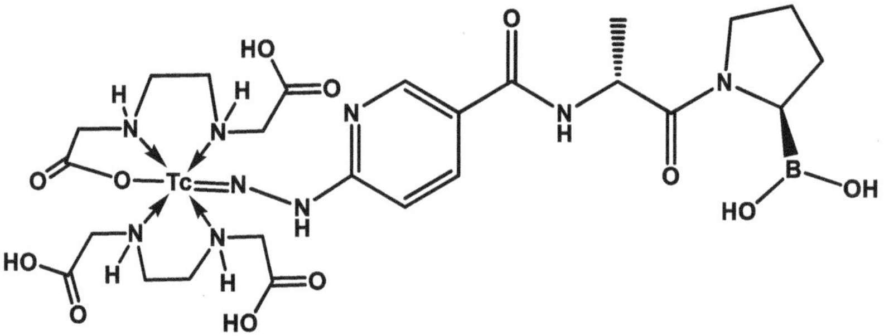

Figure 9. The proposed [Tc(V)]EDDA/HYNIC core structure in the [^{99m}Tc]Tc-iFAP radioligand. The advantage that HYNIC-iFAP presents with respect to previously reported quinolinoyl-cyanopyrrolidine-based FAP inhibitors is the possibility of obtaining [^{99m}Tc]Tc(V)-EDDA/HYNIC stable cores from instant freeze-dried kit formulations.

The chemical characterization of the iFAP ligand included analysis by mass spectrometry (UPLC-mass), ^{1}H–NMR, UV–Vis and FT-IR. Radiochemical characterization included reversed-phase radio-HPLC and ITLC-SG (instant thin layer chromatography-silica gel) with the following mobile phases: 2-butanone, 0.1 M sodium citrate, and ammonium acetate-methanol (1:1 *v/v*), as reported in detail previously [8].

4.3. Patients

Thirty-two patients (mean $\pm$ SD age, 50.8 $\pm$ 16.7 years; 28 women and 4 men) with different types of cancer (breast cancer (n = 21), lung cancer (n = 2), adrenal cortex NETs (n = 1), colorectal cancer (n = 1), cervical cancer (n = 3) and gliomas (n = 5)) were included.

The patients were divided into two groups as follows: Group 1 (n = 5 gliomas) and Group 2 (n = 27 breast, lung, colon, renal cortex NET, and cervical cancer). The characteristics of the patients are shown in Table 1 and with a detailed clinical description in Appendix A (Table A1). All oncological diagnoses were determined histopathologically (Table 2).

The patients underwent SPECT/CT 1–3 h (with an average of 2 h) after the intravenous application of [^{99m}Tc]Tc-iFAP (735 $\pm$ 63.5MBq). In Group 1, only of the brain region, and in Group 2, the thoracoabdominal region. The tumor/background ratio is optimal for diagnostic images from 30 min post-injection [9]. However, it was decided 2 h after radiotracer administration to improve the contrast of the images (lesions vs. background). The acquisition protocol and the post-injection waiting time were the same for all types of cancer evaluated. However, in patients with cervical cancer or pelvic etiology, immediate image acquisition was performed post-micturition to reduce the artifact of radiotracer accumulation in the urine.

All patients in Group 1 had previous cranial MRI (6 $\pm$ 1 days interval) and patients in Group 2 had previous [^{18}F]FDG PET/CT studies carried out (11 $\pm$ 12.6 days interval).

This research was performed in the Department of Nuclear Medicine of the National Cancer Institute (INCan), Mexico. The patients signed an informed consent declaration, and the protocol was approved by the institutional Nuclear Medicine Ethics Committee.

4.4. Image Acquisition

[^{99m}Tc]Tc-iFAP SPECT/CT images were acquired with a dual-head gamma camera (SPECT/CT, Symbia TruePoint, Siemens, Malvern, PA, USA), with low-energy, high-resolution collimators; parameters: window at 140 keV, matrix size of 128 $\times$ 128, with dispersion correction, 90 images of 8 s, rotation of 360 degrees. For the attenuation correction map, the low-dose CT parameters were obtained. A Butterworth filter (cutoff: 0.5, 5th order) and an iterative method (8 iterations /4 subsets) were used for the reconstruction of the raw data.

SPECT/CT images were acquired 2 h after the intravenous administration of [^{99m}Tc]Tc-iFAP (735 $\pm$ 63.5 MBq). The anatomical region studied in Group 1 was only the brain and in Group 2 it was thoracoabdominal. Activity in regions of interest was quantified, via 3D imaging, as Bq/cm^3.

All patients in Group 2 had undergone a prior PET/CT (Excel 20) scan (Siemens Medical Solutions), performed at 1 h after [^{18}F]FDG administration (CT: slice thickness of 5 mm, 180 mAs and 120 kVp). Whole-body scans were obtained in 3D mode from the vertex to mid-thighs (2–3 min per bed position). PET images were reconstructed using a two-dimensional expectation algorithm of ordered subsets.

4.5. Image Analysis

Images obtained with [^{99m}Tc]Tc-iFAP and [^{18}F]FDG were examined on a Siemens VG60 multimodal workstation. Visual and semi-quantitative analyses were performed by two physicians with more than 9 years of experience in nuclear medicine and molecular imaging (workstation with processing software for volumetric analysis).

Visual analysis was performed on both groups of patients. Uptake was compared with the morphology of the corresponding lesion using CT and/or MRI, depending on the patient group. The detected lesions were divided into three categories for study: primary tumor (PT), lymph node metastases (LNm), and distant metastases (Dm). The semiquantitative analysis of lesion uptake was obtained by calculating the tumor-to-background ratio (T/B) with spherical volumes of interest (VOIs) to homogenize the data obtained with both radiopharmaceuticals. Additionally, in Group 2, the concordance of uptake between both radiotracers was compared by quantifying the number of lesions (PT, LNm, and Dm).

4.6. Tumor Tissue Samples

All patients underwent a biopsy of the primary tumor lesion. Histopathology was used to determine the existence of viable tumor tissue and to verify the diagnosis. The histopathological reports were interpreted by a certified and experienced pathologist.

4.7. Statistical Analysis

The Pearson correlation coefficient was calculated between the T/Bp [^{99m}Tc]Tc-iFAP values and the molecular subtypes of the patients with breast cancer; a value of $p < 0.05$ was considered statistically significant.

5. Conclusions

The results of this pilot study show that SPECT imaging with [^{99m}Tc]Tc-iFAP is a promising and potentially useful tool in the evaluation of the tumor microenvironment of multiple solid neoplastic entities. Within the different types of cancer that we included, we observed a potential panorama in the prognostic evaluation of recently diagnosed breast cancer, as well as its probable diagnostic superiority in peritoneal carcinomatosis in recurrent colon cancer. [^{18}F]FDG was superior to [^{99m}Tc]Tc-iFAP in the detection of LNm and Dm. However, with the analyses carried out, we can establish that the role of [^{99m}Tc]Tc-iFAP is not to displace metabolic molecular imaging, but rather that it serves as a complement for an adequate prognostic evaluation.

Further prospective [^{99m}Tc]Tc-iFAP clinical studies are needed to define the clinical impact of the non-invasive in vivo detection of FAP in newly diagnosed breast cancer patients and its implication in determining candidates for immunotherapy and target therapy combined with conventional therapies.

Author Contributions: Conceptualization, G.F.-F., P.V.-A., E.A.-V. and M.L.-G.; methodology, P.V.-A., M.L.-G., P.C.-T. and B.S.-B.; formal analysis, P.V.-A., C.S.-C., G.F.-F., F.O.G.-P., B.O.-G. and E.A.-V.; writing—original draft preparation, P.V.-A. and G.F.-F.; writing—review and editing, G.F.-F.; funding acquisition, M.L.-G. and C.S.-C. All authors have read and agreed to the published version of the manuscript.

Funding: This research was funded by the "Consejo Mexiquense de Ciencia y Tecnología" (COME-CyT), through the EDOMEX financing program for female scientists, grant number FICDTEM-2021-008.

Institutional Review Board Statement: The clinical study was conducted according to the guidelines of the Declaration of Helsinki and was approved by the Ethics Committee of "Medicina Nuclear, Instituto Nacional de Cancerología" (protocol code 021/04/MNIC, 3 May 2021).

Informed Consent Statement: Informed consent was obtained from the patients involved in the study. Written informed consent has been obtained from the healthy subjects and patients to publish this paper.

Data Availability Statement: Data are contained within the article.

Acknowledgments: This study was conducted as a component of the activities of the "Laboratorio de Investigación y Desarrollo de Radiofármacos, ININ". Paola Vallejo-Armenta is grateful for the scholarship granted to participate in this research through the program "Estancias de investigación especializadas COMECYT-EDOMEX" Grant No. EESP2021-0008.

Conflicts of Interest: The authors declare no conflict of interest.

Appendix A

Table A1. Detailed description of the disease reported in cancer patients included in this research for tumor evaluation with the [^{99m}Tc]Tc-iFAP radioligand.

No.	Age (years)	Gender	Clinical Setting	Type of Cancer	Extent of Cancer
1	69	Male	Initial staging	Lung cancer, NSCLC (adenocarcinoma).	Primary, lymph node, bone
2	51	Female	Initial staging	Lung cancer, NSCLC (adenocarcinoma).	Primary, lymph node
3	66	Female	Restaging	Cervical cancer (squamous cell carcinoma).	Lymph node
4	60	Female	Initial staging	Cervical cancer (squamous cell carcinoma).	Primary
5	91	Female	Initial staging	Cervical cancer (squamous cell carcinoma). Breast cancer (ductal carcinoma; SBR 7, G2, moderate DR, Ki67 50%). Luminal B HER2+	Primary Primary, lymph node
6	70	Female	Initial staging	Breast cancer (ductal carcinoma; SBR 8, G3, moderate DR, Ki67 70%). Triple-negative.	Primary, lymph node
7	44	Female	Initial staging	Breast cancer (ductal carcinoma; SBR 9, G3, mild DR, Ki67 70%). Triple-negative.	Primary, lymph node
8	54	Female	Initial staging	Breast cancer (ductal carcinoma; SBR 6, G2, mild DR). Luminal A	Primary, lymph node
9	49	Female	Initial staging	Breast cancer (ductal carcinoma; SBR 7, G2, mild DR, Ki67 40%). HER2+	Primary, lymph node
10	40	Female	Initial staging	Breast cancer (ductal carcinoma; SBR 8, G3, moderate DR, Ki67 70%). Triple-negative.	Primary, lymph node
11	28	Female	Initial staging	Breast cancer (ductal carcinoma; SBR 8, G3, Ki67 60%). Triple-negative.	Primary, lymph node
12	29	Female	Initial staging	Breast cancer (ductal carcinoma; SBR 8, G3, moderate DR, Ki67 60%). Luminal B	Primary, lymph node
13	60	Female	Initial staging	Breast cancer (ductal carcinoma; SBR 9, G3, moderate DR, Ki67 30%). Luminal B HER2+	Primary, lymph node
14	55	Female	Initial staging	Breast cancer (ductal carcinoma; SBR 6, G2, mild DR, Ki67 15%). Luminal B	Primary, lymph node
15	55	Female	Initial staging	Breast cancer (ductal carcinoma; SBR 5, G1, moderate DR). Luminal A.	Lymph node
16	36	Female	Initial staging	Breast cancer (ductal carcinoma; SBR 8, G3, mild DR, Ki67 80%). Triple-negative.	Primary, lymph node
17	41	Female	Initial staging	Breast cancer (ductal carcinoma; SBR 6, G2, moderate DR, Ki67 40%). Luminal B HER2+	Primary, lymph node
18	48	Female	Initial staging	Breast cancer (ductal carcinoma; SBR 8, G3, mild DR, Ki67 30%). Luminal B HER2+	Primary, lymph node, lung
19	46	Female	Initial staging	Breast cancer (ductal carcinoma; SBR 7, G2). Luminal B HER2+	Primary, lymph node, liver
20	58	Female	Initial staging	Breast cancer (ductal carcinoma; SBR 6, G2, Ki67 30%). Luminal B.	Primary, lymph node
21	63	Female	Initial staging	Breast cancer (ductal carcinoma; SBR 7, G2, mild DR, Ki67 50%). Her2+	Primary, lymph node
22	44	Female	Initial staging	Breast cancer (ductal carcinoma; SBR 9, G3, mild DR, Ki67 70%). Triple-negative.	Primary, lymph node
23	42	Female	Initial staging	Breast cancer (ductal carcinoma; SBR 8, G3, mild DR, Ki67 80%). Triple-negative.	Primary, lymph node

Table A1. *Cont.*

No.	Age (years)	Gender	Clinical Setting	Type of Cancer	Extent of Cancer
24	68	Female	Initial staging	Breast cancer (ductal carcinoma; SBR 9, G3, Ki67 50%). Triple-negative.	Primary, lymph node, lung
25	55	Female	Initial staging	Breast cancer (ductal carcinoma; SBR 9, G3, moderate DR, Ki67 60%). Triple-negative.	Primary
26	37	Female	Restaging	Glioblastoma NOS (WHO IV)	Primary
27	76	Male	Initial staging	Anaplastic astrocytoma NOS (WHO III)	Primary
28	40	Female	Initial staging	Astrocytoma NOS (WHO II)	Primary
29	32	Female	Restaging	Anaplastic astrocytoma NOS (WHO III)	Primary
30	27	Male	Restaging	Anaplastic astrocytoma NOS (WHO III)	Primary
31	47	Male	Restaging	Colorectal cancer (adenocarcinoma).	Peritoneal carcinomatosis
32	23	Female	Initial staging	Adrenal cortical neuroendocrine tumor (poorly differentiated, Ki67 30%)	Primary, lung

DR: desmoplastic reaction; NSCLC: non-small cell lung cancer; SBR: Scarff–Bloom–Richardson grading; Ki67: cell proliferation index.

References

1. Hamson, E.J.; Keane, F.M.; Tholen, S.; Schilling, O.; Gorrell, M.D. Understanding fibroblast activation protein (FAP): Substrates, activities, expression and targeting for cancer therapy. *Proteom. Clin. Appl.* **2014**, *8*, 454–463. [CrossRef] [PubMed]
2. Altmann, A.; Haberkorn, U.; Siveke, J. The Latest Developments in Imaging of Fibroblast Activation Protein. *J. Nucl. Med.* **2021**, *62*, 160–167. [CrossRef] [PubMed]
3. Lindner, T.; Altmann, A.; Krämer, S.; Kleist, C.; Loktev, A.; Kratochwil, C.; Giesel, F.; Mier, W.; Marme, F.; Debus, J.; et al. Design and Development of ^{99m}Tc-Labeled FAPI Tracers for SPECT Imaging and 188Re Therapy. *J. Nucl. Med.* **2020**, *61*, 1507–1513. [CrossRef] [PubMed]
4. Loktev, A.; Lindner, T.; Mier, W.; Debus, J.; Altmann, A.; Jäger, D.; Giesel, F.; Kratochwil, C.; Barthe, P.; Roumestand, C.; et al. A tumor-imaging method targeting cancer-associated fibroblasts. *J. Nucl. Med.* **2018**, *59*, 1423–1429. [CrossRef]
5. Lindner, T.; Loktev, A.; Altmann, A.; Giesel, F.; Kratochwil, C.; Debus, J.; Jäger, D.; Mier, W.; Haberkorn, U. Development of quinoline-based theranostic ligands for the targeting of fibroblast activation protein. *J. Nucl. Med.* **2018**, *59*, 1415–1422. [CrossRef]
6. Giesel, F.L.; Adeberg, S.; Syed, M.; Lindner, T.; Jiménez-Franco, L.D.; Mavriopoulou, E.; Staudinger, F.; Tonndorf-Martini, E.; Regnery, S.; Rieken, S.; et al. FAPI-74 PET/CT Using Either ^{18}F-AlF or Cold-Kit ^{68}Ga Labeling: Biodistribution, Radiation Dosimetry, and Tumor Delineation in Lung Cancer Patients. *J. Nucl. Med.* **2021**, *62*, 201–207. [CrossRef]
7. Baum, R.P.; Schuchardt, C.; Singh, A.; Chantadisai, M.; Robiller, F.C.; Zhang, J.; Mueller, D.; Eismant, A.; Almaguel, F.; Zboralski, D.; et al. Biodistribution, and Preliminary Dosimetry in Peptide-Targeted Radionuclide Therapy of Diverse Adenocarcinomas Using ^{177}Lu-FAP-2286: First-in-Humans Results. *J. Nucl. Med.* **2022**, *63*, 415–423. [CrossRef]
8. Trujillo-Benítez, D.; Luna-Gutiérrez, M.; Ferro-Flores, G.; Ocampo-García, B.; Santos-Cuevas, C.; Bravo-Villegas, G.; Morales-Ávila, E.; Cruz-Nova, P.; Díaz-Nieto, L.; García-Quiroz, J.; et al. Design, Synthesis and Preclinical Assessment of 99mTc-iFAP for In Vivo Fibroblast Activation Protein (FAP) Imaging. *Molecules* **2022**, *27*, 264. [CrossRef]
9. Coria-Domínguez, L.; Vallejo-Armenta, P.; Luna-Gutiérrez, M.; Ocampo-García, B.; Gibbens-Bandala, B.; García-Pérez, F.; Ramírez-Nava, G.; Santos-Cuevas, C.; Ferro-Flores, G. [^{99m}Tc]Tc-iFAP Radioligand for SPECT/CT Imaging of the Tumor Microenvironment: Kinetics, Radiation Dosimetry, and Imaging in Patients. *Pharmaceuticals* **2022**, *15*, 590. [CrossRef]
10. Röhrich, M.; Loktev, A.; Wefers, A.K.; Altmann, A.; Paech, D.; Adeberg, S.; Windisch, P.; Hielscher, T.; Flechsig, P.; Floca, R.; et al. IDH-wildtype glioblastomas and grade III/IV IDH-mutant gliomas show elevated tracer uptake in fibroblast activation protein-specific PET/CT. *Eur. J. Nucl. Med. Mol. Imaging* **2019**, *46*, 2569–2580. [CrossRef]
11. Feng, T.; Fang, F.; Zhang, C.; Li, T.; He, J.; Shen, Y.; Yu, H.; Liu, X. Fluid Shear Stress-Induced Exosomes from Liver Cancer Cells Promote Activation of Cancer-Associated Fibroblasts via IGF2-PI3K Axis. *Front. Biosci. Landmark* **2022**, *27*, 104. [CrossRef] [PubMed]
12. The Human Protein Atlas. Available online: https://www.proteinatlas.org/ENSG00000078098-FAP/pathology (accessed on 6 May 2022).
13. Kratochwil, C.; Flechsig, P.; Lindner, T.; Abderrahim, L.; Altmann, A.; Mier, W.; Adeberg, S.; Rathke, H.; Röhrich, M.; Winter, H.; et al. ^{68}Ga-FAPI PET/CT: Tracer Uptake in 28 Different Kinds of Cancer. *J. Nucl. Med.* **2019**, *60*, 801–805. [CrossRef] [PubMed]
14. Jo, H.; Lee, J.; Jeon, J.; Kim, S.Y.; Chung, J.I.; Ko, H.Y.; Lee, M.; Yun, M. The critical role of glucose deprivation in epithelial-mesenchymal transition in hepatocellular carcinoma under hypoxia. *Sci. Rep.* **2020**, *10*, 1538. [CrossRef] [PubMed]

15. Ye, X.; Brabletz, T.; Kang, Y.; Longmore, G.D.; Nieto, M.A.; Stanger, B.Z.; Yang, J.; Weinberg, R.A. Upholding a role for EMT in breast cancer metastasis. *Nature* **2017**, *547*, E1–E3. [CrossRef]
16. Arora, L.; Pal, D. Remodeling of Stromal Cells and Immune Landscape in Microenvironment During Tumor Progression. *Front. Oncol.* **2021**, *11*, 596798. [CrossRef]
17. Costa, A.; Kieffer, Y.; Scholer-Dahirel, A.; Pelon, F.; Bourachot, B.; Cardon, M.; Sirven, P.; Magagna, I.; Fuhrmann, L.; Bernard, C.; et al. Fibroblast Heterogeneity and Immunosuppressive Environment in Human Breast Cancer. *Cancer Cell* **2018**, *33*, 463–479.e10. [CrossRef]
18. Banerjee, S.; Lo, W.C.; Majumder, P.; Roy, D.; Ghorai, M.; Shaikh, N.K.; Kant, N.; Shekhawat, M.S.; Gadekar, V.S.; Ghosh, S.; et al. Multiple roles for basement membrane proteins in cancer progression and EMT. *Eur. J. Cell Biol.* **2022**, *101*, 151220. [CrossRef]
19. Javadi, S.; Zhiani, M.; Mousavi, M.A.; Fathi, M. Crosstalk between Epidermal Growth Factor Receptors (EGFR) and integrins in resistance to EGFR tyrosine kinase inhibitors (TKIs) in solid tumors. *Eur. J. Cell Biol.* **2020**, *99*, 151083. [CrossRef]
20. Wang, K.; Ma, Q.; Ren, Y.; He, J.; Zhang, Y.; Zhang, Y.; Chen, W. Geldanamycin destabilizes HER2 tyrosine kinase and suppresses Wnt/beta-catenin signaling in HER2 overexpressing human breast cancer cells. *Oncol. Rep.* **2007**, *17*, 89–96.
21. Weinberg, F.; Peckys, D.B.; de Jonge, N. EGFR Expression in HER2-Driven Breast Cancer Cells. *Int. J. Mol. Sci.* **2020**, *21*, 9008. [CrossRef]
22. Luque, M.; Sanz-Álvarez, M.; Santamaría, A.; Zazo, S.; Cristóbal, I.; de la Fuente, L.; Mínguez, P.; Eroles, P.; Rovira, A.; Albanell, J.; et al. Targeted Therapy Modulates the Secretome of Cancer-Associated Fibroblasts to Induce Resistance in HER2-Positive Breast Cancer. *Int. J. Mol. Sci.* **2021**, *22*, 13297. [CrossRef] [PubMed]
23. Pelon, F.; Bourachot, B.; Kieffer, Y.; Magagna, I.; Mermet-Meillon, F.; Bonnet, I.; Costa, A.; Givel, A.M.; Attieh, Y.; Barbazan, J.; et al. Cancer-associated fibroblast heterogeneity in axillary lymph nodes drives metastases in breast cancer through complementary mechanisms. *Nat. Commun.* **2020**, *11*, 404. [CrossRef] [PubMed]
24. Kömek, H.; Can, C.; Güzel, Y.; Oruç, Z.; Gündoğan, C.; Yildirim, Ö.A.; Kaplan, İ.; Erdur, E.; Yıldırım, M.S.; Çakabay, B. 68Ga-FAPI-04 PET/CT, a new step in breast cancer imaging: A comparative pilot study with the 18F-FDG PET/CT. *Ann. Nucl. Med.* **2021**, *35*, 744–752. [CrossRef]
25. Backhaus, P.; Burg, M.C.; Roll, W.; Büther, F.; Breyholz, H.J.; Weigel, S.; Heindel, W.; Pixberg, M.; Barth, P.; Tio, J.; et al. Simultaneous FAPI PET/MRI Targeting the Fibroblast-Activation Protein for Breast Cancer. *Radiology* **2022**, *302*, 39–47. [CrossRef] [PubMed]
26. Chen, P.Y.; Wei, W.F.; Wu, H.Z.; Fan, L.S.; Wang, W. Cancer-Associated Fibroblast Heterogeneity: A Factor That Cannot Be Ignored in Immune Microenvironment Remodeling. *Front. Immunol.* **2021**, *12*, 671595. [CrossRef]
27. Mhaidly, R.; Mechta-Grigoriou, F. Fibroblast heterogeneity in tumor micro-environment: Role in immunosuppression and new therapies. *Semin. Immunol.* **2020**, *48*, 101417. [CrossRef]
28. Kieffer, Y.; Hocine, H.R.; Gentric, G.; Pelon, F.; Bernard, C.; Bourachot, B.; Lameiras, S.; Albergante, L.; Bonneau, C.; Guyard, A.; et al. Single-Cell Analysis Reveals Fibroblast Clusters Linked to Immunotherapy Resistance in Cancer. *Cancer Discov.* **2020**, *10*, 1330–1351. [CrossRef]
29. Zhao, L.; Pang, Y.; Luo, Z.; Fu, K.; Yang, T.; Zhao, L.; Sun, L.; Wu, H.; Lin, Q.; Chen, H. Role of [68Ga]Ga-DOTA-FAPI-04 PET/CT in the evaluation of peritoneal carcinomatosis and comparison with [18F]-FDG PET/CT. *Eur. J. Nucl. Med. Mol. Imaging.* **2021**, *48*, 1944–1955. [CrossRef]

pharmaceuticals

Review

Future Prospects of Positron Emission Tomography–Magnetic Resonance Imaging Hybrid Systems and Applications in Psychiatric Disorders

Young-Don Son [1,2], Young-Bo Kim [2,3], Jong-Hoon Kim [2,4], Jeong-Hee Kim [5], Dae-Hyuk Kwon [6], Haigun Lee [7,*] and Zang-Hee Cho [6,*]

1. Department of Biomedical Engineering, College of Health Science, Gachon University, Incheon 21936, Korea; ydson@gachon.ac.kr
2. Neuroscience Research Institute, Gachon University, Incheon 21565, Korea; neurokim@gachon.ac.kr (Y.-B.K.); jhnp@chol.com (J.-H.K.)
3. Department of Neurosurgery, Gachon University College of Medicine, Gil Medical Center, Gachon University, Incheon 21565, Korea
4. Department of Psychiatry, Gachon University College of Medicine, Gil Medical Center, Gachon University, Incheon 21565, Korea
5. Biomedical Engineering Research Center, Gachon University, Incheon 21936, Korea; jhkim1104@gachon.ac.kr
6. Neuroscience Convergence Center, Institute of Green Manufacturing Technology, Korea University, Seoul 02841, Korea; davidkwon2501@gmail.com
7. Department of Materials Science and Engineering, Institute of Green Manufacturing Technology, Korea University, Seoul 02841, Korea
* Correspondence: haigunlee@korea.ac.kr (H.L.); zhcho36@gmail.com (Z.-H.C.)

Citation: Son, Y.-D.; Kim, Y.-B.; Kim, J.-H.; Kim, J.-H.; Kwon, D.-H.; Lee, H.; Cho, Z.-H. Future Prospects of Positron Emission Tomography–Magnetic Resonance Imaging Hybrid Systems and Applications in Psychiatric Disorders. *Pharmaceuticals* 2022, 15, 583. https://doi.org/10.3390/ph15050583

Academic Editor: Xuyi Yue

Received: 25 February 2022
Accepted: 4 May 2022
Published: 8 May 2022

Publisher's Note: MDPI stays neutral with regard to jurisdictional claims in published maps and institutional affiliations.

Copyright: © 2022 by the authors. Licensee MDPI, Basel, Switzerland. This article is an open access article distributed under the terms and conditions of the Creative Commons Attribution (CC BY) license (https://creativecommons.org/licenses/by/4.0/).

Abstract: A positron emission tomography (PET)–magnetic resonance imaging (MRI) hybrid system has been developed to improve the accuracy of molecular imaging with structural imaging. However, the mismatch in spatial resolution between the two systems hinders the use of the hybrid system. As the magnetic field of the MRI increased up to 7.0 tesla in the commercial system, the performance of the MRI system largely improved. Several technical attempts in terms of the detector and the software used with the PET were made to improve the performance. As a result, the high resolution of the PET–MRI fusion system enables quantitation of metabolism and molecular information in the small substructures of the brainstem, hippocampus, and thalamus. Many studies on psychiatric disorders, which are difficult to diagnose with medical imaging, have been accomplished using various radioligands, but only a few studies have been conducted using the PET–MRI fusion system. To increase the clinical usefulness of medical imaging in psychiatric disorders, a high-resolution PET–MRI fusion system can play a key role by providing important information on both molecular and structural aspects in the fine structures of the brain. The development of high-resolution PET–MR systems and their potential roles in clinical studies of psychiatric disorders were reviewed as prospective views in future diagnostics.

Keywords: positron emission tomography; magnetic resonance imaging; high resolution; hybrid imaging; psychiatric disorders

1. Introduction

The brain, the most complex organ in the human body, plays an important role in modulating physiological responses, including human behaviors and cognition. Modern imaging techniques include computed tomography (CT) [1], magnetic resonance imaging (MRI) [2], and positron emission tomography (PET) [3,4]. These medical imaging devices provide structural, functional, and molecular information regarding the human body and brain. This information is visualized in image formats or quantitated numeric values. The PET–CT system was one of the first successful hybrid models, successfully providing high-resolution anatomical and molecular images and simultaneous functional and molecular information [5]. Many clinicians expected that the PET–MRI fusion images could be more useful in the brain application rather than PET–CT fusion images because MRI can provide higher contrast between gray matter and white matter in the brain than CT [6]. However, the PET–MRI fusion system is not yet considered a successful hybrid system compared to the PET-CT system due to the various technical challenges of combining PET and MRI. In particular, because the brain is compactly populated with very small areas, if the spatial resolution of PET and MRI does not match in the PET–MRI fusion, molecular information irrelevant to the structural location is provided, making the accuracy of quantitative analysis unreliable. Therefore, matching the spatial resolution between the two images in PET–MRI fusion imaging of the brain is important. In terms of spatial resolution, the current performance of PET cannot meet the anatomical discrimination requirements of the brain. Furthermore, it is imperative to improve the spatial resolution of the PET system, possibly to approximate that of the MRI system, which has a much better resolution.

Unlike brain tumors, where the anatomical features of lesions are distinct, many mental disorders require functional and molecular imaging with the high anatomical resolution because they require precise anatomical locations and information, especially for the treatment of psychiatric disorders. In summary, improving the resolution and specificity performance of imaging technologies is crucial for clinical pharmaceutical medicine. This study will briefly review the advancements in MRI and PET imaging technology, especially the newly developed PET–MRI fusion molecular imaging system and its applications.

2. MRI-PET Fusion Imaging

2.1. MRI

MRI in recent decades has progressed mainly due to the introduction of ultra-high field (UHF) MRI. Currently, the widely used field strength of MRI for clinical use is 1.5 and 3.0 tesla (T) together with UHF, such as 7.0T MRI for research. On 12 October 2017, the United States Food and Drug Administration approved 7.0T MRI for clinical use. Images obtained from 7.0T MRI showed markedly improved images with a high signal-to-noise ratio (SNR) and other imaging modalities such as tractography [7,8]. This improvement has enabled the visualization of many structures hitherto unavailable in lower-magnetic-field systems [9–12]. This was particularly obvious in areas that require high spatial resolution, such as the brainstem, hippocampus, and thalamus, which have complicated substructures. Figure 1 shows an example of the brainstem area, which is composed of various small subnuclei, in 3.0T and 7.0T MRI [13].

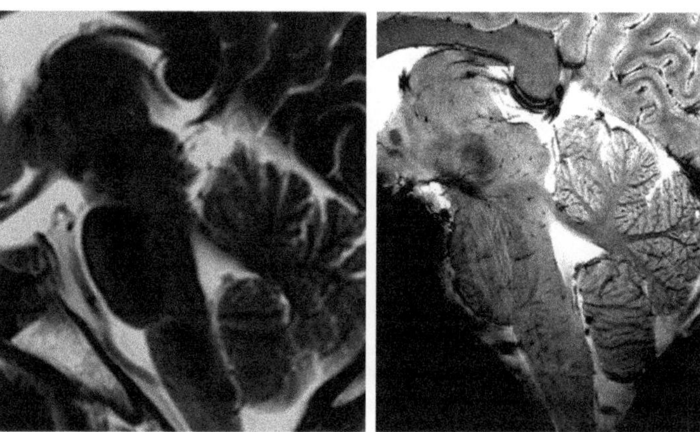

Figure 1. Comparison of 3.0T (**left**) and 7.0T (**right**) MRI in the brainstem area. As shown above, markedly different detail is visualized on the 7.0T MRI, such as the details of the thalamic and brainstem areas, including the midbrain and pons. Reprinted with permission from Cho et al., 2014, Elsevier [13].

Depending on the region of interest (ROI), different MRI modalities can be selected. For imaging the subnuclei of the thalamus, T1-weighted MRI with inversion recovery appears to be more suitable for maximizing the contrast of the subnuclei by nullifying the signal of a specific nucleus [14–17]. T2*-weighted imaging with a gradient echo pulse sequence can achieve better contrast between the white and gray matter. Further discrimination of white and gray matter can also be delineated using other techniques, such as diffusion-weighted imaging [18–20].

With an increase in the magnetic field, a high SNR and spatial resolution were achieved, as well as the uniformity and linearity of the magnetic field. However, in a high field such as susceptibility, it is higher than in a lower field. These magnetic field disturbances often result in image distortion and the mislocation of anatomical structures. The uniformity problem is related to the uniformity of the main magnetic field and the local linearity of the gradient magnetic field generated by the gradient coil. These issues can be particularly highlighted and challenged in the development of PET–MRI systems in which the PET module is inserted and integrated into the bore of the MRI [21].

A higher gradient field is also required to achieve a higher spatial resolution. However, this leads to a lower SNR and longer acquisition time due to the decrease in voxel size and the increased number of voxels. To increase the SNR under a given magnetic field, the development of radiofrequency (RF) coil technology, such as a multichannel phased-array coil, is also important. The sensitivity of the MR signal can be increased by optimizing the transmission and reception of the RF coil system [22]. The multichannel transmission/reception RF coil facilitates the reduction of the acquisition time by combining various recently introduced data sampling strategies, including compressed sensing technology.

2.2. PET

The PET system was designed to provide functional and molecular imaging of human organs, including the brain. This information depends on the properties of the radiopharmaceuticals used for PET imaging. Radiopharmaceuticals are synthesized using positron-emitting radionuclides and chemical compounds with an affinity for target molecules. The emitted positron is annihilated by an electron, and two gamma photons are generated. The PET system collects gamma photons for position determination using detectors (scintillators) with high atomic numbers. Photomultiplier tubes (PMTs) and silicon photomultipliers (SiPMs) are the two main devices used to convert scintillation light into a detectable electrical signal. SiPMs have a high gain similar to that revealed by PMTs and resistance to the high magnetic field, making them an acceptable alternative to PMTs in the PET–MRI system [23]. The new digital SiPMs provide good timing, energy, spatial resolution, and temperature stability, making them a promising candidate for MR compatibility [24–26]. After the signals are digitized, the pulses and addresses encoding the position of the detectors are processed to generate list-mode data containing pulse arrival time, energy, and position information.

Although PET typically provides sensitivity in the nM to pM range, the spatial resolution of PET images is hampered by the physical properties of detectors and gamma photons [27,28], such as detector size, positron range [29,30], penetration effect, and non-collinearity. In PET, the effect of these physical factors appears as a blurring of the image. To improve the spatial resolution, technical developments have been made in the PET detector system, particularly a decrease in the detector size to improve the intrinsic resolution of the system. Most PET systems designed today have the smallest detector size possible, but this approach reduces detection efficiency and increases the penetration effects on adjacent detectors, thereby limiting the further reduction in detector size [31,32]. The depth of interaction (DOI) design of the detector can improve the spatial resolution by decreasing the penetration effect. However, it is often impractical to employ in a real-world system.

High-resolution research tomograph (HRRT) (Siemens Healthcare, USA) is a leading high-resolution PET system among many brain-dedicated PET systems [33]. Unlike other commercial PET systems, the HRRT was designed for research purposes with 119,808 detectors, each of which has a width as small as 2.3 mm. Dual-layer lutetium oxyorthosilicate and lutetium-yttrium oxyorthosilicate (LYSO) scintillators—PMTs were used to design the DOI detector. Due to these PET designs, the spatial resolution of HRRT is as high as 2.5 mm full width at half-maximum (FWHM), which is still the highest resolution. Figure 2 shows an example of the image resolution of HRRT-PET.

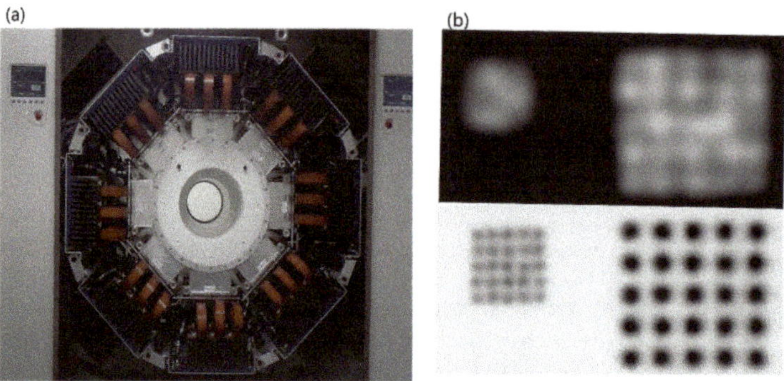

Figure 2. (**a**) HRRT-PET system. (**b**) The resolution of the phantom image obtained from conventional PET–CT (upper) and HRRT-PET (lower) with a 2 mm diameter radioisotope bar separated by 4 mm. Conventional PET–CT has a resolution of <5 mm FWHM [34], while HRRT images have a resolution of 2.5 mm FWHM.

Recently, our group proposed a novel PET system for high-resolution imaging and developed a prototype machine, as shown in Figure 3 [34]. The blurring of the PET image due to physical properties between the detectors and gamma photons that degrade the spatial resolution of PET can be measured as a point spread function or a line spread function (LSF) [35]. In the prototype system, LYSO scintillators with a 3.85 mm detector width—SiPMs were used to design the 1:1 coupling. We used the wobbling mode [36,37] with linear interpolation and reconstruction of the LSF deconvolution and achieved a substantially higher spatial resolution than conventional PET systems [34]. By combining LSF deconvolution reconstruction and wobbling sampling, we achieved a 1.56 mm FWHM transaxial resolution compared with 2.47 mm FWHM of HRRT, which was the highest among PET systems developed for both research and clinical applications. Ideally, the proposed PET system can resolve the submillimeter resolution currently available for small PET detectors using an oversampling technique.

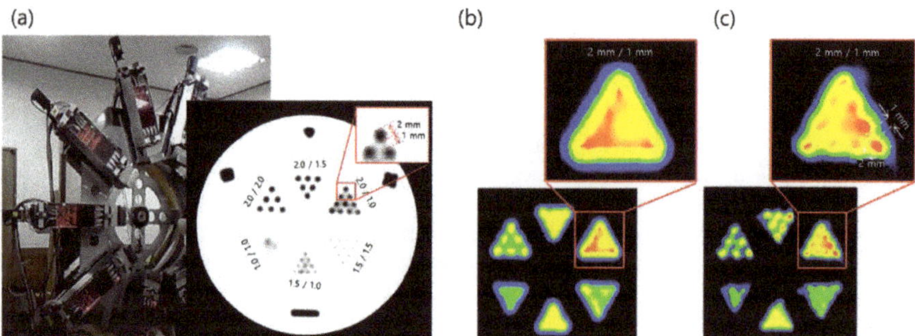

Figure 3. (**a**) Prototype high-resolution wobbling and zooming PET system developed by our group. (**b**) Images of the spatial phantom obtained via HRRT-PET and (**c**) by the prototype wobble-PET system. Reprinted with permission from Cho et al., 2019, IEEE [34].

2.3. PET–MRI Fusion Imaging

The hybrid system provides more clinically useful information than a stand-alone system [38,39]. Recently, the emerging PET–MRI fusion system is being actively researched and developed because it can overcome problems such as excessive radiation exposure and low soft-tissue contrast of CT with PET–CT system [6,40]. There are several types of combining PET and MRI into a hybrid system [21]. The technique of inserting a PET gantry inside the bore of the MRI scanner [41,42] or integrating the PET and MRI scanners in a single gantry [43,44] has the technical issues associated with the mutual interference between the two imaging modalities, including image artifacts related to the homogeneity of MRI fields in the presence of the PET detector, RF shielding for PET, and attenuation correction of PET data [45–48]. On the other hand, the tandem hybrid system in which PET and MR images are sequentially acquired by moving a patient bed between the two scanners could greatly reduce the mutual interference issues [49,50].

We developed a PET–MRI fusion system that combines HRRT-PET and 7.0T MRI for functional and molecular imaging with structural imaging for molecular imaging [49]. As we know, the spatial resolution of PET images compared to MRI is low. The resolution of MRI is approximately $1.5 \times 1.5 \times 4$ mm^3 for 1.5–3.0 T scanners and up to $80 \times 80 \times 200$ μm^3 for a UHF, 11.7T research MRI [51]. Conversely, the resolution of conventional whole-body PET is less than 5 mm, and the resolution of brain-dedicated PET is about 1.5 to 4 mm [34]. Therefore, to match the resolution mismatch of PET and MR images, we developed a PET–MRI fusion imaging system with a precision shuttle between 7.0T MRI and HRRT-PET, which resulted in image matching precision down to 0.05 mm by precision mechanical alignment. Although it is not as ideal as single-unit fusion PET–MRI, it is currently the most technologically feasible PET–MRI system. For the first time, it demonstrated precision PET–MRI fusion imaging, especially with high-resolution 7.0T MRI and HRRT-PET, by coupling the two with the shuttle concept, as shown in Figure 4.

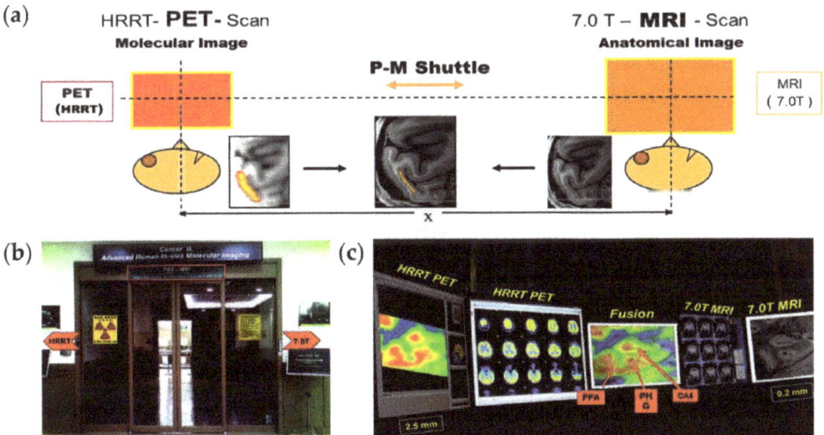

Figure 4. (**a**) Conceptual diagram of the high-resolution PET–MRI fusion imaging system and the developed PET–MRI system with HRRT and 7.0T MRI. (**b**) A gate of the PET–MRI fusion system using HRRT-PET and 7.0T MRI. The shuttle is visualized in the middle of the chamber. (**c**) Results are displayed on the PET–MRI console showing the PET images on the left and the MR images on the right. In the middle, a PET–MRI fusion image is shown. P-M; PET-MRI, FFA; fusiform face areas, PHG; parahippocampal gyrus, CA4; cornu ammonis 4. Reprinted with permission from Cho et al., 2008, JOHN WILEY AND SONS [49].

2.4. Applications of PET–MRI

A fused image of the hippocampus is displayed in the middle of the display console. The hippocampus is an area of interest to many neuroscientists because it is related to memory and human cognition. Although the hippocampus is small, it contains several structurally important subregions that are closely connected to other brain regions. Using high-resolution 7.0T MRI, the subregions of the hippocampus can be structurally visualized and compared with those defined in previous postmortem studies. Cell density, microvessels, and small gaps in the hippocampal sulcus and dentate gyrus can be clearly visualized in MRI of the hippocampus. When we combined and fused the PET image with the high-resolution MRI with a precision of 0.05 mm, the functional and molecular imaging data could be precisely located at the desired anatomical location [52]. Because PET images alone have a spatial resolution that is insufficient to localize in the proper hippocampal subregions, PET–MRI fusion can provide accurate information for the successful localization and quantification of glucose metabolism in its small substructures, including the CA1, CA2, CA3, CA4, and dentate gyrus. Using this subregional mapping of glucose metabolism within hippocampal areas, some clinical cases of encephalitis were studied, and abnormal glucose metabolism was identified. Structural atrophy and deformation in hippocampal encephalitis showed a significant decrease in glucose uptake, especially in the CA4 and proper hippocampal regions. Glucose metabolism has also been studied and compared in the hippocampal subdivisions of patients with mild Alzheimer's disease (AD) and healthy controls. High-resolution PET–MRI fusion scans were performed in nine patients with early-stage AD and ten healthy individuals [53]. MRI were acquired using a two-dimensional T2*-weighted gradient echo sequence with the following scan parameters: repetition time (TR) = 750 ms, echo time (TE) = 21 ms, flip angle (FA) = 30°, resolution = 0.2 × 0.2 × 0.2 mm^3, imaging orientation = coronal, and the number of slices = 17, and the [^{18}F]FDG PET images were obtained with a voxel size of 1.22 × 1.22 × 1.22 mm^3. Patients with early-stage AD exhibited significantly lower glucose metabolism in the posterior CA2/3 region of the left hippocampal body than the healthy controls (Figure 5).

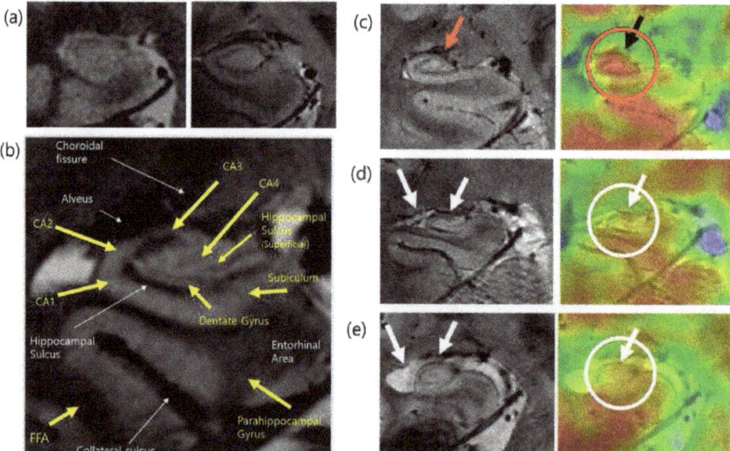

Figure 5. (**a**) Comparison of hippocampus imaging in 1.5T (left) and 7.0T (right) MRI. (**b**) Anatomical labeling of the hippocampus substructure in the 7.0T MR image. This research was originally published in JNM [52]. © SNMMI. (**c**) A 7.0T structural MRI and corresponding [^{18}F]FDG PET–MRI fusion image of the hippocampus of healthy individuals (**d**) and (**e**) hippocampal atrophy and deformation in the 7.0T MR images and the corresponding glucose metabolism in the [^{18}F]FDG PET–MRI of patients with AD. FFA; fusiform face areas, CA1-4; cornu ammonis 1-4. Reprinted with permission from Cho et al., 2014, Elsevier [13].

The brainstem is another region where several nuclei modulate neuronal signals by controlling various neurotransmitters such as serotonin and dopamine (DA). Serotonin-producing nuclei and raphe nuclei are widely dispersed along the brainstem among neurotransmitters. One of the first high-resolution images of serotonergic raphe nuclei obtained using PET–MRI is shown in Figure 6 [54]. This high-resolution molecular imaging of glucose and serotonin transporter (SERT) is visible in PET–MRI fusion images. Kim et al. examined the relationship between self-transcendence and SERT availability in the brainstem raphe nuclei of 16 healthy individuals [55]. A high-resolution MRI of these nuclei was acquired using a T1-weighted three-dimensional magnetization-prepared rapid gradient echo sequence with the following scan parameters: TR = 751 ms, TE = 21 ms, FA = 30°, resolution = 0.18 × 0.18 × 1.5 mm^3, and the number of slices = 17, and the HRRT-PET image of the SERT was obtained with a voxel size of 1.25 × 1.25 × 1.25 mm^3. This study analyzed the total self-transcendence score, which showed a significant negative correlation with SERT binding potential (BP_{ND}) in the caudal raphe. The subscale score for spiritual acceptance was significantly negatively correlated with SERT BP_{ND} in the median raphe nucleus.

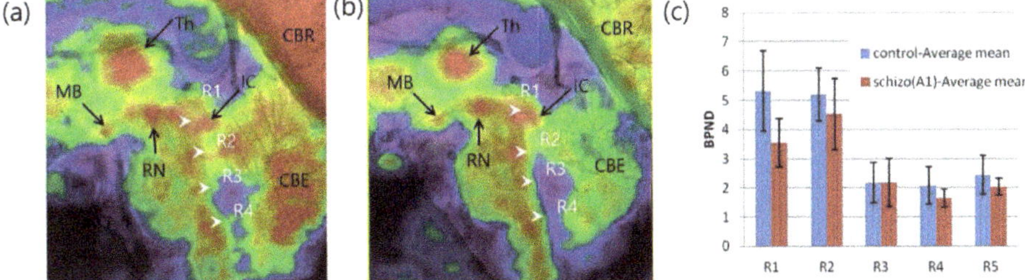

Figure 6. Raphe nuclei were identified in the brainstem by PET–MRI using (**a**) [^{18}F]FDG and (**b**) [^{11}C]DASB PET. (**c**) Group differences in the BP_{ND} of the SERT of healthy controls and schizophrenic patients observed by PET–MRI. Reprinted with permission from Son et al., 2012, Elsevier [54]. Th; thalamus, MB; mammillary body, RN; red nucleus, IC; inferior colliculus, R1-R4; raphe nuclei, CBR; cerebral cortex, CBE; cerebellum, schizo(A1); patients with acute schizophrenia.

3. Molecular Brain Imaging in Psychiatric Disorders

Molecular brain imaging studies have examined potential neurochemical biomarkers involved in various psychiatric disorders and have shown evidence of alterations in many other neurochemical systems in these diseases. Herein, we have reviewed the main findings across various neurochemical systems in the brainstem, hippocampus, and thalamus in the field of psychiatric disorders, including schizophrenia and major depressive disorder (MDD). Furthermore, the main findings in the striatum and globus pallidus were reviewed, considering the pathophysiology and symptomatology of these diseases. For these reviews, the investigators searched for clinical articles published between 1 January 2011 and 30 September 2021, in PubMed, using keywords such as "PET" AND "schizophrenia" OR "major depressive disorder" AND "serotonin" OR "dopamine" OR "glutamate" OR "norepinephrine" AND "brainstem" OR "thalamus" OR "hippocampus" OR "amygdala" OR "striatum". These reviews included full-text articles showing the results of between-group comparisons in ROIs but excluded abstracts, methodological articles, review articles, duplicate articles, and conference proceedings. A total of 37 articles were identified in PubMed, and 15 articles met the inclusion criteria. Details of the radiotracers reviewed in this section are listed in Table 1.

Table 1. Tracer targets and radiotracers reviewed in Sections 3 and 4.

Tracer target	Radiotracer	Name: Synonyms	Study
Serotonin 5-HT$_{1A}$ receptor	[^{11}C]WAY100635	N-[2-[4-(2-methoxyphenyl)-1-piperazinyl]ethyl]-N-(2-pyridinyl)cyclohexanecarboxamide trihydrochloride	Anisman et al. [56]
Serotonin 5-HT$_{1B}$ receptor	[^{11}C]AZ10419369	5-methyl-8-(4-[^{11}C]methyl-piperazin-1-yl)-4-oxo-4H-chromene-2-carboxylic acid (4-morpholin-4-yl-phenyl)-amide	Tiger et al. [57]
	[^{11}C]P943	R-1-[4-(2-methoxy-isopropyl)-phenyl]-3-[2-(4-methyl-piperazin-1-yl) benzyl]-pyrrolidin-2-one	Murrough et al. [58]
Serotonin 5-HT$_6$ receptor	[^{11}C]GSK215083	[^{11}C]-[N-methyl]3-[(3-fluorophenyl)sulfonyl]-8-(4-methyl-1-piperazi nyl)quinoline	Radhakrishnan et al. [59]
SERT	[^{11}C]DASB	[^{11}C]-3-amino-4-(2-dimethylaminomethyl-phenylsulfanyl)-benzonitrile	Kim et al. [60]
	4-[^{18}F]-ADAM	N,N-dimethyl-2-(2-amino-4-[^{18}F]fluorophenylthio)benzylamine	Yeh et al. [61]
DA D$_2$ receptor	[^{11}C]raclopride	3,5-dichloro-N-[[(2S)-1-ethylpyrrolidin-2-yl]methyl]-2-hydroxy-6-[^{11}C]methoxy-benzamide	Hamilton et al. [62]
DA D$_{2/3}$ receptor	[^{11}C]PHNO	[^{11}C]-(+)-4-propyl-9-hydroxynaphthoxazine	Caravaggio et al. [63]
	[^{18}F]fallypride	5-(3-[^{18}F]fluoropropyl)-2,3-dimethoxy-N-[(2S)-1-prop-2-enylpyrrolidin-2-yl]methyl]benzamide	Veselinović et al. [64]
DA transporter (DAT)	[^{11}C]-altropane	2β-carbomethoxy-3β-(4-fluorophenyl)-N-((E)-3-iodo-prop-2-enyl)tropane	Pizzagalli et al. [65]
	[^{18}F]FE-PE2I	N-(3-iodoprop-2E-enyl)-2β-carbo-[^{18}F]fluoroethoxy-3β-[4-methylphenyl]-nortropane	Moriya et al. [66]
	[^{11}C]PE2I	[^{11}C]N-(3-iodoprop-2E-enyl)-2β-carbomethoxy-3β-(4-methylphenyl)nortropane	Artiges et al. [67]
DA synthesis capacity	[^{18}F]-FDOPA	[^{18}F]-6-L-fluoro-L-3,4-dihydroxyphenylalanine	Guerra et al. [68]
Metabotropic glutamate receptor 5	[^{11}C]ABP688	3-(6-methyl-pyridin-2-ylethynyl)-cyclohex-2-enone-O-[^{11}C]-methyl-oxime	Guerra et al. [68]
Norepinephrine transporter (NET)	(S,S)-[^{18}F]FMeNER-D2	(S,S)-2-(α-(2-[^{18}F]fluoro[^{2}H$_2$]methoxyphenoxy) benzyl)morpholine	Moriguchi et al. [69] Arakawa et al. [70]
Phosphodiesterase 10A (PDE10A)	[^{11}C]IMA107	5-[(3R)-3-fluoropyrrolidin-1-yl]-N-[^{11}C]methyl-2-(3-methylquinoxalin-2-yl)-N-tetrahydropyran-4-yl-pyrazolo[1,5-α]pyrimidin-7-amine	Marques et al. [71]
Glucose metabolism	[^{18}F]FDG	2-deoxy-2-[^{18}F]fluoro-D-glucose	Kim et al. [60]

3.1. Schizophrenia

Recent major findings in schizophrenia have been reported in all ROIs, excluding the brainstem, in the serotonergic and dopaminergic systems.

Two PET studies in patients with schizophrenia showed serotonergic and dopaminergic dysfunction in the hippocampus [60,67]. The first study by Kim et al. investigated the interregional correlation patterns between SERT availability and glucose metabolism using 7.0T MRI and HRRT-PET with [^{11}C]DASB and [^{18}F]FDG in 19 antipsychotic-free patients with schizophrenia and 18 healthy controls to observe abnormal functional connectivity in schizophrenia [60]. In particular, Kim et al. evaluated the differences in SERT availability and glucose metabolism in all brain regions between groups, including the anterior and posterior hippocampi, based on previous studies suggesting functional segregation in hippocampal subregions [72–74]. They reported decreased SERT availability in the anterior hippocampus in patients with schizophrenia compared with healthy controls but not in glucose metabolism at the threshold of two-tailed $p < 0.01$ [60]. Moreover, this study revealed no significant correlation between SERT availability and glucose metabolism in each group's whole hippocampus or hippocampal subregions [60]. Kim et al. reported significant differences between groups in the correlations between SERT availability in the parietal and temporal cortices and glucose metabolism in the posterior cingulate cortex. These results suggest altered functional circuitry related to the posterior cingulate gyrus in the pathophysiology of schizophrenia. Figure 7 shows mean [^{11}C]DASB PET, [^{18}F]FDG PET, T1 MRI, and PET–MRI fusion images of healthy control subjects and patients with schizophrenia obtained in the PET study reported by Kim et al. [60].

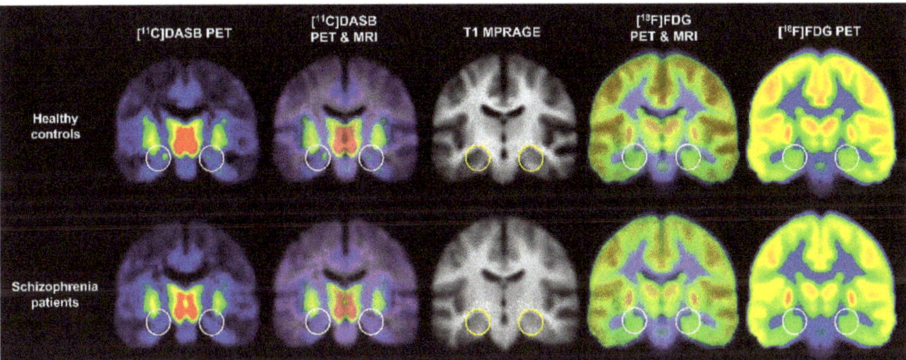

Figure 7. Mean [^{11}C]DASB PET, [^{18}F]FDG PET, T1 MRI, and PET–MRI fusion images of healthy controls and patients with schizophrenia. Circles indicate the location of the bilateral anterior hippocampus.

The second study by Artiges et al. explored striatal and extrastriatal DA dysfunction in schizophrenia using HRRT-PET with [^{11}C]PE2I in 21 male patients with chronic schizophrenia and 30 healthy male controls [67]. Artiges et al. reported an increase in DAT availability in the hippocampus in schizophrenia and significant positive correlations of DAT availability in the hippocampus with hallucinations and unusual thought content in schizophrenia. This study also showed that patients with schizophrenia have a higher DAT availability in the left thalamus. Artiges et al. reported that the high resolution and sensitivity of HRRT-PET enabled the evaluation of increased DAT availability in the hippocampus of patients with schizophrenia. Furthermore, this study demonstrated that these results were consistent with previous PET studies that suggested presynaptic DA hyperactivity in schizophrenia and that striatal and extrastriatal DA dysfunction are involved in positive psychotic symptoms. PDE10A is an enzyme present in DA neurons that degrades intracellular secondary messengers triggered by DA signaling. PDE10A inhibitors are of interest in clinical studies and the pharmaceutical industry related to psychiatric disorders because they have an antipsychotic-like effect in preclinical studies [75–79]. With the recent development of [^{11}C]IMA107, a selective PET radiotracer for PDE10A, Marques et al. investigated PDE10A availability in 12 patients with chronic schizophrenia and

12 healthy controls using [^{11}C]IMA107 PET [71]. This study revealed that PDE10A availability in the thalamus did not differ between groups and was not significantly correlated with the severity of psychotic symptoms or antipsychotic dosage. This study demonstrated that patients with schizophrenia had normal PDE10A availability in the thalamus, which is consistent with the results of previous preclinical studies showing no change in PDE10A availability in animal models of schizophrenia [75,77,78,80–82].

Most PET studies on schizophrenia reviewed in this study revealed alterations in the serotonergic and dopaminergic systems of the striatum [59,63,64,67,71] and globus pallidus [71]. A previous PET study examined the availability of 5-HT$_6$ receptors in the striatum and 5-HT$_{2A}$ receptors in the cortex following treatment with olanzapine, risperidone, aripiprazole, and quetiapine in nine male patients with schizophrenia and nine healthy male controls using PET with [^{11}C]GSK215083 [59]. All patients underwent [^{11}C]GSK215083 PET scans at the presumed steady-state trough level (trough scan) and peak serum level (peak scan) after seven days of antipsychotic treatment. This study revealed that patients treated with olanzapine showed lower availability of 5-HT$_6$ and 5-HT$_{2A}$ receptors (range: 53–95%) in the ventral striatum, caudate, putamen, and frontal cortex at both trough and peak scans than healthy controls. Furthermore, patients treated with quetiapine had a lower availability of 5-HT$_6$ receptors in the putamen at the trough scan (34%) and peak scan (45%) relative to controls. This study suggests that different antipsychotic treatments alter 5-HT$_6$ and 5-HT$_{2A}$ availability.

Caravaggio et al. estimated the D$_{2/3}$ receptor BP$_{ND}$ before and after DA depletion in three male patients with schizophrenia treated with olanzapine and ten healthy controls (six males and four females) using [^{11}C]PHNO PET to explore the changes in endogenous DA in the striatum of medicated schizophrenia [63]. This study revealed that patients treated with olanzapine showed greater ΔBP$_{ND}$ (i.e., the fractional increase in D$_{2/3}$ receptor BP$_{ND}$ after DA depletion) in the caudate and putamen than healthy controls. These findings suggest that relapse in clinical symptoms can occur when antipsychotic medications are administered. A PET study by Veselinović et al. investigated the relationship between cognitive function and dopaminergic transmission in the striatum in 15 medication-free patients with schizophrenia and 11 controls, using PET with [^{18}F]fallypride and neurocognitive assessment [64]. In this study, cognitive function was evaluated in all participants using the trail making test (TMT), measuring complex visual scanning with a motor component, motor speed, and agility in parts A and B [83], digit-symbol-substitution task (DSST) quantifying the speed of mental processing [84], verbal fluency task ("Regensburger Wortflüssigkeits-test") assessing phonemic and semantic fluency and the ability to change categories [85], and Letter-Number Span evaluating working memory performance [86]. This PET study revealed an approximately 10% higher D$_{2/3}$ receptor availability in the caudate and putamen of patients than those in controls, but the difference was not statistically significant [64].

Furthermore, cognitive performance of the TMT in individual patients was significantly negatively correlated with D$_{2/3}$ receptor availability in both the caudate and putamen. Moreover, cognitive performance in the DSST in each patient was significantly positively associated with D$_{2/3}$ receptor availability in the caudate nucleus. These results corroborated that D$_{2/3}$ receptor signaling was more involved in specific cognitive functions in patients with schizophrenia than in controls and suggested that relatively lower basal occupancy by endogenous DA in these patients favors better sparing of cognitive function.

In addition to the increased availability of DAT in the hippocampus of patients with schizophrenia, Artiges et al. reported that the availability of DAT in the left caudate head/nucleus accumbens and putamen increased in patients with schizophrenia compared with controls [67]. Furthermore, DAT availability in the hippocampus, putamen, and globus pallidus was significantly correlated with hallucinations and suspiciousness/persecution.

In addition to the thalamus, as mentioned above, a PET study by Marques et al. confirmed the differences between groups in PDE10A availability in the striatum [71]. In this study, no significant changes in the availability of PDE10A were found in the caudate, putamen, and globus pallidus in patients with schizophrenia compared to controls (% change in patients, -1–6%). In addition, the level of PDE10A binding in these regions was not significantly related to the severity of psychotic symptoms. Based on a study of intracellular signaling pathways demonstrating the effect of PDE10A inhibitors on overall signaling in the therapeutic direction of schizophrenia [76] and several studies demonstrating that PDE10A inhibitors produce behavioral effects that predict antipsychotic activity, similar to D_2 antagonists [75,77,78,80–82], this study suggests that intracellular signaling pathways affected by PDE10A inhibitors in schizophrenia should be considered compensatory pathways, rather than pathological mediators.

3.2. Major Depressive Disorder

Recent major findings of ROIs, excluding the brainstem in MDD, have emerged in various neurotransmitter systems such as serotonin, DA, and norepinephrine.

A PET study by Tiger et al. investigated the binding of 5-HT$_{1B}$ receptors in brain regions associated with MDD pathophysiology in ten drug-free recurrent patients and 10 controls using PET with [^{11}C]AZ10419369 [57]. Tiger et al. reported lower 5-HT$_{1B}$ receptor BP$_{ND}$ in the hippocampus in patients with recurrent MDD than that in controls (32% between-group difference). This study suggested that the result was in line with the decreased binding of the 5-HT$_{1A}$ receptor in this region, as reported in previous MDD studies using [^{11}C]WAY100635 PET [56]. However, this study further suggested that additional observations are needed for a more detailed understanding of serotonergic innervation in the hippocampus, as 5-HT$_{1A}$ and 5-HT$_{1B}$ receptors are functionally similar, but postsynaptic and presynaptic autoreceptors, respectively, and in different cortical layers [57].

Two PET studies investigated the NET levels in patients with MDD and healthy controls using (S,S)-[^{18}F]FMeNER-D2 PET [69,70]. The first study by Moriguchi et al. was conducted on 19 patients with MDD and 19 healthy controls to evaluate the availability of NET and its role in the clinical symptoms of MDD [69]. This study reported that patients with MDD had higher NET availability in the thalamus and thalamic subregions anatomically connected to the prefrontal cortex than controls. Furthermore, Moriguchi et al. found that NET availability in the thalamus was negatively correlated with reaction time in the TMT(A) in patients with MDD. This study suggests that increased norepinephrine transmission in patients with MDD could be associated with preserved visual attention, as estimated by reaction time in the TMT(A). The second study by Arakawa et al. investigated NET occupancy of clinically relevant doses of venlafaxine extended release (ER) in 12 patients with MDD who responded to venlafaxine ER and nine healthy controls [70]. Arakawa et al. reported that NET BP$_{ND}$ in the thalamus was significantly lower in patients treated with 150~300 mg/d venlafaxine than in controls. However, it was not significant in patients treated with 37.5~75 mg/d venlafaxine, suggesting that the NET BP$_{ND}$ in the thalamus was inversely related to the dose of venlafaxine ER. Furthermore, the NET occupancy increased in a dose- and plasma-concentration-dependent manner, although no significant differences were observed above 150 mg/d. This study revealed that clinically relevant doses of venlafaxine ER block NETs formation in the brains of patients with MDD. A PET study by Yeh et al. reported altered SERT availability in the thalamus of patients with MDD [61]. This study examined the role of SERT in MDD and suicidal behavior in 17 antidepressant-naïve patients with MDD (i.e., eight depressed suicidal and nine depressed non-suicidal patients) and 17 healthy controls using PET with 4-[^{18}F]-ADAM. This study found that SERT availability in the thalamus was significantly reduced in the MDD and depressed suicidal groups. However, this study also reported that the reduced availability of SERT in the thalamus in the depressed suicidal group was inconsistent with the results of a previous study by Ney et al. [87]. In most of the PET studies reviewed, patients with MDD showed serotonergic or dopaminergic dysfunction in the striatum [58,61–66].

In the serotonergic system, a previous PET study investigated the availability of the 5-HT$_{1B}$ receptor in the ventral striatum/ventral pallidum in ten patients with MDD in a current major depressive episode and 10 control individuals using HRRT-PET with [^{11}C]P943 [58]. This study reported lower availability of 5-HT$_{1B}$ receptors in the ventral striatum/ventral pallidum in patients with MDD than in controls. This study demonstrated results that were consistent with those of previous preclinical and postmortem studies [88–90] and suggested that abnormal 5-HT$_{1B}$ heteroreceptor function may be associated with dysfunctional reward signaling in the striatum through interactions with the DA, γ-aminobutyric acid, or glutamate systems [58]. In addition to the reduced availability of SERT in the thalamus in patients with MDD, Yeh et al. reported that SERT availability in the striatum was decreased in all patients with MDD and suicidal patients with depression [61]. Based on previous findings that the striatum and corticobasal ganglia pathways play an important role in the neuropathology of affective disorders and subsequent suicidal behaviors [91] and are associated with reward prediction involved in decision-making [92], one study suggested that the reduced availability of SERT in the striatum in patients with MDD and depression may contribute to the development of suicidal action [61].

Hamilton et al. investigated deficits in the stimulation of striatal DA receptors in MDD by decrementing the propagation of information along the cortico-striatal-pallido-thalamic (CSPT) circuit [62]. In this study, 16 MDD patients and 14 healthy controls were scanned using [^{11}C]raclopride PET and functional MRI. Hamilton et al. found an increase in the D$_2$ receptor BP$_{ND}$ in both the dorsal striatum and ventral striatum in patients with MDD compared to controls. This study reported significant negative correlations between altered D$_2$ receptor BP$_{ND}$ in the dorsal striatum and ventral striatum and connectivity in each default-mode network and salience network. Hamilton et al. suggested that the reduced striatal D$_2$ receptor BP$_{ND}$ in MDD could be partly associated with the failure of information transfer to the CSPT circuit in the pathophysiology of this disorder.

Furthermore, two PET studies reported alterations in DAT levels in patients with MDD [65,66]. Pizzagalli et al. investigated 25 patients with MDD and 23 healthy controls using [^{11}C]-altropane PET and reported that patients with MDD had lower availability of DAT in the bilateral putamen than controls [65]. This study suggests that MDD is characterized by reduced DAT expression in the striatal region. Moriya et al. estimated DAT availability in 11 geriatric patients with severe MDD and 27 healthy controls using PET with [^{18}F]FE-PE2I, assuming that anhedonia, a clinical feature of geriatric patients with MDD, is associated with reduced DA neurotransmission in the reward system [66]. Moriya et al. reported that geriatric patients with severe MDD had significantly lower DAT availability in the putamen than in the healthy controls, suggesting a link between dopaminergic neuronal dysfunction and dysregulation of the reward system.

4. PET–MRI Applications in Psychiatric Disorders

For this section, we searched for clinical articles published between 1 January 2015 and 30 September 2021 in PubMed using keywords such as "PET–MRI" AND "psychiatric disorders" OR "schizophrenia" OR "major depressive disorder". This review includes full-text articles on the use of PET–MRI fusion imaging techniques in psychiatric disorders, including schizophrenia and MDD, but excludes abstracts, methodological articles, review articles, duplicate articles, and conference proceedings. Two articles were identified in PubMed, and one article met the inclusion criteria. Details of the radiotracers reviewed in this section are listed in Table 1.

PET–MRI studies have rarely focused on patients that have psychiatric disorders. A novel fusion imaging tool, the PET–MRI-electroencephalography (EEG) system, was used in a study on schizophrenia [68]. To observe the changes in the dopaminergic system in schizophrenia, this study was performed using a whole-body mMR Biograph PET–MRI scanner (Siemens AG Healthcare, Erlangen, Germany) with [^{18}F]-FDOPA in 12 patients with schizophrenia and 13 healthy controls. In this study, patients with schizophrenia showed increased DA synthesis capacity in the nucleus accumbens and functional limbic region (which largely overlaps with the anatomical region of the nucleus accumbens) compared to healthy controls. These results suggest that PET–MRI fusion imaging is replicable in detecting significant group differences in the dopaminergic system and provides excellent anatomical-functional coregistration in some small regions, such as the nucleus accumbens and functional limbic region.

Moreover, this study was conducted using a PET–MRI–EEG system with [^{11}C]ABP688 during a mismatch negativity (MMN) task [93] in a schizophrenic patient and a healthy control to explore the feasibility of the trimodal acquisition protocol. In this experiment, schizophrenia patients had significantly reduced functional connectivity in both auditory and salience networks than healthy control. During the MMN task, the average [^{11}C]ABP688 BP$_{ND}$ values exhibited changes in the precuneus, posterior cingulum, hippocampus, parahippocampus, nucleus accumbens, and middle frontal cortex, and inferior frontal cortex in a patient with schizophrenia compared with healthy control. In addition, the loudness dependence of auditory evoked potential EEG data showed a significant group difference at a single trial level during the MMN task. These results suggest the potential use of the PET–MRI–EEG system as a fusion imaging tool for detecting biomarkers of schizophrenia. However, this is the result from a single schizophrenia patient, and this study suggests that these results need to be replicated with a much larger sample.

5. Conclusions

The brain is the most complex system in the functional domain. Within a few micrometers to millimeters, brain function differs. As shown in recent studies of psychiatric disorders, various regions of the brain, including the striatum, thalamus, and hippocampus, are involved in the alterations of neurotransmitters. Subdividing these regions into high-resolution regions can help differentiate brain diseases, especially psychiatric disorders.

Until now, a majority of PET radioligands targeting neurotransmitter receptors, transporters, and enzymes are labeled with carbon-11 [94,95]. Therefore, in the realm of psychiatry, the routine clinical use of radioligand PET imaging has limitations due to the requirement of an onsite cyclotron facility. However, PET research with 11-C radioligands probing neurotransmitter receptors and transporters has significantly contributed to unraveling the complex pathophysiology of psychiatric disorders such as schizophrenia (DA hypothesis) and depression (serotonin and catecholamine hypothesis) [94,96]. Therefore, PET imaging studies using 11-C radioligands in psychiatric disorders were mainly reviewed in this paper. The research studies can also play a significant role in the new drug development processes in the early phase using proof of concept trials [97] and in the late development phase using drug occupancy studies [98].

Regarding the hybrid PET–MRI system, high spatial resolution is particularly important in brain imaging, where structures and functions differ by a few millimeters. Therefore, considering that conventional whole-body PET has a resolution of <5 mm, it is essential to improve the resolution of PET to combine the different information between the two modalities. Achieving a spatial resolution of a PET image of <1 mm is required to make the PET–MRI fusion system truly useful for research and clinical applications. A new PET–MRI fusion system with "wobble mode" PET in conjunction with 7.0T is another future close-coupled PET–MRI fusion system with which submillimeter PET–MRI imaging can be achieved.

Over the last several decades, we have witnessed numerous medical imaging devices and improvements in their performance, which has led to immense progress in medical sciences, especially in conjunction with pharmaceuticals. As representative examples, medical imaging systems, specifically CT, MRI, and PET, have changed the basic methodology for diagnosing diseases. In particular, the emerging PET–MRI fusion system is an advanced medical imaging technique that can overcome problems such as excessive radiation exposure and low soft-tissue contrast in CT with fusion PET–CT devices, which are currently the most widely used in clinical practice. Hybrid PET–MR imaging is expected to contribute more accurate and useful information than PET–CT for the development of new pharmaceuticals. When measuring the occupancy of pharmaceuticals for the specific target neuroreceptor, delineating the anatomical structures with high resolution will improve the accuracy of measurements and the specificity of the binding region and, therefore, the care provided.

Author Contributions: Conceptualization, Z.-H.C., H.L. and Y.-D.S.; writing—original draft preparation, Y.-D.S., Y.-B.K., J.-H.K. (Jong-Hoon Kim), J.-H.K. (Jeong-Hee Kim) and D.-H.K.; writing—review and editing, Z.-H.C. and H.L.; visualization, Y.-D.S., Y.-B.K., J.-H.K. (Jong-Hoon Kim), J.-H.K. (Jeong-Hee Kim) and D.-H.K.; supervision, Z.-H.C. and H.L.; and funding acquisition, Y.-D.S. All authors have read and agreed to the published version of the manuscript.

Funding: This research was supported by the National Research Foundation (NRF) of Korea grant funded by the Korean government (MSIT) (grant numbers NRF-2017M3C7A1049026 and NRF-2020R1A4A1019623).

Institutional Review Board Statement: Not applicable.

Informed Consent Statement: Not applicable.

Data Availability Statement: Data sharing not applicable.

Conflicts of Interest: The authors declare no conflict of interest.

References

1. Cho, Z.H. General Views on 3-D Image Reconstruction and Computerized Transverse Axial Tomography. *IEEE Trans. Nucl. Sci.* **1974**, *21*, 44–71. [CrossRef]
2. Lauterbur, P.C. The Classic: Image Formation by Induced Local Interactions: Examples Employing Nuclear Magnetic Resonance. *Clin. Orthop. Relat. Res.* **1989**, *244*, 3–6. [CrossRef]
3. Phelps, M.E.; Hoffman, E.J.; Mullani, N.A.; Ter-Pogossian, M.M. Application of Annihilation Coincidence Detection to Transaxial Reconstruction Tomography. *J. Nucl. Med.* **1975**, *16*, 210–224. [PubMed]
4. Cho, Z.H.; Chan, J.K.; Eriksson, L. Circular Ring Transverse Axial Positron Camera for 3-Dimensional Reconstruction of Radionuclides Distribution. *IEEE Trans. Nucl. Sci.* **1976**, *23*, 613–622. [CrossRef]
5. Townsend, D.W.; Beyer, T.; Kinahan, P.E.; Brun, T.; Roddy, R.; Nutt, R.; Byars, L.G. The SMART Scanner: A Combined PET/CT Tomograph for Clinical Oncology. In Proceedings of the 1998 IEEE Nuclear Science Symposium Conference Record, 1998 IEEE Nuclear Science Symposium and Medical Imaging Conference (Cat. No.98CH36255), Toronto, ON, Canada, 8–14 November 1998; Volume 2, pp. 1170–1174.
6. Musafargani, S.; Ghosh, K.K.; Mishra, S.; Mahalakshmi, P.; Padmanabhan, P.; Gulyas, B. PET/MRI: A frontier in era of complementary hybrid imaging. *Eur. J. Hybrid Imag.* **2018**, *2*, 12. [CrossRef]
7. van der Kolk, A.G.; Hendrikse, J.; Zwanenburg, J.J.M.; Visser, F.; Luijten, P.R. Clinical Applications of 7 T MRI in the Brain. *Eur. J. Radiol.* **2013**, *82*, 708–718. [CrossRef]
8. Cho, Z.-H.; Calamante, F.; Chi, J.-G. *7.0 Tesla MRI Brain White Matter Atlas*, 2nd ed.; Springer: Berlin/Heidelberg, Germany, 2015; ISBN 978-3-642-54392-0.
9. Cho, Z.-H.; Kang, C.-K.; Han, J.-Y.; Kim, S.-H.; Kim, K.-N.; Hong, S.-M.; Park, C.-W.; Kim, Y.-B. Observation of the Lenticulostriate Arteries in the Human Brain In Vivo Using 7.0T MR Angiography. *Stroke* **2008**, *39*, 1604–1606. [CrossRef]
10. Cho, Z.-H.; Min, H.-K.; Oh, S.-H.; Han, J.-Y.; Park, C.-W.; Chi, J.-G.; Kim, Y.-B.; Paek, S.H.; Lozano, A.M.; Lee, K.H. Direct Visualization of Deep Brain Stimulation Targets in Parkinson Disease with the Use of 7-Tesla Magnetic Resonance Imaging: Clinical Article. *J. Neurosurg.* **2010**, *113*, 639–647. [CrossRef]
11. Kwon, D.-H.; Kim, J.-M.; Oh, S.-H.; Jeong, H.-J.; Park, S.-Y.; Oh, E.-S.; Chi, J.-G.; Kim, Y.-B.; Jeon, B.S.; Cho, Z.-H. Seven-Tesla Magnetic Resonance Images of the Substantia Nigra in Parkinson Disease. *Ann. Neurol.* **2012**, *71*, 267–277. [CrossRef]
12. Jones, S.E.; Lee, J.; Law, M. Neuroimaging at 3T vs. 7T: Is It Really Worth It? *Magn. Reason. Imaging Clin. N. Am.* **2021**, *29*, 1–12. [CrossRef]
13. Cho, Z.-H.; Kang, C.-K.; Son, Y.-D.; Choi, S.-H.; Lee, Y.-B.; Paek, S.H.; Park, C.-W.; Chi, J.-G.; Calamante, F.; Law, M.; et al. Pictorial Review of In Vivo Human Brain: From Anatomy to Molecular Imaging. *World Neurosurg.* **2014**, *82*, 72–95. [CrossRef] [PubMed]
14. Cho, Z.-H.; Son, Y.-D.; Kim, H.-K.; Kim, N.-B.; Choi, E.-J.; Lee, S.-Y.; Chi, J.-G.; Park, C.-W.; Kim, Y.-B.; Ogawa, S. Observation of Glucose Metabolism in the Thalamic Nuclei by Fusion PET/MRI. *J. Nucl. Med.* **2011**, *52*, 401–404. [CrossRef] [PubMed]
15. Tourdias, T.; Saranathan, M.; Levesque, I.R.; Su, J.; Rutt, B.K. Visualization of Intra-Thalamic Nuclei with Optimized White-Matter-Nulled MPRAGE at 7T. *Neuroimage* **2014**, *84*, 534–545. [CrossRef] [PubMed]
16. Kanowski, M.; Voges, J.; Buentjen, L.; Stadler, J.; Heinze, H.-J.; Tempelmann, C. Direct Visualization of Anatomic Subfields within the Superior Aspect of the Human Lateral Thalamus by MRI at 7T. *AJNR Am. J. Neuroradiol.* **2014**, *35*, 1721–1727. [CrossRef] [PubMed]
17. Brun, G.; Testud, B.; Girard, O.M.; Lehmann, P.; de Rochefort, L.; Besson, P.; Massire, A.; Ridley, B.; Girard, N.; Guye, M.; et al. Automatic Segmentation of Deep Grey Nuclei Using a High-Resolution 7T Magnetic Resonance Imaging Atlas-Quantification of T1 Values in Healthy Volunteers. *Eur. J. Neurosci.* **2022**, *55*, 438–460. [CrossRef] [PubMed]

18. Calamante, F.; Oh, S.-H.; Tournier, J.-D.; Park, S.-Y.; Son, Y.-D.; Chung, J.-Y.; Chi, J.-G.; Jackson, G.D.; Park, C.-W.; Kim, Y.-B.; et al. Super-Resolution Track-Density Imaging of Thalamic Substructures: Comparison with High-Resolution Anatomical Magnetic Resonance Imaging at 7.0T. *Hum. Brain Mapp.* **2013**, *34*, 2538–2548. [CrossRef] [PubMed]

19. Basile, G.A.; Bertino, S.; Bramanti, A.; Ciurleo, R.; Anastasi, G.P.; Milardi, D.; Cacciola, A. In Vivo Super-Resolution Track-Density Imaging for Thalamic Nuclei Identification. *Cereb. Cortex* **2021**, *31*, 5613–5636. [CrossRef]

20. Kwon, D.-H.; Paek, S.H.; Kim, Y.-B.; Lee, H.; Cho, Z.-H. In Vivo 3D Reconstruction of the Human Pallidothalamic and Nigrothalamic Pathways With Super-Resolution 7T MR Track Density Imaging and Fiber Tractography. *Front. Neuroanat.* **2021**, *15*, 86. [CrossRef]

21. Vandenberghe, S.; Marsden, P.K. PET–MRI: A Review of Challenges and Solutions in the Development of Integrated Multimodality Imaging. *Phys. Med. Biol.* **2015**, *60*, R115–R154. [CrossRef]

22. Hernandez, D.; Kim, K.-N. A Review on the RF Coil Designs and Trends for Ultra High Field Magnetic Resonance Imaging. *Investig. Magn. Reson. Imaging* **2020**, *24*, 95–122. [CrossRef]

23. Khalil, M. PET/MR: Basics and New Developments. In *Basic Science of PET Imaging*; Springer: Cham, Switzerland, 2017; pp. 199–228. ISBN 978-3-319-40068-6.

24. Schug, D.; Lerche, C.; Weissler, B.; Gebhardt, P.; Goldschmidt, B.; Wehner, J.; Dueppenbecker, P.M.; Salomon, A.; Hallen, P.; Kiessling, F.; et al. Initial PET Performance Evaluation of a Preclinical Insert for PET/MRI with Digital SiPM Technology. *Phys. Med. Biol.* **2016**, *61*, 2851–2878. [CrossRef] [PubMed]

25. Pratte, J.-F.; Nolet, F.; Parent, S.; Vachon, F.; Roy, N.; Rossignol, T.; Deslandes, K.; Dautet, H.; Fontaine, R.; Charlebois, S.A. 3D Photon-to-Digital Converter for Radiation Instrumentation: Motivation and Future Works. *Sensors* **2021**, *21*, 598. [CrossRef] [PubMed]

26. Lecoq, P.; Gundacker, S. SiPM Applications in Positron Emission Tomography: Toward Ultimate PET Time-of-Flight Resolution. *Eur. Phys. J. Plus* **2021**, *136*, 292. [CrossRef]

27. Moses, W.W. Fundamental Limits of Spatial Resolution in PET. Nuclear Instruments and Methods in Physics Research Section A: Accelerators, Spectrometers, Detectors and Associated Equipment. *Nucl. Instrum. Methods Phys. Res.* **2011**, *648*, S236–S240. [CrossRef] [PubMed]

28. Wang, K. Feasibility of High Spatial Resolution Working Modes for Clinical PET Scanner. *Int. J. Med. Phys. Clin. Eng. Radiat. Oncol.* **2018**, *7*, 539–552. [CrossRef]

29. Cho, Z.H.; Chan, J.K.; Ericksson, L.; Singh, M.; Graham, S.; MacDonald, N.S.; Yano, Y. Positron Ranges Obtained from Biomedically Important Positron-Emitting Radionuclides. *J. Nucl. Med.* **1975**, *16*, 1174–1176.

30. Levin, C.S.; Hoffman, E.J. Calculation of Positron Range and Its Effect on the Fundamental Limit of Positron Emission Tomography System Spatial Resolution. *Phys. Med. Biol.* **1999**, *44*, 781–799. [CrossRef]

31. Cho, Z.H.; Juh, S.C.; Friedenberg, R.M.; Bunney, W.; Buchsbaum, M.; Wong, E. A New Approach to Very High Resolution Mini-Brain PET Using a Small Number of Large Detectors. *IEEE Trans. Nucl. Sci.* **1990**, *37*, 842–851. [CrossRef]

32. Lowdon, M.; Martin, P.G.; Hubbard, M.W.J.; Taggart, M.P.; Connor, D.T.; Verbelen, Y.; Sellin, P.J.; Scott, T.B. Evaluation of Scintillator Detection Materials for Application within Airborne Environmental Radiation Monitoring. *Sensors* **2019**, *19*, 3828. [CrossRef]

33. Wienhard, K.; Schmand, M.; Casey, M.E.; Baker, K.; Bao, J.; Eriksson, L.; Jones, W.F.; Knoess, C.; Lenox, M.; Lercher, M.; et al. The ECAT HRRT: Performance and First Clinical Application of the New High Resolution Research Tomograph. *IEEE Trans. Nucl. Sci.* **2002**, *49*, 104–110. [CrossRef]

34. Cho, Z.-H.; Son, Y.-D.; Kim, H.-K.; Kwon, D.-H.; Joo, Y.-H.; Ra, J.B.; Choi, Y.; Kim, Y.-B. Development of Positron Emission Tomography With Wobbling and Zooming for High Sensitivity and High-Resolution Molecular Imaging. *IEEE Trans. Med. Imaging* **2019**, *38*, 2875–2882. [CrossRef] [PubMed]

35. Kim, H.-K.; Son, Y.-D.; Kwon, D.-H.; Joo, Y.; Cho, Z.-H. Wobbling and LSF-Based Maximum Likelihood Expectation Maximization Reconstruction for Wobbling PET. *Radiat. Phys. Chem.* **2016**, *121*, 1–9. [CrossRef]

36. Ter-Pogossian, M.M.; Mullani, N.A.; Hood, J.T.; Higgins, C.S.; Ficke, D.C. Design Considerations for a Positron Emission Transverse Tomograph (PETT V) for Imaging of the Brain. *J. Comput. Assist. Tomogr.* **1978**, *2*, 539–544. [CrossRef] [PubMed]

37. Brooks, R.A.; Sank, V.J.; Talbert, A.J.; Di Chiro, G. Sampling Requirements and Detector Motion for Positron Emission Tomography. *IEEE Trans. Nucl. Sci.* **1979**, *26*, 2760–2763. [CrossRef]

38. Zheng, Y.; Sun, X.; Wang, J.; Zhang, L.; Di, X.; Xu, Y. FDG-PET/CT Imaging for Tumor Staging and Definition of Tumor Volumes in Radiation Treatment Planning in Non-Small Cell Lung Cancer. *Oncol. Lett.* **2014**, *7*, 1015–1020. [CrossRef]

39. Ming, Y.; Wu, N.; Qian, T.Y.; Li, X.; Wan, D.Q.; Li, C.Y.; Li, Y.L.; Wu, Z.H.; Wang, X.; Liu, J.Q.; et al. Progress and Future Trends in PET/CT and PET/MRI Molecular Imaging Approaches for Breast Cancer. *Front. Oncol.* **2020**, *10*, 1301. [CrossRef]

40. Ehman, E.C.; Johnson, G.B.; Villanueva-Meyer, J.E.; Cha, S.; Leynes, A.P.; Larson, P.E.Z.; Hope, T.A. PET/MRI: Where Might It Replace PET/CT? *J. Magn. Reson. Imaging* **2017**, *46*, 1247–1262. [CrossRef]

41. Grant, A.M.; Lee, B.J.; Chang, C.M.; Levin, C.S. Simultaneous PET/MR imaging with a radio frequency- penetrable PET insert. *Med. Phys.* **2017**, *44*, 112–120. [CrossRef]

42. Gonzalez, A.J.; Gonzalez-Montoro, A.; Vidal, L.F.; Barbera, J.; Aussenhofer, S.; Hernandez, L.; Moliner, L.; Sanchez, F.; Correcher, C.; Pincay, E.J.; et al. Initial Results of the MINDView PET Insert Inside the 3T mMR. *IEEE Trans. Radiat. Plasma* **2019**, *3*, 343–351. [CrossRef]

43. Delso, G.; Furst, S.; Jakoby, B.; Ladebeck, R.; Ganter, C.; Nekolla, S.G.; Schwaiger, M.; Ziegler, S.I. Performance Measurements of the Siemens mMR Integrated Whole-Body PET/MR Scanner. *J. Nucl. Med.* **2011**, *52*, 1914–1922. [CrossRef]

44. Chen, S.G.; Gu, Y.S.; Yu, H.J.; Chen, X.; Cao, T.Y.; Hu, L.Z.; Shi, H.C. NEMA NU2-2012 performance measurements of the United Imaging uPMR790: An integrated PET/MR system. *Eur. J. Nucl. Med. Mol. Imaging* **2021**, *48*, 1726–1735. [CrossRef]

45. Disselhorst, J.A.; Bezrukov, I.; Kolb, A.; Parl, C.; Pichler, B.J. Principles of PET/MR Imaging. *J. Nucl. Med.* **2014**, *55*, 2s–10s. [CrossRef] [PubMed]

46. Muzic, R.F., Jr.; DiFilippo, F.P. Positron emission tomography-magnetic resonance imaging: Technical review. *Semin. Roentgenol.* **2014**, *49*, 242–254. [CrossRef] [PubMed]

47. Mehranian, A.; Arabi, H.; Zaidi, H. Vision 20/20: Magnetic resonance imaging-guided attenuation correction in PET/MRI: Challenges, solutions, and opportunities. *Med. Phys.* **2016**, *43*, 1130–1155. [CrossRef] [PubMed]

48. Catana, C. Principles of Simultaneous PET/MR Imaging. *Magn. Reason. Imaging Clin. N. Am.* **2017**, *25*, 231–243. [CrossRef] [PubMed]

49. Cho, Z.H.; Son, Y.D.; Kim, H.K.; Kim, K.N.; Oh, S.H.; Han, J.Y.; Hong, I.K.; Kim, Y.B. A fusion PET–MRI system with a high-resolution research tomograph-PET and ultra-high field 7.0 T-MRI for the molecular-genetic imaging of the brain. *Proteomics* **2008**, *8*, 1302–1323. [CrossRef]

50. Zaidi, H.; Ojha, N.; Morich, M.; Griesmer, J.; Hu, Z.; Maniawski, P.; Ratib, O.; Izquierdo-Garcia, D.; Fayad, Z.A.; Shao, L. Design and performance evaluation of a whole-body Ingenuity TF PET–MRI system. *Phys. Med. Biol.* **2011**, *56*, 3091–3106. [CrossRef]

51. Van Reeth, E.; Tham, I.W.K.; Tan, C.H.; Poh, C.L. Super-resolution in magnetic resonance imaging: A review. *Concept. Magn. Reson. A* **2012**, *40A*, 306–325. [CrossRef]

52. Cho, Z.-H.; Son, Y.-D.; Kim, H.-K.; Kim, S.-T.; Lee, S.-Y.; Chi, J.-G.; Park, C.-W.; Kim, Y.-B. Substructural Hippocampal Glucose Metabolism Observed on PET/MRI. *J. Nucl. Med.* **2010**, *51*, 1545–1548. [CrossRef]

53. Choi, E.-J.; Son, Y.-D.; Noh, Y.; Lee, H.; Kim, Y.-B.; Park, K.H. Glucose Hypometabolism in Hippocampal Subdivisions in Alzheimer's Disease: A Pilot Study Using High-Resolution 18F-FDG PET and 7.0-T MRI. *J. Clin. Neurol* **2018**, *14*, 158–164. [CrossRef]

54. Son, Y.-D.; Cho, Z.-H.; Kim, H.-K.; Choi, E.-J.; Lee, S.-Y.; Chi, J.-G.; Park, C.-W.; Kim, Y.-B. Glucose Metabolism of the Midline Nuclei Raphe in the Brainstem Observed by PET–MRI Fusion Imaging. *NeuroImage* **2012**, *59*, 1094–1097. [CrossRef] [PubMed]

55. Kim, J.-H.; Son, Y.-D.; Kim, J.-H.; Choi, E.-J.; Lee, S.-Y.; Joo, Y.-H.; Kim, Y.-B.; Cho, Z.-H. Self-Transcendence Trait and Its Relationship with in Vivo Serotonin Transporter Availability in Brainstem Raphe Nuclei: An Ultra-High Resolution PET–MRI Study. *Brain Res.* **2015**, *1629*, 63–71. [CrossRef] [PubMed]

56. Anisman, H.; Du, L.; Palkovits, M.; Faludi, G.; Kovacs, G.G.; Szontagh-Kishazi, P.; Merali, Z.; Poulter, M.O. Serotonin Receptor Subtype and P11 MRNA Expression in Stress-Relevant Brain Regions of Suicide and Control Subjects. *J. Psychiatry Neurosci.* **2008**, *33*, 131–141. [PubMed]

57. Tiger, M.; Farde, L.; Rück, C.; Varrone, A.; Forsberg, A.; Lindefors, N.; Halldin, C.; Lundberg, J. Low Serotonin1B Receptor Binding Potential in the Anterior Cingulate Cortex in Drug-Free Patients with Recurrent Major Depressive Disorder. *Psychiatry Res. Neuroimaging* **2016**, *253*, 36–42. [CrossRef]

58. Murrough, J.W.; Henry, S.; Hu, J.; Gallezot, J.-D.; Planeta-Wilson, B.; Neumaier, J.F.; Neumeister, A. Reduced Ventral Striatal/Ventral Pallidal Serotonin1B Receptor Binding Potential in Major Depressive Disorder. *Psychopharmacology* **2011**, *213*, 547–553. [CrossRef]

59. Radhakrishnan, R.; Matuskey, D.; Nabulsi, N.; Gaiser, E.; Gallezot, J.-D.; Henry, S.; Planeta, B.; Lin, S.; Ropchan, J.; Huang, Y.; et al. In Vivo 5-HT6 and 5-HT2A Receptor Availability in Antipsychotic Treated Schizophrenia Patients vs. Unmedicated Healthy Humans Measured with [11C]GSK215083 PET. *Psychiatry Res. Neuroimaging* **2020**, *295*, 111007. [CrossRef]

60. Kim, J.-H.; Kim, J.-H.; Son, Y.-D.; Joo, Y.-H.; Lee, S.-Y.; Kim, H.-K.; Woo, M.-K. Altered Interregional Correlations between Serotonin Transporter Availability and Cerebral Glucose Metabolism in Schizophrenia: A High-Resolution PET Study Using [11C]DASB and [18F]FDG. *Schizophr. Res.* **2017**, *182*, 55–65. [CrossRef]

61. Yeh, Y.-W.; Ho, P.-S.; Chen, C.-Y.; Kuo, S.-C.; Liang, C.-S.; Ma, K.-H.; Shiue, C.-Y.; Huang, W.-S.; Cheng, C.-Y.; Wang, T.-Y.; et al. Incongruent Reduction of Serotonin Transporter Associated with Suicide Attempts in Patients with Major Depressive Disorder: A Positron Emission Tomography Study with 4-[18F]-ADAM. *Int. J. Neuropsychopharmacol.* **2015**, *18*, pyu065. [CrossRef]

62. Hamilton, J.P.; Sacchet, M.D.; Hjørnevik, T.; Chin, F.T.; Shen, B.; Kämpe, R.; Park, J.H.; Knutson, B.D.; Williams, L.M.; Borg, N.; et al. Striatal Dopamine Deficits Predict Reductions in Striatal Functional Connectivity in Major Depression: A Concurrent 11C-Raclopride Positron Emission Tomography and Functional Magnetic Resonance Imaging Investigation. *Transl. Psychiatry* **2018**, *8*, 264. [CrossRef]

63. Caravaggio, F.; Borlido, C.; Wilson, A.; Graff-Guerrero, A. Examining Endogenous Dopamine in Treated Schizophrenia Using [11C]-(+)-PHNO Positron Emission Tomography: A Pilot Study. *Clin. Chim. Acta* **2015**, *449*, 60–62. [CrossRef]

64. Veselinović, T.; Vernaleken, I.; Janouschek, H.; Cumming, P.; Paulzen, M.; Mottaghy, F.M.; Gründer, G. The Role of Striatal Dopamine D2/3 Receptors in Cognitive Performance in Drug-Free Patients with Schizophrenia. *Psychopharmacology* **2018**, *235*, 2221–2232. [CrossRef] [PubMed]

65. Pizzagalli, D.A.; Berretta, S.; Wooten, D.; Goer, F.; Pilobello, K.T.; Kumar, P.; Murray, L.; Beltzer, M.; Boyer-Boiteau, A.; Alpert, N.; et al. Assessment of Striatal Dopamine Transporter Binding in Individuals With Major Depressive Disorder: In Vivo Positron Emission Tomography and Postmortem Evidence. *JAMA Psychiatry* **2019**, *76*, 854–861. [CrossRef] [PubMed]

66. Moriya, H.; Tiger, M.; Tateno, A.; Sakayori, T.; Masuoka, T.; Kim, W.; Arakawa, R.; Okubo, Y. Low Dopamine Transporter Binding in the Nucleus Accumbens in Geriatric Patients with Severe Depression. *Psychiatry Clin. Neurosci.* **2020**, *74*, 424–430. [CrossRef] [PubMed]

67. Artiges, E.; Leroy, C.; Dubol, M.; Prat, M.; Pepin, A.; Mabondo, A.; de Beaurepaire, R.; Beaufils, B.; Korwin, J.-P.; Galinowski, A.; et al. Striatal and Extrastriatal Dopamine Transporter Availability in Schizophrenia and Its Clinical Correlates: A Voxel-Based and High-Resolution PET Study. *Schizophr. Bull.* **2017**, *43*, 1134–1142. [CrossRef]

68. Guerra, A.D.; Ahmad, S.; Avram, M.; Belcari, N.; Berneking, A.; Biagi, L.; Bisogni, M.G.; Brandl, F.; Cabello, J.; Camarlinghi, N.; et al. TRIMAGE: A Dedicated Trimodality (PET/MR/EEG) Imaging Tool for Schizophrenia. *Eur. Psychiatry* **2018**, *50*, 7–20. [CrossRef]

69. Moriguchi, S.; Yamada, M.; Takano, H.; Nagashima, T.; Takahata, K.; Yokokawa, K.; Ito, T.; Ishii, T.; Kimura, Y.; Zhang, M.-R.; et al. Norepinephrine Transporter in Major Depressive Disorder: A PET Study. *Am. J. Psychiatry* **2017**, *174*, 36–41. [CrossRef]

70. Arakawa, R.; Stenkrona, P.; Takano, A.; Svensson, J.; Andersson, M.; Nag, S.; Asami, Y.; Hirano, Y.; Halldin, C.; Lundberg, J. Venlafaxine ER Blocks the Norepinephrine Transporter in the Brain of Patients with Major Depressive Disorder: A PET Study Using [18F]FMeNER-D2. *Int. J. Neuropsychopharmacol.* **2019**, *22*, 278–285. [CrossRef]

71. Marques, T.R.; Natesan, S.; Niccolini, F.; Politis, M.; Gunn, R.N.; Searle, G.E.; Howes, O.; Rabiner, E.A.; Kapur, S. Phosphodiesterase 10A in Schizophrenia: A PET Study Using [11C]IMA107. *Am. J. Psychiatry* **2016**, *173*, 714–721. [CrossRef]

72. Lepage, M.; Habib, R.; Tulving, E. Hippocampal PET Activations of Memory Encoding and Retrieval: The HIPER Model. *Hippocampus* **1998**, *8*, 313–322. [CrossRef]

73. Schacter, D.L.; Wagner, A.D. Medial temporal lobe activations in fMRI and PET studies of episodic encoding and retrieval. *Hippocampus* **1999**, *9*, 7–24. [CrossRef]

74. Strange, B.A.; Fletcher, P.C.; Henson, R.N.A.; Friston, K.J.; Dolan, R.J. Segregating the Functions of Human Hippocampus. *Proc. Natl. Acad. Sci. USA* **1999**, *96*, 4034–4039. [CrossRef] [PubMed]

75. Siuciak, J.A. The Role of Phosphodiesterases in Schizophrenia: Therapeutic Implications. *CNS Drugs* **2008**, *22*, 983–993. [CrossRef] [PubMed]

76. Siuciak, J.A.; McCarthy, S.A.; Chapin, D.S.; Fujiwara, R.A.; James, L.C.; Williams, R.D.; Stock, J.L.; McNeish, J.D.; Strick, C.A.; Menniti, F.S.; et al. Genetic Deletion of the Striatum-Enriched Phosphodiesterase PDE10A: Evidence for Altered Striatal Function. *Neuropharmacology* **2006**, *51*, 374–385. [CrossRef] [PubMed]

77. Schmidt, C.J.; Chapin, D.S.; Cianfrogna, J.; Corman, M.L.; Hajos, M.; Harms, J.F.; Hoffman, W.E.; Lebel, L.A.; McCarthy, S.A.; Nelson, F.R.; et al. Preclinical Characterization of Selective Phosphodiesterase 10A Inhibitors: A New Therapeutic Approach to the Treatment of Schizophrenia. *J. Pharmacol. Exp. Ther.* **2008**, *325*, 681–690. [CrossRef]

78. Grauer, S.M.; Pulito, V.L.; Navarra, R.L.; Kelly, M.P.; Kelley, C.; Graf, R.; Langen, B.; Logue, S.; Brennan, J.; Jiang, L.; et al. Phosphodiesterase 10A Inhibitor Activity in Preclinical Models of the Positive, Cognitive, and Negative Symptoms of Schizophrenia. *J. Pharmacol. Exp. Ther.* **2009**, *331*, 574–590. [CrossRef]

79. Kehler, J.; Nielsen, J. PDE10A Inhibitors: Novel Therapeutic Drugs for Schizophrenia. *Curr. Pharm. Des.* **2011**, *17*, 137–150. [CrossRef]

80. Uthayathas, S.; Masilamoni, G.J.; Shaffer, C.L.; Schmidt, C.J.; Menniti, F.S.; Papa, S.M. Phosphodiesterase 10A Inhibitor MP-10 Effects in Primates: Comparison with Risperidone and Mechanistic Implications. *Neuropharmacology* **2014**, *77*, 257–267. [CrossRef]

81. Piccart, E.; De Backer, J.-F.; Gall, D.; Lambot, L.; Raes, A.; Vanhoof, G.; Schiffmann, S.; D'Hooge, R. Genetic Deletion of PDE10A Selectively Impairs Incentive Salience Attribution and Decreases Medium Spiny Neuron Excitability. *Behav. Brain Res.* **2014**, *268*, 48–54. [CrossRef]

82. Sano, H.; Nagai, Y.; Miyakawa, T.; Shigemoto, R.; Yokoi, M. Increased Social Interaction in Mice Deficient of the Striatal Medium Spiny Neuron-Specific Phosphodiesterase 10A2. *J. Neurochem.* **2008**, *105*, 546–556. [CrossRef]

83. Kortte, K.B.; Horner, M.D.; Windham, W.K. The Trail Making Test, Part B: Cognitive Flexibility or Ability to Maintain Set? *Appl. Neuropsychol.* **2002**, *9*, 106–109. [CrossRef]

84. Wechsler, D. *WAIS-III: Administration and Scoring Manual: Wechsler Adult Intelligence Scale*; Psychological Corporation: San Antonio, TX, USA, 1997; ISBN 978-0-15-898103-1.

85. Harth, S.; Müller, S.V.; Aschenbrenner, S.; Tucha, O.; Lange, K.W. Regensburger Wortflüssigkeits-Test (RWT). *Z. Neuropsychol.* **2004**, *15*, 315–321. [CrossRef]

86. Gold, J.M.; Carpenter, C.; Randolph, C.; Goldberg, T.E.; Weinberger, D.R. Auditory Working Memory and Wisconsin Card Sorting Test Performance in Schizophrenia. *Arch. Gen. Psychiatry* **1997**, *54*, 159–165. [CrossRef] [PubMed]

87. Nye, J.A.; Purselle, D.; Plisson, C.; Voll, R.J.; Stehouwer, J.S.; Votaw, J.R.; Kilts, C.D.; Goodman, M.M.; Nemeroff, C.B. Decreased Brainstem and Putamen Sert Binding Potential in Depressed Suicide Attempters Using [11c]-Zient Pet Imaging. *Depress. Anxiety* **2013**, *30*, 902–907. [CrossRef] [PubMed]

88. Svenningsson, P.; Chergui, K.; Rachleff, I.; Flajolet, M.; Zhang, X.; Yacoubi, M.E.; Vaugeois, J.-M.; Nomikos, G.G.; Greengard, P. Alterations in 5-HT1B Receptor Function by P11 in Depression-Like States. *Science* **2006**, *311*, 77–80. [CrossRef]

89. Tatarczynska, E.; Klodzinska, A.; Stachowicz, K.; Chojnacka-Wójcik, E. Effects of a Selective 5-HT1B Receptor Agonist and Antagonists in Animal Models of Anxiety and Depression. *Behav. Pharmacol.* **2004**, *15*, 523–534. [CrossRef]

90. Chenu, F.; David, D.J.P.; Leroux-Nicollet, I.; Maître, E.L.; Gardier, A.M.; Bourin, M. Serotonin1B Heteroreceptor Activation Induces an Antidepressant-like Effect in Mice with an Alteration of the Serotonergic System. *J. Psychiatry Neurosci.* **2008**, *33*, 541–550.

91. Marchand, W.R.; Lee, J.N.; Johnson, S.; Thatcher, J.; Gale, P.; Wood, N.; Jeong, E.-K. Striatal and Cortical Midline Circuits in Major Depression: Implications for Suicide and Symptom Expression. *Prog. Neuro-Psychopharmacol. Biol. Psychiatry* **2012**, *36*, 290–299. [CrossRef]

92. Tanaka, S.C.; Doya, K.; Okada, G.; Ueda, K.; Okamoto, Y.; Yamawaki, S. Prediction of Immediate and Future Rewards Differentially Recruits Cortico-Basal Ganglia Loops. *Nat. Neurosci.* **2004**, *7*, 887–893. [CrossRef]

93. Takahashi, H.; Rissling, A.J.; Pascual-Marqui, R.; Kirihara, K.; Pela, M.; Sprock, J.; Braff, D.L.; Light, G.A. Neural Substrates of Normal and Impaired Preattentive Sensory Discrimination in Large Cohorts of Nonpsychiatric Subjects and Schizophrenia Patients as Indexed by MMN and P3a Change Detection Responses. *NeuroImage* **2013**, *66*, 594–603. [CrossRef]

94. Sørensen, A.; Ruhé, H.G.; Munkholm, K. The Relationship between Dose and Serotonin Transporter Occupancy of Antidepressants—A Systematic Review. *Mol. Psychiatry* **2022**, *27*, 192–201. [CrossRef]

95. Cumming, P.; Abi-Dargham, A.; Gründer, G. Molecular Imaging of Schizophrenia: Neurochemical Findings in a Heterogeneous and Evolving Disorder. *Behav. Brain Res.* **2021**, *398*, 113004. [CrossRef] [PubMed]

96. Weinstein, J.J.; Chohan, M.O.; Slifstein, M.; Kegeles, L.S.; Moore, H.; Abi-Dargham, A. Pathway-Specific Dopamine Abnormalities in Schizophrenia. *Biol. Psychiatry* **2017**, *81*, 31–42. [CrossRef] [PubMed]

97. Tauscher, J.; Kielbasa, W.; Iyengar, S.; Vandenhende, F.; Peng, X.; Mozley, D.; Gehlert, D.R.; Marek, G. Development of the 2nd Generation Neurokinin-1 Receptor Antagonist LY686017 for Social Anxiety Disorder. *Eur. Neuropsychopharmacol.* **2010**, *20*, 80–87. [CrossRef] [PubMed]

98. Nakajima, S.; Uchida, H.; Bies, R.R.; Caravaggio, F.; Suzuki, T.; Plitman, E.; Mar, W.; Gerretsen, P.; Pollock, B.G.; Mulsant, B.H.; et al. Dopamine D2/3 Receptor Occupancy Following Dose Reduction Is Predictable With Minimal Plasma Antipsychotic Concentrations: An Open-Label Clinical Trial. *Schizophr. Bull.* **2016**, *42*, 212–219. [CrossRef]

 *pharmaceuticals*

Review

Refining Glioblastoma Surgery through the Use of Intra-Operative Fluorescence Imaging Agents

Oluwakanyinsolami Netufo [1,†], Kate Connor [1,†], Liam P. Shiels [1,†], Kieron J. Sweeney [1,2], Dan Wu [3], Donal F. O'Shea [3], Annette T. Byrne [1,4,†] and Ian S. Miller [1,4,*,†]

1. Precision Cancer Medicine Group, Department of Physiology and Medical Physics, Royal College of Surgeons in Ireland, 2, D02 YN77 Dublin, Ireland; kanyisolanetufo@rcsi.ie (O.N.); kateconnor@rcsi.com (K.C.); liamshiels@rcsi.com (L.P.S.); kieronsweeney@beaumont.ie (K.J.S.); annettebyrne@rcsi.com (A.T.B.)
2. National Centre for Neurosurgery, Beaumont Hospital, 9, D09 V2N0 Dublin, Ireland
3. Department of Chemistry, Royal College of Surgeons in Ireland (RCSI), 2, D02 YN77 Dublin, Ireland; danwu@rcsi.com (D.W.); donalfoshea@rcsi.com (D.F.O.)
4. National Pre-Clinical Imaging Centre (NPIC), 2, D02 YN77 Dublin, Ireland
* Correspondence: ianmiller@rcsi.com
† These authors contributed equally to this work.

Abstract: Glioblastoma (GBM) is the most aggressive adult brain tumour with a dismal 2-year survival rate of 26–33%. Maximal safe resection plays a crucial role in improving patient progression-free survival (PFS). Neurosurgeons have the significant challenge of delineating normal tissue from brain tumour to achieve the optimal extent of resection (EOR), with 5-Aminolevulinic Acid (5-ALA) the only clinically approved intra-operative fluorophore for GBM. This review aims to highlight the requirement for improved intra-operative imaging techniques, focusing on fluorescence-guided imaging (FGS) and the use of novel dyes with the potential to overcome the limitations of current FGS. The review was performed based on articles found in PubMed an.d Google Scholar, as well as articles identified in searched bibliographies between 2001 and 2022. Key words for searches included 'Glioblastoma' + 'Fluorophore'+ 'Novel' + 'Fluorescence Guided Surgery'. Current literature has favoured the approach of using targeted fluorophores to achieve specific accumulation in the tumour microenvironment, with biological conjugates leading the way. These conjugates target specific parts overexpressed in the tumour. The positive results in breast, ovarian and colorectal tissue are promising and may, therefore, be applied to intracranial neoplasms. Therefore, this design has the potential to produce favourable results in GBM by reducing the residual tumour, which translates to decreased tumour recurrence, morbidity and ultimately, mortality in GBM patients. Several preclinical studies have shown positive results with targeted dyes in distinguishing GBM cells from normal brain parenchyma, and targeted dyes in the Near-Infrared (NIR) emission range offer promising results, which may be valuable future alternatives.

Keywords: glioblastoma; fluorescence guided surgery; 5-ALA; fluorescein; NIR-AZA

Citation: Netufo, O.; Connor, K.; Shiels, L.P.; Sweeney, K.J.; Wu, D.; O'Shea, D.F.; Byrne, A.T.; Miller, I.S. Refining Glioblastoma Surgery through the Use of Intra-Operative Fluorescence Imaging Agents. *Pharmaceuticals* **2022**, *15*, 550. https://doi.org/10.3390/ph15050550

Academic Editor: Xuyi Yue

Received: 1 February 2022
Accepted: 25 April 2022
Published: 29 April 2022

Publisher's Note: MDPI stays neutral with regard to jurisdictional claims in published maps and institutional affiliations.

Copyright: © 2022 by the authors. Licensee MDPI, Basel, Switzerland. This article is an open access article distributed under the terms and conditions of the Creative Commons Attribution (CC BY) license (https://creativecommons.org/licenses/by/4.0/).

1. Introduction

Glioblastoma (GBM) is the most common primary malignant brain tumour in adults [1,2]. The current standard of care (SOC) comprises maximum surgical resection followed by radiotherapy, with concomitant adjuvant Temozolomide (TMZ) chemotherapy [3]. Despite multimodal SOC, diagnosed patients have a dismal 2-year survival rate of just 26–33% [2]. Tumour resection serves a vital role to improve patient outcome via tumour debulking, cytoreduction and reduction of mass effect, and has been proven to significantly increase PFS [4,5]. The extent of resection (EOR) is the major determinant of surgical success [2], with complete resection of the detectable tumour (CRDT) the primary goal [4,5]. Complete resection (CR) of the contrast-enhancing tumour is associated with significantly improved overall survival (OS) among GBM patients (4.1 months OS vs. 1.8 months OS with partial resection) [3,6]. However, CR is virtually impossible. Thus, maximal safe surgical resection is a complex goal [4].

Several studies have assessed the impact of gross total resection (GTR) of the contrast-enhancing tumour on OS of GBM patients, while considering other predictive variables such as age, Karnofsky Performance Scale (KPS) score and absence of necrosis [4,7–9]. Additionally, factors such as the Ki-67 proliferation index (>20%) and high EGFR expression have been shown to be associated with poor overall survival [10–12]. An early study of $N = 416$ GBM patients sought to determine an association between EOR and survival time. Here, Lacroix et al. demonstrated that resection of 89% of enhancing tumours identified on T1 weighted MRI significantly improved OS (median OS 10.9 months) [8]. Indeed, the greatest improvement was observed in cases where >98% of contrast enhancing tumour was removed, resulting in a median OS of 13 months [8]. More recently, two additional studies [7,9] have demonstrated a significant survival benefit associated with extensive EOR. Sanai et al. showed that in $N = 500$ GBM patients, the subtotal EOR of 78% results in a significantly improved OS in newly diagnosed GBM, with aggressive EOR of >96% resulting in a further improved median OS of 12.2 months [7]. Finally, in a study of $N = 1229$ GBM patients, Li et al. demonstrated a 5.4 month increase in median OS, coupled with a 6% decrease in postoperative complications in patients who underwent CR of the T1 contrast-enhancing tumour [9]. Recently, several studies have shown that metabolic positron emission tomography (PET) may also support pre-operative planning, and aid in maximising EOR [13–15]. For example, a recent trial (NCT00006353) has implemented ^{11}C-methionine (MET) to aid in the definition of tumour volume and support improved RT planning [13]. Here, ^{11}C-MET identified tumour regions that were likely to recur, moreover the ^{11}C-MET enhanced regions indicated where greater margins of resection would be beneficial. A further study, which compared MRI contrast enhancement with ^{18}F-FDG and ^{11}C-MET as applied to surgical planning, found that the use of pre-operative PET was associated with an increased survival in GBM patients compared with tumour resection based on MRI alone [14].

Nevertheless, the surgical resection of GBM is not curative [7,16,17]. Indeed, it remains an ongoing clinical challenge to intra-operatively delineate gliomas from normal brain tissue [16], where in some cases, resection of GBM tumours is regrettably associated with significant neurological deficit [6,16,18]. The use of stereotactic pre-operative and intraoperative imaging assists in the delineation of the brain tumour interface. Currently, neuro-navigation, a computerised technique used in the localising of tumour material in the brain, is vital in pre-operative planning for surgical resection. However, these images become invalidated as brain tissue shifts during resection and debulking [17]. Neuro-navigation systems may, however, be improved via implementation of intra-operative MRI (iMRI), intra-operative Ultrasound (iOUS), and Contrast Enhanced Ultrasound (CE-US) [4,5,19]. Indeed, these technologies remain essential in intra-operative guidance today notwithstanding limitations in their capacity to accurately identify all residual and invasive tumour material. In this context, tumour recurrence is usually inevitable.

Notwithstanding, there exists a significant need for improved intra-operative guidance to maximise the extent of resection, and ultimately improving patient prognosis [16]. To address this, fluorescence probes may be introduced to illuminate tumour margins where traditional white light imaging fails to delineate tumour-normal brain margins [4,6,18]. Indeed, fluorescence-guided surgery (FGS) aims to improve visualisation of tumour cells within the surgical field, particularly in the neuro-oncology context where diffuse and invasive tumour margins persist [18]. Overall, FGS provides the opportunity to improve EOR and realise the benefits of resection beyond the limitations of equivocal margins, either radiographically or under white light [4,16].

To date, several fluorophores have reached clinical trial (summarised in [Table 1]). Many of these fluorophores may also be used in combination, with improvements in GTR shown when using multiple fluorescence imaging agents in unison [4]. Additional benefits of FGS include affordable price, wider accessibility, and an absence of ionising radiation [19]. The majority of fluorescence imaging agents emit light in the visible spectrum (400–700 nm) [19]. However, the more favourable wavelength for in vivo use is the near infrared (NIR) range (700–900 nm) [19], enabling deeper imaging and photo penetration of up to 8 mm through tissue [20,21]. Indeed, despite positive outcomes afforded by these agents, many agents afford only a 2-dimensional image, and the signal can frequently be obscured by overhanging tissue, blood, and haemostatic agents. Fluorophores may also be photobleached or destroyed by coagulation [22]. Nevertheless, utilisation of fluorophores may significantly improve EOR and ultimately patient outcome [12].

In the current review, we discuss the mechanisms and historical use of established fluorescent agents including Fluorescein Sodium, 5-Aminolevulinic Acid (5-ALA), Indocyanine Green, IRDye 800 CW, and Alkylphosphocholine analogues (APCs) (Figure 1). We further explore the evidence underpinning the development of novel fluorophores, which harbour potential to overcome limitations of current FGS. We describe the properties of these probes and reference pre-clinical trials that have yielded positive results. Building on previous reviews by Craig et al. (2016), Belykh et al. (2016) and Sun (2021), we expand on extensively researched probes such as Folate-targeted FGS, hypericin, and RGD Conjugated agents. We also highlight a novel class of switching fluorophores, NIR-AZA, and discuss their potential to overcome the shortcomings associated with currently established fluorophores. We postulate that the adaptation of these probes for use in GBM surgery is a promising area of translational research.

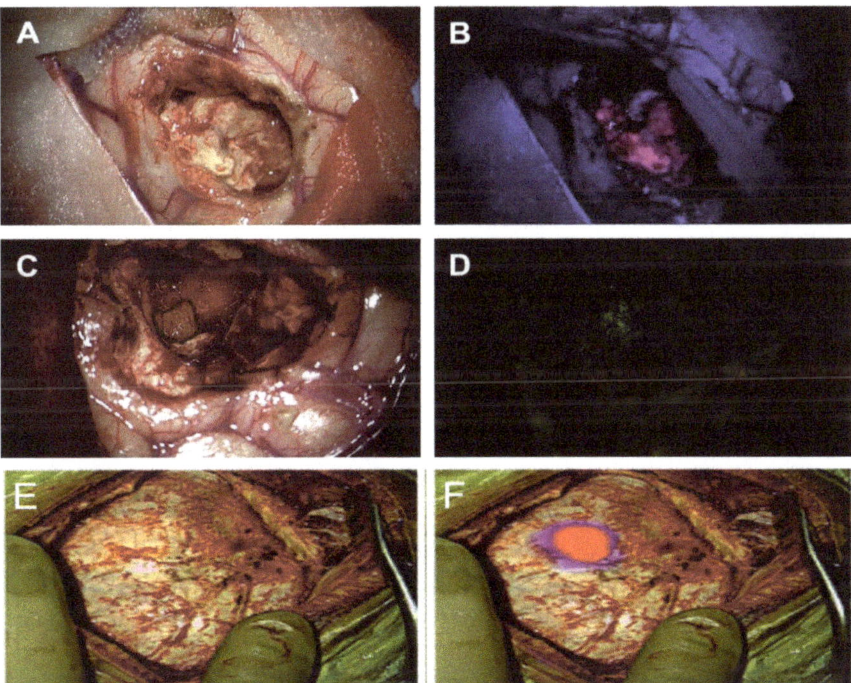

Figure 1. Examples of current fluorophores used in clinical practice. Representative images of fluorescence-guided resection of glioblastoma using (**A,B**) 5-ALA, (**C,D**) Fluorescein and (**E,F**) SWIG using white light (**A,C,E**), fluorescing light (**B,D**) or a white-light + NIR overlay (**F**). Images (**A–D**) reproduced with permission from Stummer et al. 2017. Fluorescence Imaging/Agents in tumour resection. *Neurosurg Clin. N. Am.* **2017**, *28*, 569–583. Images (**E + F**) reproduced with permission from Teng et al. 2021. *Neurosurg Focus.* **2021**, *50*, E4.

Table 1. Status of Fluorophores in clinical trials.

Fluorophore	Chemical Family	Excitation Wavelength (nm)	Emission Peak (nm)	Mode of Action	Trial Number	Tumour	Aim/Result	Reference
Fluorescein	Fluorescein	460–500	510–525	Passive	NCT03752203	Paediatric Neurosurgical Tumours	Determine EOR of Intracranial and spinal lesions using Fluorescein Sodium	https://clinicaltrials.gov/ct2/show/NCT03752203 (accessed on 16 February 2022)
					NCT02691923 Phase 2	High grade glioma	Determine the diagnostic potential of Fluorescein through an operating microscope relate to (1) contrast enhancement on co-registered preoperative MR scans, (2) intra-operative ALA-induced PpIX fluorescence and (3) gold-standard histology obtained from biopsy sampling during the procedure.	https://clinicaltrials.gov/ct2/show/NCT02691923 (accessed on 16 February 2022)
5-ALA	Endogenous non-proteinogenic amino acid	400–410	635–710	Metabolic	NCT00241670 Phase 3	Malignant Glioma	29% more Complete Resections 6-month higher PFS	[6]
					NCT02755142 Phase 1/2	Malignant Glioma	100% Positive Predictive Value 10-fold increase in dose led to 4-fold increase in contrast between tumour and brain 20 mg/kg gave the strongest fluorescence	[23]
					NCT00752323 Phase 2	Malignant Astrocytoma	Determine the optimum dose and administration time of 5-ALA	https://clinicaltrials.gov/ct2/show/NCT00752323 (accessed on 16 February 2022)
					NCT02379572	Glioblastoma	Comparison of iMRI and 5-ALA on number of complete resections	
					NCT01128218 Phase 1,2	Malignant Glioma	Determine specificity and sensitivity of 5-ALA fluorescence	https://clinicaltrials.gov/ct2/show/NCT01128218 (accessed on 16 February 2022)
					NCT02191488 Phase 1	Low, and high grade gliomas, Menihgiomas, or metastases	Red-light excitation of PpIX revealed tumour up to 5mm below resection bed in 22 of 24 tumours already visualised with blue-light.	[24]

Table 1. *Cont.*

Fluorophore	Chemical Family	Excitation Wavelength (nm)	Emission Peak (nm)	Mode of Action	Trial Number	Tumour	Aim/Result	Reference
5-ALA	Endogenous non-proteinogenic amino acid	400–410	635–710	Metabolic	NCT00870779 Phase 1	Low, and high grade gliomas, Menihgiomas, or metastases pituitary adenoma or metastasis	Determine degree of spatial correlation between local fluorescence recorded intra-operatively and co-registered conventional imaging obtained preoperatively via MRI and intra-operatively via ultrasound and operating microscope stereovision	https://clinicaltrials.gov/ct2/show/NCT00870779 (accessed on 16 February 2022)
					NCT01502280 Phase 3	Low-grade Gliomas	Intra-operative confocal microscopy identified 5-ALA tumour fluorescence at a cellular level in 10 consecutive patients.	[25]
					NCT01116661 Phase 2	Glioma	Mean CPpIX was higher in fluorescing samples than nonfluorescing samples. Visible fluorescence can be used in line with Quantitative PpIX analysis	[26]
					NCT02155452	Malignant Glioma	Study the heterogeneity of fluorescence within malignant gliomas by sampling tissues from variable areas within the same tumour	https://clinicaltrials.gov/ct2/show/NCT02155452 (accessed on 16 February 2022)
					NCT02119338	Recurrent glioma	Correlation of 5-ALA fluorescence in tumour tissue with pathological findings	https://clinicaltrials.gov/ct2/show/NCT02119338 (accessed on 16 February 2022)
					NCT02050243 Phase 1/2	CNS Tumour, Paediatric	Determine sensitivity of CNS in identifying paediatric CNS tumours and number of patients with associated side effects	https://clinicaltrials.gov/ct2/show/NCT02050243 (accessed on 16 February 2022)
ICG	Cyanine	780	800–830	Passive	NCT03262636 Phase 1	Primary and Recurrent Brain Tumour	Determine the sensitivity of ICG uptake and expression in identifying autonomic nervous system tumours	[27]

Table 1. *Cont.*

Fluorophore	Chemical Family	Excitation Wavelength (nm)	Emission Peak (nm)	Mode of Action	Trial Number	Tumour	Aim/Result	Reference
BLZ-100	Chlorotoxin peptide + ICG	730–785	760–841	Targeted	NCT02234297 Phase 1	Glioma	Determine safety of BLZ-100 in adult patients with glioma undergoing surgery.	[28]
					NCT02462629 Phase 1	Central Nervous System (CNS) Tumours	Determine safety of BLZ-100 in paediatric patients with CNS Tumours	https://clinicaltrials.gov/ct2/show/NCT02462629 (accessed on 16 February 2022)
Panitumumab-IRDye 800 CW	IRDye 800 CW	775	789–795	Targeted	NCT04085887 Phase 1/2	Paediatric brain neoplasms	Determine the safety and efficacy of Panitumumab-IRDye 800 CW in removing suspected tumours in paediatric patients	https://clinicaltrials.gov/ct2/show/NCT04085887 (accessed on 16 February 2022)
ABY-029	IRDye 800 CW	775	789–795	Targeted	NCT02901925 Phase 1	Recurrent Glioma	Determine if microdoses of ABY-029 lead to detectable signals in sampled tissues with an EGFR pathology score $\geq$ 1 based on histological staining.	https://clinicaltrials.gov/ct2/show/NCT02901925 (accessed on 16 February 2022)
LUM015	Cy5	633–647	675	Metabolic	NCT03717142	Low grade glioma, Glioblastoma	Determine the safety and efficacy of LUM015 for imaging low grade gliomas, GBM and tumour metastasis to the brain	https://clinicaltrials.gov/ct2/show/NCT03717142 (accessed on 16 February 2022)
Demeclocycline	Demeclocycline	402	535	passive	NCT02740933	Brain Tumour	Determine if fluorescence is observable via confocal microscopy.	https://clinicaltrials.gov/ct2/show/NCT02740933 (accessed on 16 February 2022)
BBN-IRDye 800 CW	**IRDye 800 CW**	775	789	Targeted	NCT02910804	Glioblastoma	Determine the efficacy of BBN-IRDye800 CW in GBM patients	https://clinicaltrials.gov/ct2/show/NCT02910804 (accessed on 16 February 2022)
					NCT03407781	Lower grade Glioma	Determine the efficacy of BBN-IRDye800 CW in lower grade glioma patients	https://clinicaltrials.gov/ct2/show/NCT03407781 (accessed on 16 February 2022)

Each fluorophore was categorised into passive (non-selective accumulation in tissue), targeted (selective binding to a specific molecule in the tissue) and metabolic (requires metabolic process for activation) n/a: not applicable, results not published.

2. Established Fluorescent Agent Utilised in Surgery

2.1. Fluorescein Sodium

Fluorophores were first used in surgery in 1948, with Fluorescein sodium (FS), the sodium salt of the fluorescent organic dye Fluorescein [19], the first agent introduced to improve the identification of intra-operative brain tumours [4,22]. Accumulation of FS depends on a leaky blood-brain-barrier (BBB), and the regions in the brain of fluorescein accumulation correspond to those established by MRI contrast-enhancement [4,5,18]. During tumour resection, the BBB is disrupted, allowing extravasation of fluorescein [29] and results in fluorescence in non-tumour areas, for instance, following surgical manipulation, in surrounding oedematous but fundamentally healthy tissue. As a result, non-specific signals are frequently observed [18,30,31]. This non-tumour fluorescence results in fluorescein's inability to serve as a tumour-specific marker and prevents clear-cut tumour resection [4,22]. Though an inexpensive method for intra-operative imaging, some studies have not shown an improvement in resection outcomes or improved survival of GBM patients using this agent [5,22]. In a study of $N = 12$ patients with high-grade glioma (HGG) who underwent FS-guided surgery, the fluorescein margins corresponded with that of gadolinium enhancement on MRI. Biopsy samples were taken and FS showed a sensitivity of 82.2% and specificity of 90% in distinguishing tumour cells from normal cortical tissue. The GTR of the enhancing tumour, as assessed by postoperative MRI, was achieved in every case. Infiltrating edges accumulate the least FS due to minimal BBB disruption [30], limiting its usefulness in enhancing tumour in resection surgeries. Moreover, as accumulation occurs in the extracellular space, the fluorescence emitted in dense tumours is restricted [29,30].

Nevertheless, in contrast to the above studies, the efficacy of FS for use in GBM resection has shown to be improved with the use of a surgical filter and this has yielded favourable results as seen in a phase II trial of 12 patients [31]. Of $N = 20$ biopsies performed at the resection margin ($N = 5$ patients), a sensitivity and specificity of 91% and 100% respectively of FS identifying tumour tissue was reported [31]. Indeed, in a recent study of 106 patients with GBM [32], GTR was seen in 84% of patients, thus displaying a great improvement compared to non-fluorescent guided surgery. Currently use of this probe for fluorescence-guided resection of glioma is still under consideration by the FDA [33].

2.2. 5-Aminolevulinic Acid (5-ALA)

The orally administered prodrug 5-ALA fluoresces slightly below the NIR spectrum and is currently the most widely used fluorophore in the clinic with peak fluorescence reached 6 h post administration [19]. The prodrug 5-ALA endogenously occurs in the heme synthesis pathway, whereby it is metabolised to Protoporphyrin IX (PpIX) in the mitochondria. Due to the reduced ferrochelatase activity of malignant cells, PpIX uptake is significantly increased in malignant tissue compared to normal tissue [4,5,18,19,34]. The 5-ALA fluoresces under blue light appearing red in bulk tumour areas, pink around the margins, and disappears completely with diminishing tumour density, thus giving room for a high degree of ambiguity [16,34]. The use of 5-ALA allows for a more precise resection procedure and does not depend on damage to the BBB [16] in order to reach the tumour site [6]. Rather, it is dependent on upregulated cellular transport mechanisms leading to intracellular accumulation and is dependent on cellular metabolism and specific tumour microenvironment [35].

In GBM, use of 5-ALA is associated with an increase of 29% in CR and a significant reduction in residual contrast-enhancing tumour on postoperative MRI (a predictor for recurrence), when compared to white light only resections [6,19]. This translates into longer PFS, a reduced need for re-interventions, and a 3.05 month increase in OS [4–6,22,36]. Moreover, fluorescence can be detected beyond the margins of the contrast-enhancing tumour on MRI, therefore suggesting its use in visualising non-enhancing tumour [35]. Recently, several groups have begun to explore the use of 5-ALA with photodynamic therapy (PDT) in the management of GBM [37]. The PpIX (thusly 5-AlA) is not only fluorescent but has been shown to be phototoxic. Due to its high specificity of accumulation in the tumour, it has been suggested that PDT of the surgical resection cavity may further enhance survival in GBM patients [38]. Intra-operative PDT can target the tumour in infiltrating margins following 5-ALA FGS surgical resection to ensure CS [38]. A pilot clinical trial (NCT03048240), which has implemented this approach, has been initiated by University Hospital Lille, in collaboration with the Institut National de la Santé Et de la Recherche Médicale (INSERM, Paris, France). Briefly, 10 patients with GBM with complete surgical removal received 5-ALA FGS and intra-operative PDT in combination with current SOC postoperatively. After iMRI to assess the extent of surgical resection, the PDT was delivered at five fractions of 5 J/cm^2. At an interim analysis of the patients, the 12-months progression-free survival (PFS) rate was 60% (median 17.1 months), and the actuarial 12-months OS rate was 80% (median 23.1 months), suggesting that 5-ALA PDT may help to decrease the recurrence risk by targeting residual tumour cells in the resection cavity [38].

Notwithstanding the ostensible successes of 5-ALA, there are drawbacks. Currently, the available microscopes to view fluorescence from 5-ALA operate under dark-field conditions [4] resulting in an inability to identify important neurological structures. Another shortcoming of 5-ALA is photobleaching, which is the reduced intensity of emission light with prolonged exposure to the activating light [4,22]. This is overcome under normal circumstances as new tissue is continually re-exposed throughout the procedure [22]. Furthermore, 5-ALA produces a 2D image meaning fluorescence can be missed by overhanging tissue, but this limitation, however, can be overcome by dissecting the tumour margin [22]. Skin sensitisation, moreover, requires the patient to avoid sunlight or direct artificial light for up to 24 h post-administration [19,22]. However, the use of 5-ALA in combination with other dyes or modalities has the potential to overcome its shortcomings. The 5-ALA may also be used in combination with FS to better illuminate tumour tissues. This combination appears orange-to-red in the parts where tumour is present, and green in normal tissue thus increasing sensitivity and specificity of tumour and yielding improved surgical resection margins [4,22]. Finally, iMRI, in addition to FGS with 5-ALA, can produce a GTR of up to 100% of contrast-enhancing tumour detailing the complimentary use of both modalities of imaging in GBM tumour resection [4]. In trials assessing the safety of 5-ALA, preoperative and postoperative KPS score, neurological status, hepatobiliary enzyme levels and blood count were generally unchanged [6,39]. This safety margin, combined with its efficacy in delineating tumour cells in high-grade gliomas has led to its approval for use in intra-operative imaging by the U.S. Food and Drug Administration (FDA) [35,40].

2.3. Indocyanine Green

Indocyanine Green (ICG) is a hydrophobic dye [4] which attaches to plasma proteins within blood vessels [19], and serves as a tool for both observing blood flow, and an aid for surgical guidance. It has an emission peak of 820 nm so is considered a near-infrared (NIR) agent [19,41]. The use of ICG results in lower tissue autofluorescence and deeper tissue penetration than 5-ALA. It is the only clinically approved NIR fluorophore [42] and is currently FDA-approved for use in ophthalmologic angiography and hepatic function assessment [43]. The ICG is excreted exclusively in bile and along with its non-specific nature and short in vivo half-life of 4 min [42], the current use in FGS of tumours is limited to hepatocellular carcinoma (HCC) [44].

Recent clinical trial data has shown its potential for intra-operative tumour/normal tissue classifications when used in conjunction with AI image analysis of tissue perfusion profiles [45]. For example, Cahill et al. (2021) demonstrated in 24 patients (11 with colorectal cancer CRC) that the wash-in kinetics of ICG (as analysed by AI) of normal and of cancerous tissue was significantly different and was able to determine the patients with CRC with a specificity in tumour detection of 95% and a sensitivity of 92%. A novel emerging technique for use in glioma surgeries is second-window ICG (SWIG) [46,47]. This approach involves a high-dose of ICG administered 24 h preoperatively and employs the enhanced permeability and retention (EPR) effect observed in tumours [46]. The EPR effect results from the abnormal vasculature and inadequate lymphatic drainage of cancers [48]. In a recent study [46], SWIG resulted in increased accumulation in HGGs (96% sensitivity) and rapid clearance from normal brain tissue. Nevertheless, the use of ICG in FGS is costly, and requires alternating use of white light and NIR, which requires separate display monitors to overlay tumour tissue fluorescence with conventional light [29]. Another limitation of ICG is the detection of false-positives by producing signals in areas of necrosis and inflammation [47,49] due to corresponding fluorescence patterns with Gadolinium-enhancing tissue [49]. Future studies should implement SWIG in the resection of GBM as this approach has the potential to overcome the shortcomings in current ICG use.

2.4. IRDye 800 CW

The NIR dye IRDye 800 CW emits at 805 nm. It is often conjugated with specific molecules aimed to specifically target cancer tissues. Cetuximab, an FDA approved monoclonal antibody that inhibits Epidermal Growth Factor Receptor (EGFR), is an example of a molecule that can be conjugated to IRDye 800 CW [4,18,30]. The EGFR is highly expressed in 50–70% [50] of GBMs and is, therefore, a promising target for diagnosis of GBM [22,29]. A first in-human trial of cetuximab conjugated IRDye 800 CW was conducted by Miller et al. (2018) [50]. Therein, two patients, with preoperative MRI scans showing contrast-enhanced tumours ranging from 1.5–8 cm in diameter, were enrolled in the trial. One patient received a low dose of the cetuximab-conjugate (50 mg), while the second patient received a high dose cetuximab-conjugate (100 mg). Viable tumour tissue was identifiable in the low dose patient with a sensitivity of 73.0% and a specificity of 66.3% (tumour vs. normal tissue). In the high dose patient, a sensitivity of 98.2% was achieved, with a specificity of 69.8% (CI 64.3–74.9). One of the major benefits of the NIR conjugate systems is the wide availability in many major surgery centres of the imaging infrastructure necessary to utilise this technology. This eliminates the need to train surgeons or purchase new imaging devices while rapidly implementing the technology.

The use of IRDye 800 CW conjugated to an affibody (ABY-029) in glioma surgery has also been investigated. Affibody molecules are synthetic peptides and non-immunoglobulin proteins, which can be modified for use in radioactive labelling. Therein, Samkoe et al. utilised the affibody ABY-029, which serves as an EGFR inhibitor [51] and a targeted fluorophore for intra-operative imaging. In a recent preclinical study [51], in the first hour post-administration in GBM tissue of mice, there was an 8- to 16-fold average increase in fluorescence visualised in the tumour relative to normal brain with fluorescence still present after 48 h [52]. The mean half-life for cetuximab is 4.7 days [53], and it was demonstrated that when in used conjugation with the NIR dye, this half-life was notably reduced to an average of 27 h across cohorts [54]. A long half-life translates to maximum uptake in tissue [54], however it also requires lengthened clearance from surrounding tissue [19,44], and an improved balance is therefore reached with the use of dyes in the NIR spectrum. Future studies should focus on reducing the prolonged accumulation in normal brain parenchyma while optimising imaging protocols.

3. Novel Dyes in Pre-Clinical Development

Due to the limitations associated with currently available fluorophores, as outlined above, there is a continuing need to develop new strategies for fluorescence imaging as applied in neuro-oncology surgical applications. These strategies include the development of novel agents, as well as the adaption of new variants of existing probes, which may be implemented either alone or in combination with existing approaches (Figure 2). One approach to designing novel agents is to fluorescently tag known tumour targets, such as EGFR, HER2, and VEGF. These targeted fluorescence probes may then be employed for visualisation of tumours [55]. Table 2 shows novel fluorophores still undergoing preclinical trials. As discussed below, these probes aim to address the shortcomings of established fluorophores, such as non-specific fluorescence and unwanted adverse effects.

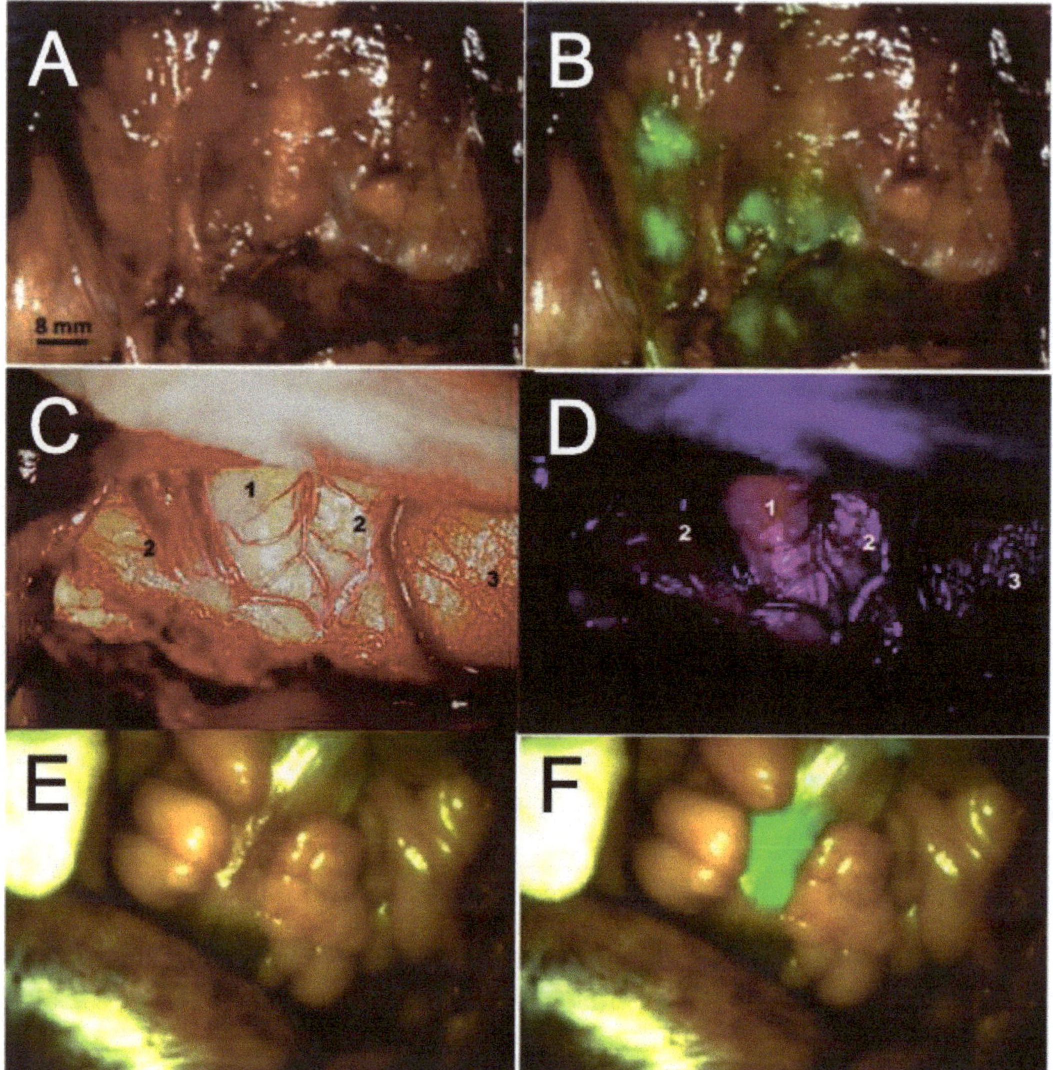

Figure 2. Pre-clinical assessment of novel dyes for Glioblastoma Fluorescence Guided Surgery (FGS). Representative images of novel dyes currently in pre-clinical assessment for use in Glioblastoma Fluorescence Guided Surgery (FGS). (**A,B**) Intra-operative detection of ovarian metastasis using a Folate-Targeted fluorescent probe. (**C,D**) Intra-operative detection of malignant glioma following IV injection of Hypericin. (**E,F**) Intra-operative fluorescence of a palpable colorectal tumour using an RGD conjugated agent. Images are shown under either white light (**A,C,E**), white light and fluorescence overlay (**B,F**), or under blue fluorescence. Images (**A**) + (**B**) reproduced with permission from Hoogstins et al. *Clin Cancer Res.* **2016**, *22*, 2929–2938. Images (**C**) + (**D**) reproduced with permission from Ritz et al. *Eur. J. Surg. Oncol.* (EJSO). **2012**, *38*, 352–360. Images (**E**) + (**F**) reproduced with permission from de Valk et al. *Ann. Surg. Oncology.* **2021**, *28*, 1832–1844.

Table 2. Fluorophores undergoing preclinical evaluation.

Fluorophore	Chemical Family	Excitation Wavelength (nm)	Emission Peak (nm)	Mode of Action	Tissue Type	Result	Reference
CF8 − DiR	CSP + F8 + DiR	750	782	Targeted	Glioma in mice	Folate-targeted CF8-DiR showed a significantly higher accumulation than CSP-DiR. Free DiR dye remained localised in injection point showing accumulation was due to conjugation with CF8.	[56]
Hypericin		510–550	590–650	Passive	Glioma in rats	Tumour Background Ratio (TBR)s of 6 and 1.4	[18,57]
Cetuximab-IRDye 800 CW	IRDye 800 CW	775	789–795	Targeted	Orthotopic mice GBM	87% luciferase signal reduction compared to 41% with white light.	[58]
Panitumumab-IRDye 800 CW	IRDye 800 CW	775	789–795	Targeted	GBM in mice	30% higher TBR when using Panitumumab-IRDye 800 CW than 5-ALA	[59]
IRDye 800 CW-RGD	RGD Conjugate + IRDye 800 CW	775	789–795	Targeted	Mice Glioblastoma (GBM)	Renal clearance of IRDye 800 CW-RGD. The dye selectively binds to Integrin receptors on GBM tissue. TBR of 79.7 ± 6.9 in GBM	[60]
Cyclic-RGD-PLGC (Me)AG-ACPP	RGD Conjugate + Matrix Metalloproteinase (MMP-2)	620	670	Targeted	GBM Cells	Dual targeting improved uptake compared to either cRGD or MMP-2 alone. TBR of 7.8 ± 1.6 in GBM	[61]
cRGD-ZW800-1	RGD Conjugate	750–785	800	Targeted	GBM cell lines	36% more fluorescence signal recorded in comparison to unlabelled cRGD	[21]
CLR1502	Alkylphosphocholine (APCs) Analogues	760	778	Metabolic	Glioma in mice	TBR of 9.28 ± 1.08)	[62]
CLR1501	Alkylphosphocholine (APCs) Analogues	500	517	Metabolic	Glioma in mice	TBR of 3.51 ± 0.44 on confocal imaging; 7.23 ± 1.63 on IVIS imaging	[62]
Chlorotoxin:Cy5.5	Cyanine5.5	633	694	Targeted	Glioma-bearing mice	Mice injected with Chlorotoxin: Cy5.5 a 15-fold higher TBR at day 1 in comparison to mice with Cy5.5 alone	[63]

Table 2. *Cont.*

Fluorophore	Chemical Family	Excitation Wavelength (nm)	Emission Peak (nm)	Mode of Action	Tissue Type	Result	Reference
Angiopep-2-Cy5.5	Cyanine5.5	660–680	694	Targeted	GBM in mice	Tumour to normal fluorescence ratio (TNR) of 1.6 and 63% higher intracerebral uptake than PEG-Cy5.5, tumour margin was delineated non-invasively in vivo	[64]
PEG-Cy5.5	Cyanine5.5	650	665	Passive	GBM in mice	TNR of 1.1	[64]
DA364	RGD Conjugate	675	694–720	Targeted	GBM in mice	TBR of 5.14	[65]
Methylene Blue		642	688–700	Passive/Metabolic	Patient samples of Gliomas	Sensitivity and specificity of 95% and 100% respectively. Dye-enhanced multimodal confocal microscopy shows architectural and morphological features with similar quality to haematoxylin and eosin (H & E)	[66]
PARPi-FL	Inhibitor of the DNA repair enzyme PARP1	503	525	Targeted	GBM in mice	PARPi-FL showed low toxicity, high stability in vivo, and accumulates selectively in glioblastomas due to high PARP1 expression	[67]
CH1055	NIR-II	750	1055	Passive	Brain tumours in mice	Tumour was detected at depths of 4 mm.	[68]
Anti-EGFR Affibody-IRDye-800 CW	IRDye 800 CW	720	730–900	Targeted	GBM cell in mice	The small (6.7 kDa) protein Anti-EGFR Affibody was observed at high levels in outer edges of the tumour	[69]
SDF-1-IRDye-800 CW	IRDye 800 CW	685 and 785	702 or 789	Targeted	GBM cells	Fluorescence persisted for up to 4 days in-vivo	[70]
IRDye800 CW-AE344 (uPAR)	IRDye 800 CW	740 nm	850 nm	Targeted	Orthotopic GBM in mice	TBR above 4.5 between 1 to 12 h post injection	[71]
VEGF labelled IRDye-800 CW	IRDye 800 CW	675 and 745 nm	800 nm	Targeted	Mouse models of ovarian, breast and gastric cancers	TBR of 1.93 $\pm$ 0.40 on day 6 post administration	[72]
EGFR2 labelled IRDye-800 CW	IRDye 800 CW	675 and 745 nm	800 nm	Targeted	Mouse models of ovarian, breast and gastric cancers	TBR of 2.92 $\pm$ 0.29 on day 6 post administration	[72]

Each fluorophore was categorised into passive (non-selective accumulation in tissue), targeted (selective binding to a specific molecule in the tissue) and metabolic (requires metabolic process for activation).

3.1. Folate-Targeted FGS

The folate receptor (FR) is highly expressed in neoplastic cells, and can serve as a useful target for fluorescence probing [19]. Fluorescein iso-thiocyanate (Folate-FITC) has an emission wavelength of 520 nm, and is an example of one such FR-targeting fluorophore [19]. Once Folate-FITC binds to the folate receptor, endocytosis occurs slowly with the fluorescent conjugation persisting within the cell after 2 h [19]. Indeed, GBM tumours display a high expression of FR-α, rendering this receptor a potential target for brain tumour-specific fluorescence imaging [56,73]. Effective targeting of GBM via FR requires the consideration of other FR expressing components of the tumour microenvironment (TME), such as tumour-associated macrophages (TAMs) [56]. The TAMs constitute up to 50% of tumour bulk in GBM and have been shown to play an important role in tumour maintenance and progression [74]. However, FR is only moderately expressed within TAMs, presenting a challenge to target them specifically. It has been suggested to overcome this challenge through the use of recently developed Carbon nanosphere technology [56]. Carbon nanospheres (CSPs) are distinguished by their ability to cross the BBB and may improve the targeting of FRs specifically on GBM cells [56]. The CSPs along with an FR-targeting agent (F8) form CF8 [56], which has the combined ability to target FR-expressing cancer cells and TAMs across the BBB [56]. The CF8 may, therefore, serve as a selective target to FR-expressing glioma cells. Therefore, this BBB-crossing feature of CSPs overcomes a limitation of conventional FR-targeted delivery systems [56]. In this study investigating the use of CF8 as a dual drug delivery system to glioma cells and TAMs, the dye 1,1-dioctadecyl-3,3,3,3- tetramethylindotricarbocyanine iodide (DiR)-labelled CF8 was injected into glioma-bearing mice. There was increased accumulation of CF8-DiR in glioma tumours when compared to CSP-DiR or free DiR dye alone, and fluorescence was observed using an in vivo imaging system [56]. The findings from this preclinical study revealed clear benefits of using this dual targeting strategy for simultaneously targeting both FR-expressing tumour cells and TAMs in the tumour microenvironment. Indeed, this strategy may lead to improved tumour resection and increased patient survival.

3.2. Hypericin

Hypericin, a naphthodianthrone, is a strong lipophilic fluorophore [75] with an emission peak of 650 nm [18]. This agent has the potential to improve glioma diagnosis and treatment due to its higher photostability and penetration depth in comparison to 5-ALA [75]. In a pilot study to investigate its use in identifying High Grade Gliomas (HGG) [75], tumour tissue was clearly apparent from normal brain tissue. Hypericin appears red in areas where it is strongly fluorescent, as in the core of the tumour bulk, and appears pink in weakly fluorescent areas towards the margins where the tumour density is decreased, while normal tissue appears blue [75]. In a further study employing implanted C6 glioma cells in rats, hypericin selectively accumulated intracerebrally and maximum uptake was recorded 24 h following administration [57]. Sensitivity and specificity in differentiating tumour from healthy tissue ranged between 90–100% and 91–94%, respectively [75]. The performance of hypericin has the potential to be highly beneficial to patients with tumour recurrence. This is because clear tumour delineation is difficult, particularly at tumour margins, due to the infiltrative nature of GBM [75]. Nevertheless, the decision on what is considered strongly and weakly fluorescent is user-dependent [75], thus giving variable results, which can affect the EOR.

3.3. RGD Conjugated Agents

The NIR fluorophores such as ICG are limited in their use due to their non-specific accumulation in tissue and their short in vivo half-life [44]. Nevertheless, more recent NIR bio-conjugated fluorophores have been developed to specifically target certain tumour types such as GBM [44]. Additionally, bio-conjugated fluorophores have a large molecular weight and, therefore, have a longer half-life. This greatly enhances their therapeutic uses in treating cancer, as this leads to prolonged accumulation in GBM tissue. However, to allow for clearance from normal tissue, there is an increased time delay experienced between administration of the fluorophore and the commencement of imaging [44]. To allow for more accurate intra-operative imaging at earlier timepoints, it will be necessary to find other approaches that enhance the target-to-background signal ratio. A plausible approach to achieve this would be using mechanisms of selective fluorescence quenching in background tissue. By first establishing the emitting potential of the fluorophore within the region of interest (ROI), and then quenching fluorescence in background areas, it will allow the issue of background clearance to be overcome [44].

Integrins are a potential target for bio-conjugated fluorophores with $\alpha v \beta 3$ and $\alpha v \beta 5$ highly expressed in GBM [21,60,76]. Integrins such as $\alpha v \beta 3$, play a role in tumour angiogenesis and their upregulation is also associated with increased cancer growth and metastasis [44]. The tripeptide arginine-glycine aspartic acid (RGD) sequence can recognise and bind $\alpha v \beta 3$ and $\alpha v \beta 5$, and promotes cellular internalisation. Due to these characteristics, conjugates of the more stable cyclic variant c(RGDfk) are being investigated as selectivity enhancers for tumour therapies and diagnostics [44]. In a study investigating therapeutic targeting of integrins in cancer [76], cRGD has shown affinity towards integrins $\alpha v \beta 3$ and $\alpha v \beta 5$ in GBM [21,60,76]. Like cRGD, the peptide iRDG has affinity for high levels of αv integrins on the surface of tumour vessels. The iRDG peptide binds to $\alpha v \beta 3$ and $\alpha v \beta 5$ and is then proteolytically cleaved within the tumour to produce CRGDK/R [44,77]. The iRGD peptide also has an affinity for neuropilin-1 (NRP-1). Binding of the iRGD peptide to NRP-1 results in tumour tissue penetration and uptake, which is useful in drug delivery [77]. This tumour-specific targeted approach of RGD conjugates and the success in targeting cells for drug delivery can be translated for use in FGS of GBM as is done with RGD-conjugated fluorophores like IRDye 800 CW-RGD and cRGD-ZW800-1. In a transgenic GBM mouse model, the IRDye 800 CW-RGD dye specifically bound to the integrin receptors harboured on GBM cells and gave a detectable fluorescence [60]. In this study, conjugation of this NIR fluorophore (IRDye 800 CW) to the RGD motif did not impede the fluorescence activity of IRDye 800 CW, nor the integrin affinity of RGD [60]. Furthermore, the study was performed with the IRDye 800 CW dye, which is of 800 nm emission wavelength simultaneously with the 700 nm of the dye and reports showed no interference on the images [60]. This demonstrates the potential for combining dyes with varying emission wavelengths in future research to improve EOR in GBM surgery [60]. There were also no reports of effects from photobleaching.

cRGD-ZW800-1 is a cyclic pentapeptide RGD conjugated to the 800 nm NIR fluorophore, ZW800-1 [21]. Delineation of tumour tissue from normal brain was observed in a GBM-mouse model investigating with cRGD-ZW800-1 and when compared to IRDye 800 CW [21], the conjugate had a lower non-specific uptake in normal tissues. An increased dose of 30 nmol of cRGD-ZW800-1 resulted in no non-specific uptake in surrounding tissue [21]. This, therefore, has the potential to be the optimal dose of cRGD-ZW800-1 to achieve maximal safe resection in GBM surgery and future research needs to investigate the benefits of resection at this dose. Additionally, a TBR of 17.2 was achieved within 4 h in comparison to 5.1 with IRDye 800 CW [60]. Optimal fluorescence is achieved within hours of administration as opposed to days when compared to antibody labelled fluorophores, and overall targeting RGD-binding integrins has shown to be well-tolerated in preclinical studies [21]. Ultimately, these RDG conjugated compounds have promising potential to vastly improve FGS in GBM as they overcome tumour specificity issues, toxicity, and photobleaching. These compounds represent a new chapter for intra-operative agents in GBM, which will ultimately lead to improved surgical outcomes for patients.

3.4. Alkylphosphocholine Analogues (APCs)

The APCs are synthetic phospholipid ether molecules that selectively accumulate [62] in overexpressed lipid grafts of various tumours such as GBM [29]. The APCs are resistant to catabolic breakdown, which prolongs time in the tumour microenvironment, and translates to prolonged fluorescence in targeted tissue during surgical resection [29]. In a preclinical study comparing the selective accumulation of 5-ALA with fluorophore labelled APCs (CLR1501 and CLR1502), in five preclinical models of GBM (U251, two patient derived xenografts (PDXs) and three GBM stem cell PDX lines) [62], CLR1501 showed an accumulation and fluorescence profile similar to 5-ALA, while CLR1502 showed a clearer and more defined tumour to brain distinction using IVIS spectrum imaging, Fluobeam detection, and commercially available operative microscopes with appropriate fluorescence detection attachments. These data agree with previous studies, as CLR1502 contains a NIR cyanine fluorophore and light in this spectrum can effectively decrease light absorption in targeted tissue and the false positives of tissue autofluorescence [62]. The tumour-specific uptake of APC-conjugated fluorophores enables greater precision during GBM resection and a heightened potential for preservation of normal brain tissue. Nevertheless, it is noteworthy that practical limitations for these NIR compounds remain. Specifically, the use of NIR agents in surgery requires the tumour to be visualised on a separate monitor in a darkened operating theatre. As a result, surgeons cannot simultaneously visualise the resection field and instruments directly as they would in 5-ALA-guided surgeries.

3.5. NIR-AZA Compounds

BF2-azadipyrromethene (NIR-AZA) is an NIR fluorophore class, developed by the authors, which has shown promise in FGS. Our research efforts have focused on the BF2-azadipyrromethene (NIR-AZA) fluorophore class due to excellent photophysical characteristics including substituent determined emission maxima between 675 and 800 nm, high quantum yields and exceptional photostability [20,78]. Furthermore, linking poly(ethyleneglycol) (PEG) units to NIR-AZA enhanced fluorescence emission. Photostability was also improved, with no light activation of O_2 and no in vivo toxicity associated with the doses recommended for imaging [79]. The PEG plays a vital role in drug delivery by solubilizing and protecting the paired agent, which in this case is the fluorophore. It protects the fluorophore from the aqueous environment while extending the half-life [32]. Additionally, the bio-responsive NIR-AZA has off/on fluorescence switching controlled by the in vivo interconversion of phenol to phenolate [34,56]. Bio-responsive NIR-AZA can be used for real-time continuous imaging of cellular processes such as endocytosis, lysosomal trafficking, and cellular efflux [56]. With the photostability and long half-life, and the specific nature of bio-responsive NIR-AZA compounds, NIR-AZA compounds have the potential to be applied for use in targeted GBM imaging as they address the necessary shortcomings in established fluorophores currently in use for surgical resection of GBM. In recent unpublished work investigating the efficacy of the NIR-AZA agents in a mouse model of GBM by the authors (Figure 3), we demonstrate that following IV injection of the novel fluorophore, activity is only observed in the tumour bearing region of the brain. The implanted tumour (Nfpp10a mouse GBM cell line [80]) is additionally fluorescently labelled with green fluorescent protein (GFP) which can be detected via optical imaging. Figure 3B highlights the activity of the novel fluorophore which overlaps with the GFP fluorescence of the tumour, detailing the specificity of this novel compound. Further work is now required to investigate this agent as a robust intra-operative imaging agent.

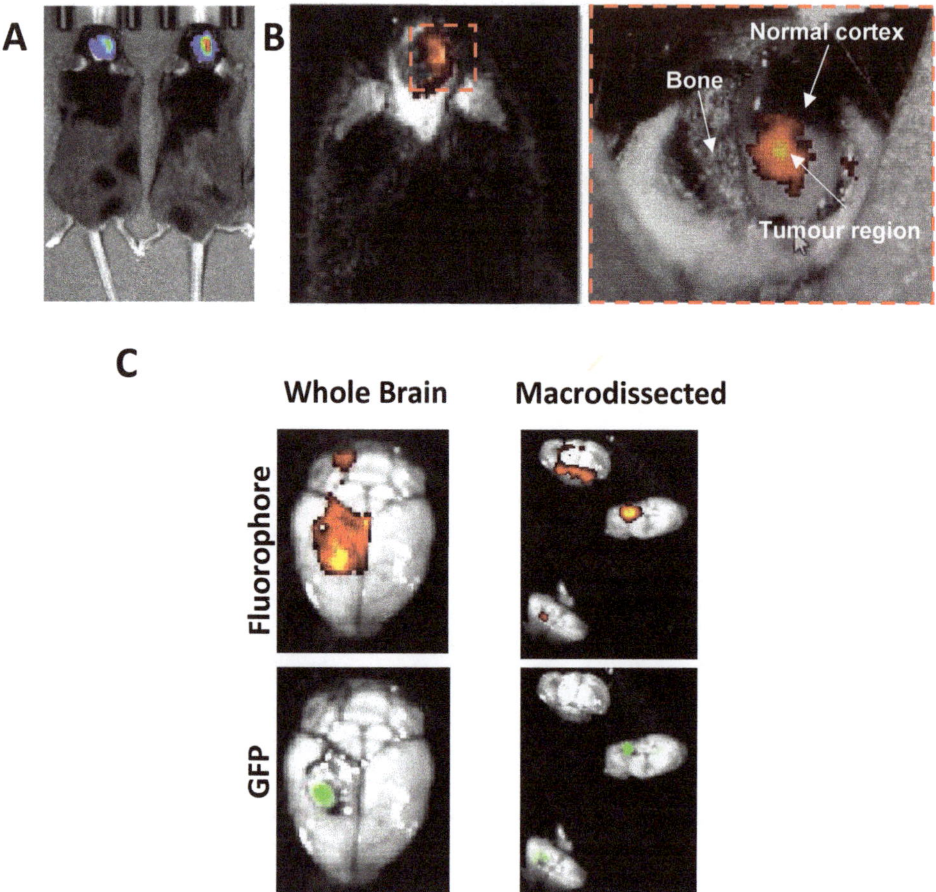

Figure 3. Preclinical assessment of a novel NIR-AZA Fluorophore. Mice were implanted with 2×10^5 NFPp10a-GFP cells, and tumours allowed to develop for 14 days. Tumour growth was monitored by bioluminescence imaging. Subsequently mice then underwent a partial craniotomy to expose tumour and normal tissue. Fluorophore was then injected IV and mice fluorescently imaged for 90 min on IVIS Spectrum. (**A**) Representative image of mice showing location of implanted GBM tumour by bioluminescence. (**B**) Representative images of mouse post partial craniotomy illustrating exposure of normal and tumour tissue after fluorophore injection. (**C**) Ex-vivo imaging of whole and macro-dissected brain. As the tumour was also tagged with Green Fluorescent Protein (GFP), fluorescence imaging was also performed to confirm tumour location [Connor, Shiels et al., Unpublished data].

4. Conclusions

Glioblastoma (GBM) remains one of the most aggressive tumours and a significant source of mortality and morbidity in adults. Current SOC aims to achieve maximum safe resection of neoplastic tissue while preserving healthy brain tissue. This is especially important in the eloquent regions where supramarginal resection is not possible without risk of neurological deficit [81]. However, delineation of normal from tumour tissue continues to be an ongoing challenge. Indeed, despite radiological tools like MRI and ultrasonography, which have proven capable of identifying tumour bulk and in some cases residual tumour pre- and post-operatively, use intra-operatively is met with several limitations, which has given rise to the use of fluorescence-guided surgery (FGS). Nonetheless, there is opportunity for improvement in surgical resection protocols with the use of FGS, specifically as many challenges persist in the development of targeted, effective, and safe fluorophores for use in this disease context.

Challenges in the development of fluorophores include a requirement for well-designed clinical trials to assess safety and efficacy. Within these trials, it is important that appropriate outcome measures are selected to assess performance. The GTR and PFS are often used as indicators of clinical outcome. However, overall survival or development of neurological symptoms may be more relevant clinical markers [18]. Further research is also required to investigate fluorophores, both alone and in combination, to determine the optimal dose and pre-imaging administration schedule to optimise their fluorescence profile and achieve maximal resection. Indeed, a combination of PET guided-surgical planning with intra-operative fluorescent agents, may provide improved resection margins compared to intra-operative agents alone. This combination has the potential to greatly improve the impact of both individual modalities on safe maximal surgical resection and, therefore, enhance the welfare and outcome of GBM patients [82]. The half-life of the fluorophore within the body must also be considered, as this will affect the clearance from the tissue surrounding the tumour. Finally, the technological advancements in neurosurgery do not obviate the need for a detailed consideration of neuroanatomy and rigorous pre-operative planning and monitoring during the procedure [83]. Critically, in the recent review by Mieog et al. [84], it was suggested that for FGS to succeed, target selection, imaging agents and their related detection machinery and their implementation in the clinic have to operate in synergy with each other. Only then will FGS truly improve patient outcomes. Furthermore it has been suggested that the favoured approach of FGS is specifically targeted fluorophores which allow specific accumulation in the tumour and its microenvironment, with biological conjugates leading the way [84]. These conjugates are molecules that target specific tumour epitopes (e.g., integrins or surface receptors). Therefore, the design of targeting tumours (tumour tuning) with a conjugated fluorophore has the potential to produce favourable results in GBM via reduction in residual tumour, which will translate to decreased tumour recurrence, GBM patient morbidity and mortality. Indeed as discussed in this review, one future direction for the development of novel fluorophores is the use of "switching fluorophores" i.e., probes that become fluorescentally active under particular tumour associated conditions, such as acidic pH of a tumour [85]. Switching fluorophores greatly enhances tumour selectivity and may increase the extent of tumour resection. Furthermore, these agents could be conjugated to tumour specific ligands like RDG peptides or other tumour specific markers (EGFRVIII for example). Overall, the field of FGS as applied in the neuro-oncology setting is rapidly evolving and expanding. Nonetheless, this FGS in neuro-oncology still needs to overcome some major hurdles before widespread implementation as a critical tool for improving surgical resection of GBM.

Author Contributions: Conceptualization: A.T.B.; writing—original draft preparation: O.N., K.C., L.P.S., D.W., I.S.M.; writing—review and editing: I.S.M., K.C., K.J.S., D.F.O., A.T.B.; supervision: A.T.B., K.C., L.P.S.; funding acquisition, A.T.B. All authors have read and agreed to the published version of the manuscript.

Funding: This research was funded by the "GLIOTRAIN" (http://www.gliotrain.eu) award, a Horizon 2020 Research and Innovation program funded under the Marie Skłodowska-Curie ETN initiative (Grant Agreement #766069). This research was also funded by Science Foundation Ireland Research infrastructure grant 18/RI/5759 "NPIC" (www.NPIC.ie). We are also grateful for financial support from the Beaumont Hospital Foundation. DFOS acknowledges funding from the Irish Government Department of Business, Enterprise and Innovation's Disruptive Technology Innovation Fund.

Institutional Review Board Statement: The in-vivo study was performed at University College Dublin (UCD) and were approved by UCD Animal Research Ethics Committee (AREC). All work was performed according to local rules and regulations and under Health Products Regulatory Authority HPRA licence number AE18982/P189.

Informed Consent Statement: Not applicable.

Data Availability Statement: No new data were created or analyzed in this study. Data sharing is not applicable to this article.

Conflicts of Interest: DFOS has a financial interest in patents filed and granted relating to NIR fluorophores, all other authors have no conflict of interest to declare.

References

1. Schupper, A.J.; Hadjipanayis, C. Use of Intraoperative Fluorophores. *Neurosurg. Clin. N. Am.* **2021**, *32*, 55–64. [CrossRef] [PubMed]
2. Fernandes, C.; Costa, A.; Osório, L.; Lago, R.C.; Linhares, P.; Carvalho, B.; Caeiro, C. Current Standards of Care in Glioblastoma Therapy. In *Glioblastoma*; De Vleeschouwer, S., Ed.; Codon Publications: Brisbane, QLD, Australia, 2017. [CrossRef]
3. Stupp, R.; Hegi, M.; Mason, W.; van den Bent, M.; Taphoorn, M.; Janzer, R.; Ludwin, S.; Allgeier, A.; Fisher, B.; Belanger, K.; et al. Effects of radiotherapy with concomitant and adjuvant temozolomide versus radiotherapy alone on survival in glioblastoma in a randomised phase III study: 5-year analysis of the EORTC-NCIC trial. *Lancet Oncol.* **2009**, *10*, 459–466. [CrossRef]
4. Orillac, C.; Stummer, W.; Orringer, D.A. Fluorescence Guidance and Intraoperative Adjuvants to Maximize Extent of Resection. *Neurosurgery* **2021**, *89*, 727–736. [CrossRef] [PubMed]
5. Youngblood, M.W.; Stupp, R.; Sonabend, A.M. Role of Resection in Glioblastoma Management. *Neurosurg. Clin. N. Am.* **2021**, *32*, 9–22. [CrossRef] [PubMed]
6. Stummer, W.; Pichlmeier, U.; Meinel, T.; Wiestler, O.D.; Zanella, F.; Reulen, H.-J. Fluorescence-guided surgery with 5-aminolevulinic acid for resection of malignant glioma: A randomised controlled multicentre phase III trial. *Lancet Oncol.* **2006**, *7*, 392–401. [CrossRef]
7. Sanai, N.; Berger, M.S. Operative techniques for gliomas and the value of extent of resection. *Neurotherapeutics* **2009**, *6*, 478–486. [CrossRef]
8. Lacroix, M.; Abi-Said, D.; Fourney, D.R.; Gokaslan, Z.L.; Shi, W.; DeMonte, F.; Lang, F.F.; McCutcheon, I.E.; Hassenbusch, S.J.; Holland, E.; et al. A multivariate analysis of 416 patients with glioblastoma multiforme: Prognosis, extent of resection, and survival. *J. Neurosurg.* **2001**, *95*, 190–198. [CrossRef]
9. Li, Y.M.; Suki, D.; Hess, K.; Sawaya, R. The influence of maximum safe resection of glioblastoma on survival in 1229 patients: Can we do better than gross-total resection? *J. Neurosurg.* **2016**, *124*, 977–988. [CrossRef]
10. Armocida, D.; Frati, A.; Salvati, M.; Santoro, A.; Pesce, A. Is Ki-67 index overexpression in IDH wild type glioblastoma a predictor of shorter Progression Free survival? A clinical and Molecular analytic investigation. *Clin. Neurol. Neurosurg.* **2020**, *198*, 106126. [CrossRef]
11. Li, J.; Liang, R.; Song, C.; Xiang, Y.; Liu, Y. Prognostic significance of epidermal growth factor receptor expression in glioma patients. *OncoTargets Ther.* **2018**, *11*, 731–742. [CrossRef]
12. Ahmadipour, Y.; Jabbarli, R.; Gembruch, O.; Pierscianek, D.; Darkwah Oppong, M.; Dammann, P.; Wrede, K.; Ozkan, N.; Muller, O.; Sure, U.; et al. Impact of Multifocality and Molecular Markers on Survival of Glioblastoma. *World Neurosurg.* **2019**, *122*, e461–e466. [CrossRef] [PubMed]
13. Pessina, F.; Navarria, P.; Clerici, E.; Bellu, L.; Franzini, A.; Milani, D.; Simonelli, M.; Persico, P.; Politi, L.S.; Casarotti, A.; et al. Role of ^{11}C Methionine Positron Emission Tomography (11CMETPET) for Surgery and Radiation Therapy Planning in Newly Diagnosed Glioblastoma Patients Enrolled into a Phase II Clinical Study. *J. Clin. Med.* **2021**, *10*, 2313. [CrossRef] [PubMed]
14. Pirotte, B.J.; Levivier, M.; Goldman, S.; Massager, N.; Wikler, D.; Dewitte, O.; Bruneau, M.; Rorive, S.; David, P.; Brotchi, J. Positron emission tomography-guided volumetric resection of supratentorial high-grade gliomas: A survival analysis in 66 consecutive patients. *Neurosurgery* **2009**, *64*, 471–481; discussion 481. [CrossRef] [PubMed]
15. Ort, J.; Hamou, H.A.; Kernbach, J.M.; Hakvoort, K.; Blume, C.; Lohmann, P.; Galldiks, N.; Heiland, D.H.; Mottaghy, F.M.; Clusmann, H.; et al. (18)F-FET-PET-guided gross total resection improves overall survival in patients with WHO grade III/IV glioma: Moving towards a multimodal imaging-guided resection. *J. Neuro-Oncol.* **2021**, *155*, 71–80. [CrossRef]
16. Liu, J.T.C.; Meza, D.; Sanai, N. Trends in fluorescence image-guided surgery for gliomas. *Neurosurgery* **2014**, *75*, 61–71. [CrossRef]
17. Gerard, I.J.; Kersten-Oertel, M.; Petrecca, K.; Sirhan, D.; Hall, J.A.; Collins, D.L. Brain shift in neuronavigation of brain tumors: A review. *Med. Image Anal.* **2017**, *35*, 403–420. [CrossRef]
18. Senders, J.T.; Muskens, I.S.; Schnoor, R.; Karhade, A.V.; Cote, D.J.; Smith, T.R.; Broekman, M.L. Agents for fluorescence-guided glioma surgery: A systematic review of preclinical and clinical results. *Acta Neurochir.* **2017**, *159*, 151–167. [CrossRef]
19. Nagaya, T.; Nakamura, Y.A.; Choyke, P.L.; Kobayashi, H. Fluorescence-Guided Surgery. *Front. Oncol.* **2017**, *7*, 314. [CrossRef]
20. Daly, H.C.; Conroy, E.; Todor, M.; Wu, D.; Gallagher, W.M.; O'Shea, D.F. An EPR Strategy for Bio-responsive Fluorescence Guided Surgery with Simulation of the Benefit for Imaging. *Theranostics* **2020**, *10*, 3064–3082. [CrossRef]
21. Handgraaf, H.J.M.; Boonstra, M.C.; Prevoo, H.; Kuil, J.; Bordo, M.W.; Boogerd, L.S.F.; Sibinga Mulder, B.G.; Sier, C.F.M.; Vinkenburg-van Slooten, M.L.; Valentijn, A.; et al. Real-time near-infrared fluorescence imaging using cRGD-ZW800-1 for intraoperative visualization of multiple cancer types. *Oncotarget* **2017**, *8*, 21054–21066. [CrossRef]
22. Stummer, W.; Suero Molina, E. Fluorescence Imaging/Agents in Tumor Resection. *Neurosurg. Clin. N. Am.* **2017**, *28*, 569–583. [CrossRef] [PubMed]

23. Stummer, W.; Stepp, H.; Wiestler, O.D.; Pichlmeier, U. Randomized, Prospective Double-Blinded Study Comparing 3 Different Doses of 5-Aminolevulinic Acid for Fluorescence-Guided Resections of Malignant Gliomas. *Neurosurgery* **2017**, *81*, 230–239. [CrossRef] [PubMed]

24. Roberts, D.W.; Olson, J.D.; Evans, L.T.; Kolste, K.K.; Kanick, S.C.; Fan, X.; Bravo, J.J.; Wilson, B.C.; Leblond, F.; Marois, M.; et al. Red-light excitation of protoporphyrin IX fluorescence for subsurface tumor detection. *J. Neurosurg.* **2018**, *128*, 1690–1697. [CrossRef] [PubMed]

25. Sanai, N.; Snyder, L.A.; Honea, N.J.; Coons, S.W.; Eschbacher, J.M.; Smith, K.A.; Spetzler, R.F. Intraoperative confocal microscopy in the visualization of 5-aminolevulinic acid fluorescence in low-grade gliomas. *J. Neurosurg.* **2011**, *115*, 740–748. [CrossRef]

26. Widhalm, G.; Olson, J.; Weller, J.; Bravo, J.; Han, S.J.; Phillips, J.; Hervey-Jumper, S.L.; Chang, S.M.; Roberts, D.W.; Berger, M.S. The value of visible 5-ALA fluorescence and quantitative protoporphyrin IX analysis for improved surgery of suspected low-grade gliomas. *J. Neurosurg.* **2019**, *133*, 79–88. [CrossRef]

27. Li, C.; Buch, L.; Cho, S.; Lee, J.Y.K. Near-infrared intraoperative molecular imaging with conventional neurosurgical microscope can be improved with narrow band "boost" excitation. *Acta Neurochir.* **2019**, *161*, 2311–2318. [CrossRef]

28. Patil, C.G.; Walker, D.G.; Miller, D.M.; Butte, P.; Morrison, B.; Kittle, D.S.; Hansen, S.J.; Nufer, K.L.; Byrnes-Blake, K.A.; Yamada, M.; et al. Phase 1 Safety, Pharmacokinetics, and Fluorescence Imaging Study of Tozuleristide (BLZ-100) in Adults With Newly Diagnosed or Recurrent Gliomas. *Neurosurgery* **2019**, *85*, E641–E649. [CrossRef]

29. Schupper, A.J.; Yong, R.L.; Hadjipanayis, C.G. The Neurosurgeon's Armamentarium for Gliomas: An Update on Intraoperative Technologies to Improve Extent of Resection. *J. Clin. Med.* **2021**, *10*, 236. [CrossRef]

30. Diaz, R.J.; Dios, R.R.; Hattab, E.M.; Burrell, K.; Rakopoulos, P.; Sabha, N.; Hawkins, C.; Zadeh, G.; Rutka, J.T.; Cohen-Gadol, A.A. Study of the biodistribution of fluorescein in glioma-infiltrated mouse brain and histopathological correlation of intraoperative findings in high-grade gliomas resected under fluorescein fluorescence guidance. *J. Neurosurg.* **2015**, *122*, 1360–1369. [CrossRef]

31. Acerbi, F.; Broggi, M.; Eoli, M.; Anghileri, E.; Cuppini, L.; Pollo, B.; Schiariti, M.; Visintini, S.; Orsi, C.; Franzini, A.; et al. Fluorescein-guided surgery for grade IV gliomas with a dedicated filter on the surgical microscope: Preliminary results in 12 cases. *Acta Neurochir.* **2013**, *155*, 1277–1286. [CrossRef]

32. Höhne, J.; Schebesch, K.-M.; de Laurentis, C.; Akçakaya, M.O.; Pedersen, C.B.; Brawanski, A.; Poulsen, F.R.; Kiris, T.; Cavallo, C.; Broggi, M.; et al. Fluorescein Sodium in the Surgical Treatment of Recurrent Glioblastoma Multiforme. *World Neurosurg.* **2019**, *125*, e158–e164. [CrossRef]

33. Okuda, T.; Yoshioka, H.; Kato, A. Fluorescence-guided surgery for glioblastoma multiforme using high-dose fluorescein sodium with excitation and barrier filters. *J. Clin. Neurosci.* **2012**, *19*, 1719–1722. [CrossRef] [PubMed]

34. Stummer, W.; Novotny, A.; Stepp, H.; Goetz, C.; Bise, K.; Reulen, H.J. Fluorescence-guided resection of glioblastoma multiforme by using 5-aminolevulinic acid-induced porphyrins: A prospective study in 52 consecutive patients. *J. Neurosurg.* **2000**, *93*, 1003–1013. [CrossRef] [PubMed]

35. Michael, A.P.; Watson, V.L.; Ryan, D.; Delfino, K.R.; Bekker, S.V.; Cozzens, J.W. Effects of 5-ALA dose on resection of glioblastoma. *J. Neurooncol.* **2019**, *141*, 523–531. [CrossRef] [PubMed]

36. Gandhi, S.; Tayebi Meybodi, A.; Belykh, E.; Cavallo, C.; Zhao, X.; Syed, M.P.; Borba Moreira, L.; Lawton, M.T.; Nakaji, P.; Preul, M.C. Survival Outcomes Among Patients With High-Grade Glioma Treated With 5-Aminolevulinic Acid-Guided Surgery: A Systematic Review and Meta-Analysis. *Front. Oncol.* **2019**, *9*, 620. [CrossRef]

37. Stepp, H.; Stummer, W. 5-ALA in the management of malignant glioma. *Lasers Surg. Med.* **2018**, *50*, 399–419. [CrossRef]

38. Vermandel, M.; Dupont, C.; Lecomte, F.; Leroy, H.A.; Tuleasca, C.; Mordon, S.; Hadjipanayis, C.G.; Reyns, N. Standardized intraoperative 5-ALA photodynamic therapy for newly diagnosed glioblastoma patients: A preliminary analysis of the INDYGO clinical trial. *J. Neurooncol.* **2021**, *152*, 501–514. [CrossRef] [PubMed]

39. Teixidor, P.; Arráez, M.; Villalba, G.; Garcia, R.; Tardáguila, M.; González, J.J.; Rimbau, J.; Vidal, X.; Montané, E. Safety and Efficacy of 5-Aminolevulinic Acid for High Grade Glioma in Usual Clinical Practice: A Prospective Cohort Study. *PLoS ONE* **2016**, *11*, e0149244. [CrossRef]

40. Hadjipanayis, C.G.; Stummer, W. 5-ALA and FDA approval for glioma surgery. *J. Neurooncol.* **2019**, *141*, 479–486. [CrossRef]

41. Hilderbrand, S.A.; Weissleder, R. Near-infrared fluorescence: Application to in vivo molecular imaging. *Curr. Opin. Chem. Biol.* **2010**, *14*, 71–79. [CrossRef]

42. Wu, D.; Daly, H.C.; Conroy, E.; Li, B.; Gallagher, W.M.; Cahill, R.A.; O'Shea, D.F. PEGylated BF(2)-Azadipyrromethene (NIR-AZA) fluorophores, for intraoperative imaging. *Eur. J. Med. Chem.* **2019**, *161*, 343–353. [CrossRef]

43. Kobayashi, H.; Ogawa, M.; Alford, R.; Choyke, P.L.; Urano, Y. New strategies for fluorescent probe design in medical diagnostic imaging. *Chem. Rev.* **2010**, *110*, 2620–2640. [CrossRef]

44. Wu, D.; Daly, H.C.; Grossi, M.; Conroy, E.; Li, B.; Gallagher, W.M.; Elmes, R.; O'Shea, D.F. RGD conjugated cell uptake off to on responsive NIR-AZA fluorophores: Applications toward intraoperative fluorescence guided surgery. *Chem. Sci.* **2019**, *10*, 6944–6956. [CrossRef]

45. Cahill, R.A.; O'Shea, D.F.; Khan, M.F.; Khokhar, H.A.; Epperlein, J.P.; Mac Aonghusa, P.G.; Nair, R.; Zhuk, S.M. Artificial intelligence indocyanine green (ICG) perfusion for colorectal cancer intra-operative tissue classification. *Br. J. Surg.* **2021**, *108*, 5–9. [CrossRef]

46. Cho, S.S.; Salinas, R.; De Ravin, E.; Teng, C.W.; Li, C.; Abdullah, K.G.; Buch, L.; Hussain, J.; Ahmed, F.; Dorsey, J.; et al. Near-Infrared Imaging with Second-Window Indocyanine Green in Newly Diagnosed High-Grade Gliomas Predicts Gadolinium Enhancement on Postoperative Magnetic Resonance Imaging. *Mol. Imaging Biol.* **2020**, *22*, 1427–1437. [CrossRef]

47. Teng, C.W.; Huang, V.; Arguelles, G.R.; Zhou, C.; Cho, S.S.; Harmsen, S.; Lee, J.Y.K. Applications of indocyanine green in brain tumor surgery: Review of clinical evidence and emerging technologies. *Neurosurg. Focus* **2021**, *50*, E4. [CrossRef]

48. Kalyane, D.; Raval, N.; Maheshwari, R.; Tambe, V.; Kalia, K.; Tekade, R.K. Employment of enhanced permeability and retention effect (EPR): Nanoparticle-based precision tools for targeting of therapeutic and diagnostic agent in cancer. *Mater. Sci. Eng. C* **2019**, *98*, 1252–1276. [CrossRef]

49. Cho, S.S.; Salinas, R.; Lee, J.Y.K. Indocyanine-Green for Fluorescence-Guided Surgery of Brain Tumors: Evidence, Techniques, and Practical Experience. *Front. Surg.* **2019**, *6*, 11. [CrossRef]

50. Miller, S.E.; Tummers, W.S.; Teraphongphom, N.; van den Berg, N.S.; Hasan, A.; Ertsey, R.D.; Nagpal, S.; Recht, L.D.; Plowey, E.D.; Vogel, H.; et al. First-in-human intraoperative near-infrared fluorescence imaging of glioblastoma using cetuximab-IRDye800. *J. Neurooncol.* **2018**, *139*, 135–143. [CrossRef]

51. Samkoe, K.S.; Gunn, J.R.; Marra, K.; Hull, S.M.; Moodie, K.L.; Feldwisch, J.; Strong, T.V.; Draney, D.R.; Hoopes, P.J.; Roberts, D.W.; et al. Toxicity and Pharmacokinetic Profile for Single-Dose Injection of ABY-029: A Fluorescent Anti-EGFR Synthetic Affibody Molecule for Human Use. *Mol. Imaging Biol.* **2017**, *19*, 512–521. [CrossRef]

52. de Souza, A.L.; Marra, K.; Gunn, J.; Samkoe, K.S.; Hoopes, P.J.; Feldwisch, J.; Paulsen, K.D.; Pogue, B.W. Fluorescent Affibody Molecule Administered In Vivo at a Microdose Level Labels EGFR Expressing Glioma Tumor Regions. *Mol. Imaging Biol.* **2017**, *19*, 41–48. [CrossRef] [PubMed]

53. Hubbard, J.M.; Alberts, S.R. Alternate dosing of cetuximab for patients with metastatic colorectal cancer. *Gastrointest. Cancer Res.* **2013**, *6*, 47–55. [PubMed]

54. Rosenthal, E.L.; Warram, J.M.; de Boer, E.; Chung, T.K.; Korb, M.L.; Brandwein-Gensler, M.; Strong, T.V.; Schmalbach, C.E.; Morlandt, A.B.; Agarwal, G.; et al. Safety and Tumor Specificity of Cetuximab-IRDye800 for Surgical Navigation in Head and Neck Cancer. *Clin. Cancer Res.* **2015**, *21*, 3658–3666. [CrossRef] [PubMed]

55. Belykh, E.; Martirosyan, N.L.; Yagmurlu, K.; Miller, E.J.; Eschbacher, J.M.; Izadyyazdanabadi, M.; Bardonova, L.A.; Byvaltsev, V.A.; Nakaji, P.; Preul, M.C. Intraoperative Fluorescence Imaging for Personalized Brain Tumor Resection: Current State and Future Directions. *Front. Surg.* **2016**, *3*, 55. [CrossRef] [PubMed]

56. Elechalawar, C.K.; Bhattacharya, D.; Ahmed, M.T.; Gora, H.; Sridharan, K.; Chaturbedy, P.; Sinha, S.H.; Jaggarapu, M.M.C.S.; Narayan, K.P.; Chakravarty, S. Dual targeting of folate receptor-expressing glioma tumor-associated macrophages and epithelial cells in the brain using a carbon nanosphere–cationic folate nanoconjugate. *Nanoscale Adv.* **2019**, *1*, 3555–3567. [CrossRef]

57. Noell, S.; Mayer, D.; Strauss, W.S.; Tatagiba, M.S.; Ritz, R. Selective enrichment of hypericin in malignant glioma: Pioneering in vivo results. *Int. J. Oncol.* **2011**, *38*, 1343–1348. [CrossRef]

58. Warram, J.M.; de Boer, E.; Korb, M.; Hartman, Y.; Kovar, J.; Markert, J.M.; Gillespie, G.Y.; Rosenthal, E.L. Fluorescence-guided resection of experimental malignant glioma using cetuximab-IRDye 800CW. *Br. J. Neurosurg.* **2015**, *29*, 850–858. [CrossRef]

59. Napier, T.S.; Udayakumar, N.; Jani, A.H.; Hartman, Y.E.; Houson, H.A.; Moore, L.; Amm, H.M.; van den Berg, N.S.; Sorace, A.G.; Warram, J.M. Comparison of Panitumumab-IRDye800CW and 5-Aminolevulinic Acid to Provide Optical Contrast in a Model of Glioblastoma Multiforme. *Mol. Cancer* **2020**, *19*, 1922–1929. [CrossRef]

60. Huang, R.; Vider, J.; Kovar, J.L.; Olive, D.M.; Mellinghoff, I.K.; Mayer-Kuckuk, P.; Kircher, M.F.; Blasberg, R.G. Integrin $\alpha v \beta 3$-targeted IRDye 800CW near-infrared imaging of glioblastoma. *Clin. Cancer Res.* **2012**, *18*, 5731–5740. [CrossRef]

61. Crisp, J.L.; Savariar, E.N.; Glasgow, H.L.; Ellies, L.G.; Whitney, M.A.; Tsien, R.Y. Dual targeting of integrin $\alpha v \beta 3$ and matrix metalloproteinase-2 for optical imaging of tumors and chemotherapeutic delivery. *Mol. Cancer* **2014**, *13*, 1514–1525. [CrossRef]

62. Swanson, K.I.; Clark, P.A.; Zhang, R.R.; Kandela, I.K.; Farhoud, M.; Weichert, J.P.; Kuo, J.S. Fluorescent cancer-selective alkylphosphocholine analogs for intraoperative glioma detection. *Neurosurgery* **2015**, *76*, 115–124. [CrossRef]

63. Veiseh, M.; Gabikian, P.; Bahrami, S.B.; Veiseh, O.; Zhang, M.; Hackman, R.C.; Ravanpay, A.C.; Stroud, M.R.; Kusuma, Y.; Hansen, S.J.; et al. Tumor paint: A chlorotoxin:Cy5.5 bioconjugate for intraoperative visualization of cancer foci. *Cancer Res.* **2007**, *67*, 6882–6888. [CrossRef] [PubMed]

64. Yan, H.; Wang, J.; Yi, P.; Lei, H.; Zhan, C.; Xie, C.; Feng, L.; Qian, J.; Zhu, J.; Lu, W.; et al. Imaging brain tumor by dendrimer-based optical/paramagnetic nanoprobe across the blood-brain barrier. *Chem. Commun.* **2011**, *47*, 8130–8132. [CrossRef] [PubMed]

65. Lanzardo, S.; Conti, L.; Brioschi, C.; Bartolomeo, M.P.; Arosio, D.; Belvisi, L.; Manzoni, L.; Maiocchi, A.; Maisano, F.; Forni, G. A new optical imaging probe targeting $\alpha V \beta 3$ integrin in glioblastoma xenografts. *Contrast Media Mol. Imaging* **2011**, *6*, 449–458. [CrossRef] [PubMed]

66. Snuderl, M.; Wirth, D.; Sheth, S.A.; Bourne, S.K.; Kwon, C.S.; Ancukiewicz, M.; Curry, W.T.; Frosch, M.P.; Yaroslavsky, A.N. Dye-enhanced multimodal confocal imaging as a novel approach to intraoperative diagnosis of brain tumors. *Brain Pathol.* **2013**, *23*, 73–81. [CrossRef] [PubMed]

67. Irwin, C.P.; Portorreal, Y.; Brand, C.; Zhang, Y.; Desai, P.; Salinas, B.; Weber, W.A.; Reiner, T. PARPi-FL—A fluorescent PARP1 inhibitor for glioblastoma imaging. *Neoplasia* **2014**, *16*, 432–440. [CrossRef] [PubMed]

68. Antaris, A.L.; Chen, H.; Cheng, K.; Sun, Y.; Hong, G.; Qu, C.; Diao, S.; Deng, Z.; Hu, X.; Zhang, B.; et al. A small-molecule dye for NIR-II imaging. *Nat. Mater.* **2016**, *15*, 235–242. [CrossRef]

69. Sexton, K.; Tichauer, K.; Samkoe, K.S.; Gunn, J.; Hoopes, P.J.; Pogue, B.W. Fluorescent affibody peptide penetration in glioma margin is superior to full antibody. *PLoS ONE* **2013**, *8*, e60390. [CrossRef]

70. Meincke, M.; Tiwari, S.; Hattermann, K.; Kalthoff, H.; Mentlein, R. Near-infrared molecular imaging of tumors via chemokine receptors CXCR4 and CXCR7. *Clin. Exp. Metastasis* **2011**, *28*, 713–720. [CrossRef]

71. Kurbegovic, S.; Juhl, K.; Sorensen, K.K.; Leth, J.; Willemoe, G.L.; Christensen, A.; Adams, Y.; Jensen, A.R.; von Buchwald, C.; Skjoth-Rasmussen, J.; et al. IRDye800CW labeled uPAR-targeting peptide for fluorescence-guided glioblastoma surgery: Preclinical studies in orthotopic xenografts. *Theranostics* **2021**, *11*, 7159–7174. [CrossRef]

72. Terwisscha van Scheltinga, A.G.; van Dam, G.M.; Nagengast, W.B.; Ntziachristos, V.; Hollema, H.; Herek, J.L.; Schroder, C.P.; Kosterink, J.G.; Lub-de Hoog, M.N.; de Vries, E.G. Intraoperative near-infrared fluorescence tumor imaging with vascular endothelial growth factor and human epidermal growth factor receptor 2 targeting antibodies. *J. Nucl. Med.* **2011**, *52*, 1778–1785. [CrossRef] [PubMed]

73. Wu, H.; Zhan, Y.; Qu, Y.; Qi, X.; Li, J.; Yu, C. Changes of folate receptor -α protein expression in human gliomas and its clinical relevance. *Zhonghua Wai Ke Za Zhi* **2014**, *52*, 202–207.

74. Landry, A.P.; Balas, M.; Alli, S.; Spears, J.; Zador, Z. Distinct regional ontogeny and activation of tumor associated macrophages in human glioblastoma. *Sci. Rep.* **2020**, *10*, 19542. [CrossRef]

75. Ritz, R.; Daniels, R.; Noell, S.; Feigl, G.C.; Schmidt, V.; Bornemann, A.; Ramina, K.; Mayer, D.; Dietz, K.; Strauss, W.S.L.; et al. Hypericin for visualization of high grade gliomas: First clinical experience. *Eur. J. Surg. Oncol. (EJSO)* **2012**, *38*, 352–360. [CrossRef]

76. Desgrosellier, J.S.; Cheresh, D.A. Integrins in cancer: Biological implications and therapeutic opportunities. *Nat. Rev. Cancer* **2010**, *10*, 9–22. [CrossRef] [PubMed]

77. Zuo, H. iRGD: A Promising Peptide for Cancer Imaging and a Potential Therapeutic Agent for Various Cancers. *J. Oncol.* **2019**, *2019*, 9367845. [CrossRef] [PubMed]

78. Daly, H.C.; Sampedro, G.; Bon, C.; Wu, D.; Ismail, G.; Cahill, R.A.; O'Shea, D.F. BF2-azadipyrromethene NIR-emissive fluorophores with research and clinical potential. *Eur. J. Med. Chem.* **2017**, *135*, 392–400. [CrossRef]

79. Curtin, N.; Wu, D.; Cahill, R.; Sarkar, A.; Aonghusa, P.M.; Zhuk, S.; Barberio, M.; Al-Taher, M.; Marescaux, J.; Diana, M.; et al. Dual Color Imaging from a Single BF2-Azadipyrromethene Fluorophore Demonstrated in vivo for Lymph Node Identification. *Int. J. Med. Sci.* **2021**, *18*, 1541–1553. [CrossRef] [PubMed]

80. Allen, E.; Jabouille, A.; Rivera, L.B.; Lodewijckx, I.; Missiaen, R.; Steri, V.; Feyen, K.; Tawney, J.; Hanahan, D.; Michael, I.P.; et al. Combined antiangiogenic and anti-PD-L1 therapy stimulates tumor immunity through HEV formation. *Sci. Transl. Med.* **2017**, *9*, eaak9679. [CrossRef]

81. Wang, L.M.; Banu, M.A.; Canoll, P.; Bruce, J.N. Rationale and Clinical Implications of Fluorescein-Guided Supramarginal Resection in Newly Diagnosed High-Grade Glioma. *Front. Oncol.* **2021**, *11*, 666734. [CrossRef]

82. Li, D.; Zhang, J.; Chi, C.; Xiao, X.; Wang, J.; Lang, L.; Ali, I.; Niu, G.; Zhang, L.; Tian, J.; et al. First-in-human study of PET and optical dual-modality image-guided surgery in glioblastoma using (68)Ga-IRDye800CW-BBN. *Theranostics* **2018**, *8*, 2508–2520. [CrossRef] [PubMed]

83. Sun, R.; Cuthbert, H.; Watts, C. Fluorescence-Guided Surgery in the Surgical Treatment of Gliomas: Past, Present and Future. *Cancers* **2021**, *13*, 3508. [CrossRef] [PubMed]

84. Mieog, J.S.D.; Achterberg, F.B.; Zlitni, A.; Hutteman, M.; Burggraaf, J.; Swijnenburg, R.J.; Gioux, S.; Vahrmeijer, A.L. Fundamentals and developments in fluorescence-guided cancer surgery. *Nat. Rev. Clin. Oncol.* **2022**, *19*, 9–22. [CrossRef] [PubMed]

85. Ge, Y.; O'Shea, D.F. RGD conjugated switch on near infrared-fluorophores for fluorescence guided cancer surgeries. *Future Oncol.* **2019**, *15*, 4123–4125. [CrossRef] [PubMed]

 *pharmaceuticals*

Review

Radiopharmaceutical Labelling for Lung Ventilation/Perfusion PET/CT Imaging: A Review of Production and Optimization Processes for Clinical Use

Frédérique Blanc-Béguin [1],*, Simon Hennebicq [1], Philippe Robin [1], Raphaël Tripier [2], Pierre-Yves Salaün [1] and Pierre-Yves Le Roux [1]

1   Univ Brest, UMR-INSERM 1304, GETBO, CHRU Brest, Médecine Nucléaire, Avenue Foch, 29200 Brest, France; simon.hennebicq@chu-brest.fr (S.H.); philippe.robin@chu-brest.fr (P.R.); pierre-yves.salaun@chu-brest.fr (P.-Y.S.); pierre-yves.leroux@chu-brest.fr (P.-Y.L.R.)
2   Univ Brest UMR-CNRS 6521 (CEMCA), IFR 148, Avenue Le Gorgeu, 29200 Brest, France; raphael.tripier@univ-brest.fr
*   Correspondence: frederique.blanc@chu-brest.fr

Abstract: Lung ventilation/perfusion (V/Q) positron emission tomography-computed tomography (PET/CT) is a promising imaging modality for regional lung function assessment. The same carrier molecules as a conventional V/Q scan (i.e., carbon nanoparticles for ventilation and macro aggregated albumin particles for perfusion) are used, but they are labeled with gallium-68 (^{68}Ga) instead of technetium-99m (^{99m}Tc). For both radiopharmaceuticals, various production processes have been proposed. This article discusses the challenges associated with the transition from ^{99m}Tc- to ^{68}Ga-labelled radiopharmaceuticals. The various production and optimization processes for both radiopharmaceuticals are reviewed and discussed for optimal clinical use.

Keywords: V/Q PET/CT; [^{68}Ga]Ga-MAA; ^{68}Ga-labelled carbon nanoparticles

Citation: Blanc-Béguin, F.; Hennebicq, S.; Robin, P.; Tripier, R.; Salaün, P.-Y.; Le Roux, P.-Y. Radiopharmaceutical Labelling for Lung Ventilation/Perfusion PET/CT Imaging: A Review of Production and Optimization Processes for Clinical Use. *Pharmaceuticals* 2022, 15, 518. https://doi.org/10.3390/ph15050518

Academic Editor: Klaus Kopka

Received: 16 March 2022
Accepted: 20 April 2022
Published: 22 April 2022

Publisher's Note: MDPI stays neutral with regard to jurisdictional claims in published maps and institutional affiliations.

Copyright: © 2022 by the authors. Licensee MDPI, Basel, Switzerland. This article is an open access article distributed under the terms and conditions of the Creative Commons Attribution (CC BY) license (https://creativecommons.org/licenses/by/4.0/).

1. Introduction

Lung ventilation-perfusion (V/Q) scintigraphy allows the regional lung function distribution of the two major components of gas exchanges, namely ventilation and perfusion, to be assessed [1]. Regional lung ventilation can be imaged after inhaling inert gases or radiolabelled aerosols that reach alveoli or terminal bronchioles. Regional lung perfusion can be assessed after intravenous injection of radiolabelled macroaggregated albumin (MAA) particles trapped during the first pass in the terminal pulmonary arterioles [2,3].

Pulmonary embolism (PE) diagnosis is the main clinical indication of lung V/Q scintigraphy in pulmonary embolism (PE) diagnosis. V/Q scanning was the first non-invasive test validated for PE diagnosis. The technique was then further improved with the introduction of single-photon emission computed tomography (SPECT) and, more recently, SPECT/computed tomography (CT) imaging [4]. There are many other clinical situations in which an accurate assessment of regional lung function may improve patient management besides PE diagnosis. This includes predicting post-operative pulmonary function in patients with lung cancer, radiotherapy planning to minimize the dose to the lung parenchyma with preserved function and reduce radiation-induced lung toxicities, or pre-surgical assessment of patients with severe emphysema undergoing a lung volume reduction surgery. However, although lung scintigraphy should play a central role in these clinical scenarios, its use has not been widely implemented in daily clinical practice [5]. One of the likely explanations could be the inherent technical limitations of SPECT imaging for the accurate delineation and quantification of regional ventilation and perfusion function [4].

Lung V/Q positron emission tomography (PET)/CT is a novel promising imaging modality for regional lung function assessment [6,7]. The technique has shown promising results in various clinical scenarios, including PE diagnosis [8], radiotherapy planning [9], or pre-surgical evaluation of lung cancer patients [10]. Several large prospective clinical trials are underway, such as (NCT04179539, NCT03569072, NCT04942275, and NCT05103670). The rationale is simple [5]. PET/CT uses the same carrier molecules as conventional V/Q scanning, i.e., carbon nanoparticles for ventilation imaging and MAA particles for perfusion imaging. Similar physiological processes are therefore assessed with SPECT or PET imaging. However, carrier molecules are labelled with gallium-68 (^{68}Ga) instead of technetium-99m (^{99m}Tc), allowing the acquisition of images with PET technology. PET has technical advantages compared with SPECT, including higher sensitivity, higher spatial and temporal resolution, superior quantitative capability and much greater access to respiratory-gated acquisition [11].

This article discusses the challenges associated with the switch from ^{99m}Tc- to ^{68}Ga-labelled V/Q radiopharmaceuticals. The various synthesis and optimization processes for both radiopharmaceuticals are reviewed and discussed, focusing on optimal clinical use.

2. Challenges of the Transition from ^{99m}Tc- to ^{68}Ga-Labelled Radiopharmaceuticals for Lung Imaging

As the chemical and physical properties of ^{99m}Tc and ^{68}Ga are different, the transition from V/Q scintigraphy to V/Q PET/CT implies some adaptation.

^{99m}Tc, a metallic radionuclide, is the most widely available isotope in diagnostic nuclear medicine. It is found in oxidation states −I to VII, but the technetium (Tc) complexes for medical applications are found mostly in oxidation state V [12]. ^{99m}Tc, which mainly decays (88%) with a half-life of 6.02 h by gamma emission (E_γ = 140 keV) to the ground state technetium-99 (^{99}Tc), is obtained as ^{99m}TcO$_4{}^-$ from a molybdenum-99 (^{99}Mo)/^{99m}Tc generator commercially available and compatible with the requirements of Good Manufacturing Practices (GMP). The Tc(VII) in ^{99m}TcO$_4{}^-$ has to be reduced to a lower oxidation state to produce a ^{99m}Tc-stable peptide complex or a reactive intermediate complex [13]. Tc(V) forms 5- or 6-coordinate complexes, always in the presence of multiple bond cores with heteroatoms such as oxygen (O), nitrogen (N) or sulfur (S), among which the most common in radiopharmaceuticals are the oxotechnetium and the nitridotechnetium cores [12].

The prevalent gallium (Ga) oxidation state in aqueous solution is +3, forming several gallate anions as gallium hydroxides Ga(OH)$_4{}^-$ at pH superior to 7. Ga(III) is a hard acid and is strongly bound to ligands featuring multiple anionic oxygen donor sites according to HSAB (hard-soft acid-base). Still, some cases have shown it to have a good affinity for thiolates [14]. Ga(III) ions can form four, five and six bonds, explaining all the possible salts or chelates. ^{68}Ga decays with a half-life of 67.71 min by positron emission (88.88%) and electron capture (11.11%) to the ground state zinc-68 (^{68}Zn). It is obtained from a commercially available germanium-68 (^{68}Ge)/^{68}Ga generator, compatible with GMP requirements.

The first challenge of the switch from ^{99m}Tc- to ^{68}Ga-labelled radiopharmaceuticals for lung V/Q imaging is to maintain the pharmacological properties of V and Q tracers. Both MAA and carbon nanoparticles labelled with ^{99m}Tc have been largely studied. They have been shown to have a biodistribution throughout the lungs that allow an accurate assessment of regional lung perfusion and ventilation function. The principle of lung V/Q PET/CT imaging is to assess similar physiological processes than with conventional V/Q scan, but with greater technology for image acquisition.

The technique needs to be easy to implement in nuclear medicine facilities to enable routine use. The preparation should be fast, simple, GMP-compliant and safe for the operators. Furthermore, radiopharmaceutical production should use unmodified commercially available kits of carrier molecules and similar equipment and devices as much as possible as those used for conventional V/Q scans.

3. Lung Perfusion Imaging

3.1. [^{99m}Tc]Tc-MAA

3.1.1. Chemical Aspects of [^{99m}Tc]Tc-MAA Particles

Among the various type of human serum albumin (HSA) available for radionuclide labeling, MAA is the most commonly used form in nuclear medicine facilities. The nature of the complex [^{99m}Tc]Tc-MAA has not been fully elucidated. It was hypothesized that the labelling of proteins with ^{99m}TcO$_4^-$ involved reduction of the anionic Tc(VII) to a cationic Tc by the tin Sn(II) contained in the commercial kit, which was then complexed with electron-donating groups [15–17]. Some authors have assumed that ^{99m}TcO$_4^-$ reduced by the Sn(II)- albumin aggregates probably formed a (Tc = O)$^{3+}$ complex with the aggregates [15]. More recently, high positive cooperativity was shown between ^{99m}Tc and MAA, although MAA particles did not seem to have binding pockets [18,19]. Moreover, it has been shown that the speed of radiolabelling increased from HAS to albumin nanocolloids (NC) to MAA due to the greater reaction surface [18]. This result agreed with the hypothesis, which assumed that in HAS labelling kits, Sn^{2+} may be enclosed in the tertiary structure of the protein and that it may take some time for the ^{99m}TcO$_4^-$ added to diffuse the site of Sn^{2+} for reduction reaction [17]. Furthermore, MAA has very complex shapes with larger surfaces than a spherical shape (as for NC) of equivalent diameter, enhancing the reactivity properties [18,19].

Whatever the exact nature of the link between ^{99m}Tc and MAA, the complex [^{99m}Tc]Tc-MAA demonstrates high stability as more than 90% of the radioactivity is still associated with the MAA after 24h of in vitro incubation in whole human blood at 37 °C [20].

3.1.2. Technical Aspects: [^{99m}Tc]Tc-MAA Preparation

[^{99m}Tc]Tc-MAA particles are manually prepared by introducing a ^{99m}Tc solution in a commercially available MAA kit. The ^{99m}Tc is obtained from a ^{99}Mo/^{99m}Tc generator as sodium pertechnetate (^{99m}TcO$_4^-$, Na$^+$). The MAA labelling with ^{99m}Tc, which occurred at pH 6, is a simple and fast (about 15 min) process, which allows the production of GMP [^{99m}Tc]Tc-MAA without heating step [18].

Before intravenous administration to the patients, the [^{99m}Tc]Tc-MAA suspensions are tested according to the standards mentioned by the kit supplier for clinical use. The radiochemical purity (RCP) is generally controlled using instant thin layer chromatography (iTLC), and the radioactivity distribution is assessed by filtration of the [^{99m}Tc]Tc-MAA suspension through a 3-μm pore size membrane. The results are obtained by measuring filter and filtrate radioactivity. The radionuclidic purity and the pH have to be controlled as well. As MAA are large particles, [^{99m}Tc], Tc-MAA must be resuspended by gentle agitation before dispensing.

3.1.3. Pharmacological Aspects

In a [^{99m}Tc]Tc-MAA suspension, the average particle size is 20–40 μm, and 90% have a size between 10 and 90 μm. There should be no particles larger than 150 μm [21]. [^{99m}Tc]Tc-MAA particles reach the lung via the pulmonary arterial circulation. Due to the size of the alveolar capillaries (5.5 μm on average), the [^{99m}Tc]Tc-MAA does not reach the alveolar capillaries but largely accumulates in the terminal pulmonary arterioles. Particles inferior to 10 μm may pass through the lungs and then phagocytose by the reticuloendothelial system [21]. According to the requirement of the MAA suppliers, the number of MAA particles injected should range from 60,000 to 700,000 to obtain uniform distribution of activity reflecting regional perfusion (for over 280 billion pulmonary capillaries and 300 million pre-capillary arterioles) [22].

Many studies have shown the suitability of the [^{99m}Tc]Tc-MAA suspension to perform pulmonary perfusion scintigraphy. In rabbits, it has been shown that more than 90% of the activity was found in the lungs within a few minutes of administration and that greater than 80% of the activity remained in the lungs over the first hour of the study [20]. Malone et al. assessed the biodistribution of MAA particles in humans [23]. A total of 98% of activity was measured in the lungs immediately after injection. The removal of activity from the lungs followed an approximately bi-exponential form with the first phase in which 56% of the components had an effective half-life of 0.88 ± 0.16 h and the second phase in which 44% of the components had a half-life of 4.56 ± 0.39 h. Moreover, 3 h after injection, the [^{99m}Tc]Tc-MAA uptake in the kidneys and the bladder was 3.6 ± 2.1% and 5.1 ± 4.0%, respectively [23].

3.2. [^{68}Ga]Ga-MAA

3.2.1. Chemical Aspects of [^{68}Ga]Ga-MAA Particles

MAA labelling with ^{68}Ga has been proposed using bifunctional chelators such as EDTA or DTPA, forming quite stable and inert chelates [24,25]. However, direct labelling was performed by most groups. Direct labelling uses a co-precipitation of ^{68}Ga(III) and albumin particles [26,27]. Mathias et al. hypothesized that ^{68}Ga was adsorbed to the surface of the MAA particles after hydrolysis to insoluble gallium hydroxide without excluding specific interactions of Ga(III) ion with ion pairs exposed at the particle surface [28]. As Ga is present as Ga(OH)$_4$$^-$ at a basic pH, ^{68}Ga does not bind to MAA at a pH above 7. The MAA behavior matches with solvent-exposed glutamate and aspartate amino acids, which should be binding sites for multivalent cations with low affinity and low cation specificity [18]. Furthermore, Jain et al. found that stannous chloride (SnCl$_2$) present in the MAA kits for the reduction of ^{99m}Tc had a strong influence on [^{68}Ga]Ga-MAA formation (radiochemical yield, mean particle diameter, serum stability), suggesting that Sn could be linked to MAA or ^{68}Ga [29]. More recently, it has been assumed that MAA has multiple affinity binding sites for ^{68}Ga [18]. Moreover, during ^{99m}Tc and ^{68}Ga competition evaluation for MAA binding sites, MAA showed no discrimination between ^{99m}Tc and ^{68}Ga coherently without a binding pocket [18].

3.2.2. Technical Aspects: [^{68}Ga]Ga-MAA Preparation

[^{99m}Tc]Tc-MAA preparation is a manual and simple process involving only 2 steps: generator elution and mixing the eluate with the MAA. In contrast, because of the chemical properties of ^{68}Ga, at least four steps are required to label MAA particles with ^{68}Ga: ^{68}Ge/^{68}Ga generator elution, mixing the ^{68}Ga eluate with the MAA, heating the reaction medium and the purification of the [^{68}Ga]Ga-MAA. The key steps of MAA labelling with ^{68}Ga are presented in Figure 1.

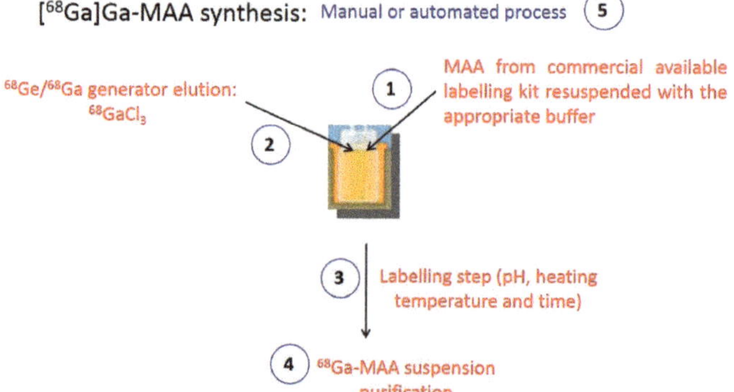

Figure 1. Key points of MAA labelling with ^{68}Ga.

Table 1 summarizes the various [^{68}Ga]Ga-MAA preparation processes described in the literature. The main diverging points of the preparations include (1) the choice of MAA particles, (2) the need for a ^{68}Ga eluate pre-purification, (3) the labelling conditions (pH, heating temperature and time), (4) the [^{68}Ga]Ga-MAA suspension purification, and (5) the automation of the process (Figure 1).

Table 1. [^{68}Ga]Ga-MAA labelling conditions of various methods described in the literature.

	Albumin Particles Labelled		^{68}Ga Eluate Pre-Purification		Labelling Conditions					Radiolabelling Yield (%)	Process Duration (min)	Manual/Automated Process
	Nature and Origin	Washed/Non-Washed	Yes/No	Method Used	pH	Buffer	Heating Time (min)	T °C				
Hnatowich et al. [26]	HSA microspheres. Commercial kit.	Non-washed	Yes	Anion exchange chromatography	2.6–3		15	40–60		65.0	40	Manual
Hayes et al. [30]	HSA microspheres. Commercial kit.	Non-washed	Yes	Anion exchange chromatography	4.8	Sodium acetate solution	10	37		97.4 ± 1.0		Manual
Maziere et al. [31]	HSA microspheres. Commercial kit.	Non-washed	Yes	Pre-concentration	4.7		20	85		93.2 ± 2.5	40	Manual
Even and Green [24]	MAA. Commercial kit.	Washed	Yes	Pre-concentration	4.7	Sodium acetate buffer	15	74 ± 1		77.0–97.0		Manual
Mathias et al. [28]	MAA. Commercial kit.	Washed			5–6	Sodium acetate solution	15	75		81.6 ± 5.3	25	Manual
Maus et al. [32]	MAA. Commercial kit.	Washed	Yes	Fractionation	4	HEPES	20	75		85.0 ± 2.0	30	Manual
Hofman et al. [6]	MAA. Commercial kit.	Washed	Yes	SCX cartridge	6.5	Sodium acetate solution	5	70		≥90.0		Manual
Ament et al. [33]	MAA. Commercial kit.	Washed	Yes	Fractionation	4	HEPES	20	75		85.0 ± 2.0		Manual
Amor-Coarasa et al. [34]	MAA. Commercial kit.	Washed	No		4.7	Sodium acetate solution	15	75		78.3 ± 3.1		Manual
			Yes	Combination of chromatographic exchange resins (cationic then anionic)	4.7	Sodium acetate solution	15	75		97.6 ± 1.5		Manual
Jain et al. [29]	MAA. Homemade with and without SnCl$_2$		No		6		15	75		77.6		Manual
Mueller et al. [35]	MAA. Commercial kit.	Washed and non-washed	Yes	SCX cartridge	4.5	Sodium acetate buffer	10	115		Not mentioned (non-washed) 75.0 (washed)	20 (manual) 14 (auto-mated)	Manual and automated
Shannehsazzadeh et al. [22]	MAA. Commercial kit.	Washed	Yes	Fractionation	4	HEPES	8	75		90.0 -95.0		Manual
Persico et al. [36]	MAA. Commercial kit.	Washed	Yes	SCX cartridge	6–6.5	Sodium acetate buffer	15	40		97.0		Manual and automated

Table 1. *Cont.*

	Albumin Particles Labelled		⁶⁸Ga Eluate Pre-Purification		Labelling Conditions				Radiolabelling Yield (%)	Process Duration (min)	Manual/Automated Process
	Nature and Origin	Washed/Non-Washed	Yes/No	Method Used	pH	Buffer	Heating Time (min)	T °C			
Gultekin et al. [37]	MAA. Commercial kit.	Washed	Yes	PSH⁺ cartridge	4–5	HEPES	7	90	80.0		Manual
Ayşe et al. [38]	MAA. Commercial kit.	Washed	Yes	PSH⁺ cartridge	4–5	HEPES	7	90	85.0 ± 3.0	16	Automated
Blanc-Béguin et al. [39]	MAA. Commercial kit.	Non washed	No		4.3	Sodium acetate solution	5	60	96.0 ± 1.7	15	Automated

- MAA

As shown in Table 1, almost all authors used commercial kits available to prepare [^{99m}Tc]Tc-MAA for labelling MAA particles with ^{68}Ga. The use of commercially available kits is an important consideration to facilitate the implementation of the technique in nuclear medicine facilities. However, commercially available MAA kits contain SnCl$_2$ and free albumin. Consequently, many groups carried out MAA labelling with washed MAA to remove the excess of free albumin and SnCl$_2$ (stannous chloride), which is usually used as a reduction component, from the kit and thus improve the labelling yields (Table 1). Ayşe et al. obtained a better final product RCP by washing the MAA particles before the labelling (RCP = 99.0 ± 0.1%) rather than not washing (RCP = 95.0 ± 0.1%) [38]. On the other hand, Mueller et al. found no significant difference in the radiolabelling yields using non-washed and pre-washed MAA (80% and 75%, respectively) [35]. In studies that used unmodified commercially available MAA kits, radiolabelling yields were consistently superior to 75.0% (Table 1) [28,35,39]. Furthermore, Jain et al. found lower radiolabelling yields using in-house synthesized MAA without SnCl$_2$ than MAA with SnCl$_2$ (49.9 ± 1.3% and 84.5 ± 5.3%, respectively). They found that stannous chloride present in the MAA kits used had a strong influence on the [^{68}Ga]Ga-MAA formation (radiochemical yield, mean particle diameter, serum stability), suggesting that Sn could be linked to MAA or ^{68}Ga [29]. Consequently, using unmodified non-washed commercially available MAA kits produced for ^{99m}Tc seems to be a suitable solution for [^{68}Ga]Ga-MAA labelling;

- ^{68}Ga eluate

^{68}Ga eluate obtained from currently available generators are contaminated with long-lived parent nuclide ^{68}Ge and cationic metal ion impurities such as titanium (Ti)$^{4+}$ from the column material, zinc (Zn)$^{2+}$ from the decay of ^{68}Ga or iron (Fe)$^{3+}$. These impurities might compete with ^{68}Ga in the complexation reaction [40,41].

Pre-purification of the eluate has been proposed by several groups, with various methods such as anion exchange chromatography, cationic cartridge, fractionation or eluate pre-concentration (Table 1) proposed to overcome this issue. Most groups performed an eluate pre-purification using an SCX cartridge or an equivalent pre-conditioned with hydrochloric acid (HCl) and water (Table 1).

In contrast, a few groups did not perform ^{68}Ga eluate pre-purification before MAA labelling and obtained ^{68}Ge impurity levels lower than 0.0001% and radiolabelling yields superior to 96.0% (Table 1) [28,29,34,39]. Based on these results and the fact that this is time-consuming, the pre-purification of the ^{68}Ga eluate does not seem mandatory and could be avoided for [^{68}Ga]Ga-MAA preparation.

- Labelling conditions

The three key points of the MAA labelling with ^{68}Ga are the pH, the heating temperature, and the reaction medium's heating time (Figure 1).

Among the various [^{68}Ga]Ga-MAA labelling processes published, the labelling pH ranges from 3 to 6.5, with the radiolabelling yield varying from 65.0 to 97.4% (decay corrected or not) (Table 1). The optimal pH range seems to be between 4 and 6.5, where ^{68}Ga is a water-soluble cation [14]. Importantly, it has been shown that Ga does not bind to albumin at a pH above 7 [19]. For assessing the optimal pH for the labelling reaction, three buffers or equivalents have been used: acetate buffer, sodium acetate solution and HEPES buffer (Table 1). HEPES and acetate buffers are biocompatible with no toxicity issue [40]. They have demonstrated better characteristics to stabilize and prevent ^{68}Ga(III) precipitation and colloid formation [40]. Nevertheless, in contrast to sodium acetate, HEPES is not approved for human use, and thus, purification and additional quality control analyses are required, resulting in further time and resource consumption [40].

The labelling temperatures reported in the literature ranged from 40 °C to 115 °C (Table 1). Whatever the labelling conditions, the radiolabelling yields were superior to 75.0% in all but one study (Table 1). Several studies have tested various labelling conditions and obtained increasing radiolabelling yield by increasing the heating temperature (from 25 to 100 °C, pH: 4–6) [29,38]. On the other hand, low heating temperature (40–70 °C) has been described to preserve MAA structure and size because high temperatures may induce the rupture of bigger macroaggregates [34,36,39]. Accordingly, the heating temperature range seems quite large, as long as MAA integrity is maintained.

Finally, the labelling heating time ranged from 5 to 20 min (Table 1). Some studies have compared various heating times (from 3 to 50 min) and found higher labelling yields with heating times between 7 and 20 min [31,38,39]. Due to the short half-life of ^{68}Ga and for practical considerations, the heating step should be as short as possible.

[68]GaGa-MAA purification

Many authors have shown that the final product RCP was improved by performing a purification step at the end of the labelling (Table 2). Various processes were used to purify ^{68}Ga]Ga-MAA suspension. The most commonly used process was to wash the labelled MAA with saline and centrifuge to isolate ^{68}Ga]Ga-MAA particles. Another process was to increase the reaction mixture to 10 mL with sterile water at the end of the heating step and to pass the suspension through a Sep-Pak C18 cartridge. Then, the cartridge was washed two times with sterile water. The RCP of the obtained final product was superior to 97% (Table 2).

Table 2. Process of ^{68}Ga]Ga-MAA purification and radiochemical purity according to the various authors.

	[68]Ga]Ga-MAA Purification Conditions		Radiochemical Purity (%)
	Process of Purification	**Manual/Automated**	
Maziere et al. [31]	Centrifugation	Manual	99.9
Even and Green [24]	Centrifugation	Manual	89.0 ± 5.0–98.4 ± 0.3
Mathias et al. [28]	Centrifugation	Manual	99.8 ± 0.1
Maus et al. [32]	Sep-Pak C_{18} cartridge	Manual	>97.0
Ament et al. [33]	Centrifugation	Manual	>97.0
Amor-Coarasa et al. [34]	Centrifugation	Manual	>95.0
Jain et al. [29]	Centrifugation	Manual	98.0 ± 0.8
Mueller et al. [35]	No purification		>95.0
Shannehsazzadeh et al. [22]	Centrifugation	Manual	100.0
Persico et al. [36]	No purification		97.0
Gultekin et al. [37]	Centrifugation	Manual	99.0
Ayşe et al. [38]	No purification		99.0
Blanc-Béguin et al. [39]	Filtration	Automated	99.0 ± 0.6

While purification processes have only been manual to date, an innovative automated procedure was recently proposed for the purification step using a low protein-binding filter. At the end of the heating step, the reaction medium was passed through the filter from the bottom. Then, ^{68}Ga]Ga-MAA was removed from the filter and transferred into the final vial by passing 10 mL of saline from the top to the bottom of the filter (see Figure 2) [39]. The RCP of the obtained ^{68}Ga]Ga-MAA suspension was $99.0 \pm 0.6\%$.

- Manual or Automated Process

Most of the ^{68}Ga]Ga-MAA preparation processes described in the literature were manual and were carried out in 20 to 40 min (See Table 1) [28,31,32]. Mueller et al. showed that an automated process reduced the preparation time from 20 to 14 min [35]. A few automated processes were described in the literature to perform the synthesis from the elution to the end of the heating steps (see Table 1). More recently, a fully automated process has been proposed, including both MAA labelling and ^{68}Ga]Ga-MAA purification, which was carried out in 15 min (see Tables 1 and 2) [39]. To the best of our knowledge, this is the only fully automated process to date.

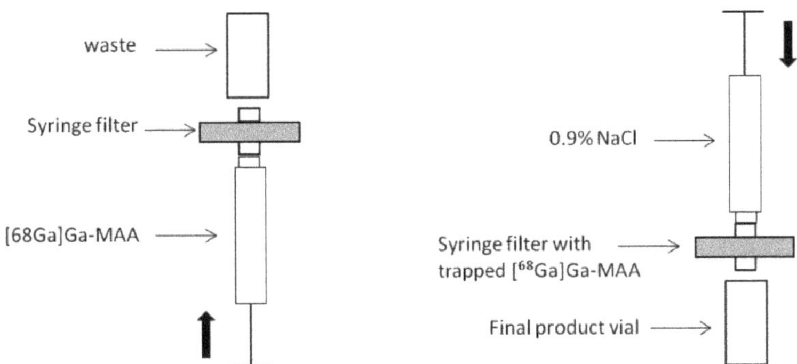

Figure 2. Schematic representation of the [^{68}Ga]Ga-MAA purification stage based on the process by Blanc-Béguin et al. Used with permission from Blanc-Béguin et al. [39].

3.2.3. Pharmacological Aspects

An important challenge of the switch from ^{99m}Tc- to ^{68}Ga-labelled MAA is maintaining the pharmacological properties of particles to ensure similar biodistribution throughout the terminal pulmonary arterioles. Accordingly, the key parameter is the particle size, which should range between 10.0 and 90.0 µm, with no particles size superior to 150.0 µm. On the other hand, particles should not be inferior to 10.0 µm because the target organs would be the reticuloendothelial system and the bones instead of the lungs [22]. Most of the literature data reported a mean diameter ranging from 10 to 90 µm (15.0–75.0 µm for Blanc-Béguin et al., 52.9 ± 15.2 for Jain et al. and 43.0–51.0 for Canziani et al.) (Table 3) [18,29,39]. Furthermore, all published data on [^{68}Ga]Ga-MAA described a preserved structure of radiolabelled particles, whatever the labelling conditions (see Table S1).

Hence, [^{99m}Tc]Tc-MAA and [^{68}Ga]Ga-MAA particles have similar sizes and structures. The number of [^{68}Ga]Ga-MAA particles injected should range from 60,000 to 700,000, no differently from [^{99m}Tc]Tc-MAA, to obtain uniform distribution of activity reflecting regional perfusion.

All studies that assessed the biodistribution of labelled MAA particles in the animals described an almost complete retainment of activity in the lungs from 5 min to 30, 45, 60 min, or 4 h after the injection of [^{68}Ga]Ga-MAA particles, and very low activity in other organs, especially in the liver [22,29,33,34]. Indeed, experiments performed with rats or mice have shown that more than 80% of the injected activity was located in the lungs from 15 min to 4 h post-injection [22,29,34]. In female wild-type rats, the peak of activity occurred 1 min and 35 min post-injection in the kidneys and bladder, respectively, whereas it was from 10 to 20 min in the lungs [22]. In Sprague-Dawley rats, less than 2% of the injected dose per organ (ID/o) activity was measured in other organs from 2 to 4 h post-[^{68}Ga]Ga-MAA administration in the tail vein [34]. It is noteworthy that, as compared with the [^{99m}Tc]Tc-MAA, [^{68}Ga]Ga-MAA exhibited better in vivo stability after intravenous injection in Sprague-Dawley rats [34]. Indeed, the percentage of decay-corrected ID/o (DC-ID/o) of [^{68}Ga]Ga-MAA located in the lung did not change over the study period, i.e., the 4 h following the injection (98.6 ± 0.7 at 2 h and 98.6 ± 0.1 at 4h), whereas the % DC-ID/o of [^{99m}Tc]Tc-MAA located in the lung decreased from 86.6 ± 0.7 at 2 h to 79.2 ± 1.5 at 4 h [34]. After injection in the rat tail vein, the main activity was extracted by the kidneys to the bladder, and the free ^{68}Ga remained in the blood after two and four hours (84.9 ± 4.5% and 63.1 ± 3.9% of DC-ID/o, respectively), presumably as ^{68}Ga native transferrin complex [22,34].

Table 3. Summary of important factors of the switch from ^{99m}Tc to ^{68}Ga.

		[^{99m}Tc]Tc-MAA	[^{68}Ga]Ga-MAA
Labelling conditions			
	pH	6	4–6.5
	Heating temperature (°C)	Room temperature	40–115 °C
	Heating time (min)	0	5–20 min
	Total labelling time	20 min	15–40 min
Size			
	µm	10.0–90.0	15.0–75.0
Labelled MAA suspension stability			
	hours	8	3
Labelled MAA in vitro serum stability			
	hours	24	1
Biodistribution		In humans	In animals
Lungs uptake	%	98.0 [23]	
	time	Immediately after injection	
	%	86.6 ± 0.7 [34]	98.6 ± 0.7 [34]
	time	2 h post-injection	
Kidney uptake	%	3.6 ± 2.1 [23]	1.6 ± 0.4 (right kidney), 1.4 ± 0.2 (left kidney) [34]
	time	3 h after injection	4 h after injection
Bladder uptake	%	5.1 ± 4.0	14 ± 1.7
	time	3 h after injection	4 h after injection [34]
Stomach uptake	%	3.5 ± 2.5	
	time	4.4 after injection	

^{68}Ga(III) has a high binding affinity to the blood serum protein transferrin (log K1 = 20.3). The main requirement for [^{68}Ga]Ga-MAA stability is thermodynamic stability towards hydrolysis and formation of Ga(OH)$_3$ [42]. As shown in Table 3, whatever the labelling conditions, the obtained [^{68}Ga]Ga-MAA suspension had at least a 45 min in vitro stability in animal serum or plasma, which is largely sufficient given that images are acquired immediately after injection and that the acquisition time is approximately 5 min [33].

To the best of our knowledge, no data were published on the biodistribution of [^{68}Ga]Ga-MAA in humans. However, Ament et al., who performed an exploratory study on five patients with clinical suspicion of PE who underwent V/Q PET/CT, have observed that perfusion imaging was homogeneous in most cases [33]. No significant retention and no visual uptake of [^{68}Ga]Ga-MAA particles in the liver were detectable [33].

However, many clinical studies reported consistent activity distribution on PET imaging in various pulmonary conditions after injection of [^{68}Ga]Ga-MAA [1,5,8,33,35,43].

4. Lung Ventilation Imaging

4.1. Aerosolized ^{99m}Tc-Labelled Carbon Nanoparticles (Technegas)

4.1.1. Physical and Chemical Aspects

^{99m}Tc-labelled carbon nanoparticles consist of primary hexagonally structured carbon nanoparticles, which can agglomerate into larger secondary aggregates. Primary nanoparticles are structured with graphite planes oriented parallel to the technetium surface to form nanoparticles with a thickness of about 5 nm [44]. Few data are available about the link between ^{99m}Tc and carbon nanoparticles. It was hypothesized that Tc^{7+} obtained from a ^{99}Mo/^{99m}Tc generator was reduced at the crucible interface, resulting in native metal Tc which co-condensates with carbon species once in the vapor phase [44].

Some authors have hypothesized that ^{99m}Tc-labelled carbon nanoparticles consisted of ^{99m}Tc atoms trapped by a structure similar to a fullerene cage [45,46]. Indeed, Mackey et al. have demonstrated the presence of fullerenes during the generation of the aerosolized ^{99m}Tc-labelled carbon nanoparticles available to form metallofullerenes with the ^{99m}Tc atom attached either exohedrally or endohedrally to the fullerene molecules [46]. However, this hypothesis was controversial especially because of the hexagonal platelet structure of the labelled carbon nanoparticles [44].

4.1.2. Technical Aspects

The ^{99m}Tc-labelled carbon nanoparticle production is a simple process that requires relatively little material: a ^{99}Mo/^{99m}Tc generator, a Technegas generator (Cyclomedica Pty Ltd., Kingsgrove, Australia) and a pure argon bottle. There are three main stages in ^{99m}Tc-labelled carbon nanoparticle production: the loading of the crucible, the simmer stage and the burning stage.

- Crucible loading

Using a syringe with a needle, 0.14 mL to 0.30 mL (140–925 MBq) of ^{99m}Tc eluate is introduced in a graphite crucible previously humidified with 99% ethanol to increase its wettability and placed between the generator electrodes [6,47–51]. As the volume of the crucible is limited to 0.14 mL or 0.30 mL according to the supplier, it is possible to perform several crucible loadings to introduce all the ^{99m}Tc eluate needed;

- Simmer stage

The simmer stage, performed immediately after the crucible loading, reduces ^{99m}Tc^{7+} to metallic ^{99m}Tc under a pure argon atmosphere [44]. The use of pure argon is a determining parameter for the structure and the physical properties of the ^{99m}Tc-labelled carbon nanoparticles [45,48,52–54]. During the simmer stage, the graphite crucible is heated for 6 min at 70 °C. However, it has been shown that increasing the number of simmers increases the median size of the ^{99m}Tc-labelled carbon nanoparticles [45];

- Burning stage

The simmer cycle is followed by the crucible heating to 2550 °C ± 50 °C for 15 s. Metallic Tc and carbon species are vaporized and co-condensed during this burning stage to obtain aerosolized ^{99m}Tc-labelled carbon nanoparticles [44]. At the end of this stage, the switching off of the Technegas generator fills the 6 L chamber with aerosolized ^{99m}Tc-labelled carbon nanoparticles ready for use via inhalation by the patient.

All process parameters (heating temperature and time) used for clinical production are fixed. The machine allows a 10 min window in which the ^{99m}Tc-labelled carbon nanoparticles may be administered to the patient. However, the longer the administration delay is, the higher the median size of the particles is [45,51].

4.1.3. Pharmacological Aspects

The size of primary carbon nanoparticles ranges from 5 to 60 nm, while the size of the aggregates is approximately 100–200 nm (See Table 4). Hence, aerosolized ^{99m}Tc-labelled carbon nanoparticles are considered an ultrafine aerosol with ventilation properties similar to radioactive gasses, such as krypton-81m (^{81m}Kr) and xenon-133 (^{133}Xe) [21,47,48,50,51,55–57]. Many authors agree on the mainly alveolar deposition of the ^{99m}Tc-labelled carbon nanoparticles and the stability of the nanoparticles in the lungs over time [21,45,49,57,58].

Table 4. ^{99m}Tc and ^{68}Ga-labelled carbon nanoparticle size, shape and structure according to the literature.

		Labelled Carbon Primary Nanoparticle Size (nm)	Labelled Carbon Secondary Aggregate Size (nm)	Count Median Diameter (nm)	Labelled Carbon Nanoparticle Thickness (nm)	Shape, Structure and Properties	Physical Properties
^{99m}Tc labelling	Burch et al. [47]	≤5.0					
	Strong et al. [58]			140.0 ± 1.5			
	Isawa et al. [49]	≤200.0					
	Lemb et al. [59]	12.5 ± 1.65 (7–23)	118.0 (60–160)			Primary hexagonally structured graphite particles	Hydrophobic properties. Inert properties
	Mackey et al. [46]					Fullerenes	
	Lloyd et al. [45]		100.0–300.0	158.0 ± 1.5			
	Senden et al. [44]	30.0–60.0			5.0	Thin hexagonal platelets with graphite planes oriented parallel to the Tc surface	Biological inertness
	Möller et al. [57]	10.0	100.0–200.0				Hygroscopic properties
	Pourchez et al. [51]	40.0 ± 2.9				Hexagonal platelets of metallic technetium closely encapsulated with a thin layer of graphitic carbon	Hygroscopic properties
	Blanc-Béguin et al. [60]	20.9 ± 7.2				Thin hexagonal platelets with graphite planes oriented parallel to the Tc surface	
^{68}Ga labelling	Blanc-Béguin et al. [60]	22.4 ± 10	Several hundreds			Hexagonal shape Layered structure	

4.2. Aerosolized ^{68}Ga-Labelled Carbon Nanoparticles

4.2.1. Physical and Chemical Aspects

The physical properties of aerosolized particles are important parameters in determining their penetration, deposition, and retention in the respiratory tract. The physical properties of ^{68}Ga-labelled carbon nanoparticles, prepared using a Technegas generator in the usual clinical way, were recently assessed [60]. ^{68}Ga-labelled carbon nanoparticles demonstrated similar properties as ^{99m}Tc-labelled carbon nanoparticles, with primary hexagonally shaped and layered structured particles [60]. Although the chemical process of labelling carbon nanoparticles with ^{68}Ga and the exact chelation structure of ^{68}Ga in carbon nanoparticles are unknown, the physical properties of ^{68}Ga-labelled carbon nanoparticles suggest a method of labelling similar to labeling with ^{99m}Tc.

4.2.2. Technical Aspects

In contrast with MAA labelling, the process for ^{99m}Tc-labelled carbon nanoparticle preparation is very similar across studies in the literature. ^{68}Ga-labelled carbon nanoparticles are produced using an unmodified Technegas generator and following the same stages as for the preparation of ^{99m}Tc-labelled carbon nanoparticles: the crucible loading with an eluate volume range from 0.14 mL to 0.30 mL, the simmer stage and the burning stage with similar heating time and temperature [6,7,33,60–62]. The only difference is the nature of the eluate, which is gallium-68 chloride (^{68}GaCl$_3$) instead of ^{99m}TcO$_4^-$, Na$^+$. Few authors performed an eluate pre-concentration using an anion exchange cartridge or by fractionating to purify and reduce the volume of the eluate [33,62]. However, ^{68}Ga-labelled carbon nanoparticles obtained using an unmodified eluate from the ^{68}Ge/^{68}Ga generator demonstrated similar physical properties as ^{99m}Tc-labelled carbon nanoparticle properties and were suitable for pulmonary ventilation PET/CT [6,7,60,61].

4.2.3. Pharmacological Aspects

From the pharmacological point of view, an important parameter of the switch from lung ventilation SPECT to PET/CT is to maintain the physical properties of aerosolized carbon nanoparticles to ensure similar alveolar deposition and stability in the lungs.

The size is a key factor in determining the degree of aerosol particle penetration in the human pulmonary tract [55]. To the best of our knowledge, only one recently published work has studied the physical properties of ^{68}Ga-labelled carbon nanoparticles, and few pharmacological data are available in the literature. However, it was reported that using an unmodified Technegas generator, the mean diameter of primary ^{68}Ga-labelled carbon nanoparticles was in the same range as primary ^{99m}Tc-labelled carbon nanoparticles (22.4 $\pm$ 10.0 nm and 20.9 $\pm$ 7.2 nm, respectively) with similar agglomeration into larger secondary aggregates measuring several hundreds of nm [60].

These results suggested similar lung distribution of ^{99m}Tc- and ^{68}Ga-labelled carbon nanoparticles, as confirmed by a study on healthy piglets [62]. Later, V/Q PET performed in Sprague-Dawley rats reported complete incorporation of ^{68}Ga-labelled carbon nanoparticles in the lungs without extrapulmonary activity (urinary bladder, abdomen, blood pool) [33]. Moreover, animal studies performed with aerosolized ^{68}Ga-labelled carbon nanoparticles have demonstrated greater differences between poorly and well-ventilated regions, suggesting higher resolution than ^{99m}Tc-labelled carbon nanoparticles [62].

Moreover, in healthy human volunteers, the activity distribution in the lungs after inhalation of ^{68}Ga-labelled carbon nanoparticles was intense, without bronchial deposit 15 min after the inhalation. Furthermore, the activity (decay corrected) over the lung was constant at 3.5 h without elimination via blood, urine (only trace radioactivity in urine bladder was observed) or feces suggesting the stability of the deposition of ^{68}Ga-labelled carbon nanoparticles over this time [7]. Another exploratory study performed on five patients with clinical suspicion of PE observed homogeneous ventilation imaging in two cases and inhomogeneous accumulation with central deposition of labelled carbon nanoparticles in three cases [33].

Finally, many clinical studies reported consistent activity distribution in the lungs on PET imaging in various pulmonary conditions after inhalation of ^{68}Ga-labelled carbon nanoparticles [5,6,8,33].

5. Practical Considerations for an Optimal Clinical Use

Lung V/Q PET/CT is a promising imaging modality for regional lung function assessment. Indeed, PET imaging has great technical advantages over SPECT imaging (higher sensitivity, spatial and temporal resolution, superior quantitative capability, easier to perform respiratory-gated acquisition). PET may also be a useful alternative to SPECT imaging in a ^{99m}Tc shortage. The success of the switch from conventional scintigraphy to PET imaging, and therefore from [99m]Tc- to ^{68}Ga-labelled radiopharmaceuticals, relies on two main factors: preserving the pharmacological properties of the labelled MAA and carbon nanoparticles, whose biodistribution is well known; and facilitating the implementation in nuclear medicine departments. In that respect, several studies have been conducted on the production of both perfusion and ventilation ^{68}Ga-labelled radiopharmaceuticals, which have led to simplification, optimization and, more recently, automation of the processes.

For lung perfusion PET/CT imaging, various processes have been used for [^{68}Ga]Ga-MAA labelling, with different options in the key steps of the preparation, including the choice of MAA particles, the need for ^{68}Ga eluate pre-purification, the labelling conditions or the [^{68}Ga]Ga-MAA suspension purification. However, simpler processes appear to be suitable for optimal clinical use. This includes using a non-modified commercially available MAA kit, with no need for a ^{68}Ga eluate pre-purification, use of an easy to use buffer such as sodium acetate solution, and a short reaction medium heating time (5 min). Automated processes have been developed to facilitate processing time and reduce the radiation dose to the operator. Thus, a simple and fast (15 min) automated GMP compliant [^{68}Ga]Ga-MAA synthesis process was proposed, using a non-modified MAA commercial kit, a ^{68}Ga eluate without pre-purification and including an innovative process for [^{68}Ga]Ga-MAA purification, which maintains the pharmacological properties of the tracer and provided labelling yields >95% [39]. Moreover, whatever the labelling conditions, the obtained [^{68}Ga]Ga-MAA suspension was described to be stable in 0.9% sodium chloride for at least one hour [35,39]. Given the radioactive concentration of [^{68}Ga], Ga-MAA obtained at the end of the synthesis (i.e., from 300MBq/10 mL to 900 MBq/10 mL according to the age of the [68]Ge/^{68}Ga generator) and the dose injected (i.e., around 50 MBq), up to 6 perfusion PET/CT scans can be performed with one synthesis [5,6,39,63].

For lung ventilation PET/CT imaging, preparing and administering aerosolized ^{68}Ga-labelled carbon nanoparticles is very straightforward. The process is very similar to the production of ^{99m}Tc-labelled carbon nanoparticles and, therefore, fairly easy to implement in nuclear medicine facilities. Indeed, adding a ^{68}Ga eluate instead of ^{99m}Tc eluate in the carbon crucible of an unmodified commercially available Technegas™ generator provides carbon nanoparticles with similar physical properties. Furthermore, recently, an automated process included a step to fractionate the ^{68}Ga eluate into two samples, one for [^{68}Ga]Ga-MAA labelling and the other for aerosolized ^{68}Ga-labelled carbon nanoparticle production, which has been developed [39].

Besides radiopharmaceutical production, many factors may facilitate the implementation of V/P PET/CT imaging in nuclear medicine facilities. [68]Ge/^{68}Ga generators are increasingly available in the nuclear medicine departments due to ^{68}Ga tracers for neuroendocrine tumors and prostate cancer imaging. PET/CT cameras are also increasingly accessible due to the development of digital PET/CT cameras and might be total-body PET/CT in the future. Most nuclear medicine facilities already have the necessary equipment to carry-out V/P PET/CT imaging, including carbon nanoparticle generators and MAA kits. Automating the MAA labelling is now possible; commercial development of ready-to-use sets for automated synthesis radiolabelling of ^{68}Ga-MAA would be of interest.

In conclusion, recent data support the ease of using well-established carrier molecules and ^{68}Ga to enable the switch from SPECT to PET imaging for regional lung function.

The technology may be easily implemented in most nuclear medicine facilities and open perspectives for the improved management of patients with lung disease.

Supplementary Materials: The following supporting information can be downloaded at: https://www.mdpi.com/article/10.3390/ph15050518/s1, Table S1: [^{68}Ga]Ga-MAA particles size and biodistribution study according to the various authors.

Author Contributions: Conceptualization, P.-Y.L.R. and F.B.-B.; formal analysis, F.B.-B.; data curation, F.B.-B.; writing—original draft preparation, F.B.-B. and P.-Y.L.R.; writing—review and editing, F.B.-B., S.H., P.R., R.T., P.-Y.S. and P.-Y.L.R.; visualization, F.B.-B. and P.-Y.L.R.; supervision, P.-Y.L.R., P.R. and P.-Y.S.; project administration, P.-Y.L.R. All authors have read and agreed to the published version of the manuscript.

Funding: This research received no external funding.

Institutional Review Board Statement: Not applicable.

Informed Consent Statement: Not applicable.

Data Availability Statement: Not applicable.

Conflicts of Interest: The authors declare no conflict of interest.

References

1. Le Roux, P.Y.; Siva, S.; Steinfort, D.P.; Callahan, J.; Eu, P.; Irving, L.B.; Hicks, R.J.; Hofman, M.S. Correlation of ^{68}Ga Ventilation-Perfusion PET/CT with Pulmonary Function Test Indices for Assessing Lung Function. *J. Nucl. Med. Off. Publ. Soc. Nucl. Med.* **2015**, *56*, 1718–1723. [CrossRef] [PubMed]
2. Bajc, M.; Schumichen, C.; Gruning, T.; Lindqvist, A.; Le Roux, P.Y.; Alatri, A.; Bauer, R.W.; Dilic, M.; Neilly, B.; Verberne, H.J.; et al. EANM guideline for ventilation/perfusion single-photon emission computed tomography (SPECT) for diagnosis of pulmonary embolism and beyond. *Eur. J. Nucl. Med. Mol. Imaging* **2019**, *46*, 2429–2451. [CrossRef]
3. Le Roux, P.Y.; Blanc-Beguin, F.; Bonnefoy, P.B.; Bourhis, D.; Camilleri, S.; Moreau-Triby, C.; Pinaquy, J.B.; Salaün, P.Y. Guide pour la re' daction de protocoles pour la scintigraphie pulmonaire. *Méd. Nucl.* **2021**, *45*, 8. [CrossRef]
4. Le Roux, P.Y.; Robin, P.; Salaun, P.Y. New developments and future challenges of nuclear medicine and molecular imaging for pulmonary embolism. *Thromb. Res.* **2018**, *163*, 236–241. [CrossRef]
5. Le Roux, P.Y.; Hicks, R.J.; Siva, S.; Hofman, M.S. PET/CT Lung Ventilation and Perfusion Scanning using Galligas and Gallium-68-MAA. *Semin. Nucl. Med.* **2019**, *49*, 71–81. [CrossRef] [PubMed]
6. Hofman, M.S.; Beauregard, J.M.; Barber, T.W.; Neels, O.C.; Eu, P.; Hicks, R.J. ^{68}Ga PET/CT ventilation-perfusion imaging for pulmonary embolism: A pilot study with comparison to conventional scintigraphy. *J. Nucl. Med. Off. Publ. Soc. Nucl. Med.* **2011**, *52*, 1513–1519. [CrossRef] [PubMed]
7. Kotzerke, J.; Andreeff, M.; Wunderlich, G. PET aerosol lung scintigraphy using Galligas. *Eur. J. Nucl. Med. Mol. Imaging* **2010**, *37*, 175–177. [CrossRef] [PubMed]
8. Le Roux, P.Y.; Iravani, A.; Callahan, J.; Burbury, K.; Eu, P.; Steinfort, D.P.; Lau, E.; Woon, B.; Salaun, P.Y.; Hicks, R.J.; et al. Independent and incremental value of ventilation/perfusion PET/CT and CT pulmonary angiography for pulmonary embolism diagnosis: Results of the PECAN pilot study. *Eur. J. Nucl. Med. Mol. Imaging* **2019**, *46*, 1596–1604. [CrossRef] [PubMed]
9. Siva, S.; Thomas, R.; Callahan, J.; Hardcastle, N.; Pham, D.; Kron, T.; Hicks, R.J.; MacManus, M.P.; Ball, D.L.; Hofman, M.S. High-resolution pulmonary ventilation and perfusion PET/CT allows for functionally adapted intensity modulated radiotherapy in lung cancer. *Radiother. Oncol. J. Eur. Soc. Ther. Radiol. Oncol.* **2015**, *115*, 157–162. [CrossRef]
10. Le Roux, P.Y.; Leong, T.L.; Barnett, S.A.; Hicks, R.J.; Callahan, J.; Eu, P.; Manser, R.; Hofman, M.S. Gallium-68 perfusion positron emission tomography/computed tomography to assess pulmonary function in lung cancer patients undergoing surgery. *Cancer Imaging Off. Publ. Int. Cancer Imaging Soc.* **2016**, *16*, 24. [CrossRef] [PubMed]
11. Hicks, R.J.; Hofman, M.S. Is there still a role for SPECT-CT in oncology in the PET-CT era? *Nat. Rev. Clin. Oncol.* **2012**, *9*, 9. [CrossRef] [PubMed]
12. Papagiannopoulou, D. Technetium-99m radiochemistry for pharmaceutical applications. *J. Label. Compd. Radiopharm.* **2017**, *60*, 502–520. [CrossRef]
13. Liu, S.; Edwards, D.S. ^{99m}Tc-Labeled Small Peptides as Diagnostic Radiopharmaceuticals. *Chem. Rev.* **1999**, *99*, 2235–2268. [CrossRef]
14. Wadas, T.J.; Wong, E.H.; Weisman, G.R.; Anderson, C.J. Coordinating radiometals of copper, gallium, indium, yttrium, and zirconium for PET and SPECT imaging of disease. *Chem. Rev.* **2010**, *110*, 2858–2902. [CrossRef] [PubMed]
15. Dewanjee, M.K. The chemistry of ^{99m}Tc-labeled radiopharmaceuticals. *Semin. Nucl. Med.* **1990**, *20*, 5–27. [CrossRef]
16. Lin, M.S.; Winchell, H.S.; Shipley, B.A. Use of Fe(II) or Sn(II) alone for technetium labeling of albumin. *J. Nucl. Med.* **1971**, *12*, 8.

17. Vanbilloen, H.P.; Verbeke, K.A.; De Roo, M.J.; Verbruggen, A.M. Technetium-99m labelled human serum albumin for ventriculography: A comparative evaluation of six labelling kits. *Eur. J. Nucl. Med.* **1993**, *20*, 465–472. [CrossRef] [PubMed]
18. Canziani, L.; Marenco, M.; Cavenaghi, G.; Manfrinato, G.; Taglietti, A.; Girella, A.; Aprile, C.; Pepe, G.; Lodola, L. Chemical and Physical Characterisation of Macroaggregated Human Serum Albumin: Strength and Specificity of Bonds with ^{99m}Tc and ^{68}Ga. *Molecules* **2022**, *27*, 404. [CrossRef] [PubMed]
19. Marenco, M.; Canziani, L.; De Matteis, G.; Cavenaghi, G.; Aprile, C.; Lodola, L. Chemical and Physical Characterisation of Human Serum Albumin Nanocolloids: Kinetics, Strength and Specificity of Bonds with ^{99m}Tc and ^{68}Ga. *Nanomaterials* **2021**, *11*, 1776. [CrossRef]
20. Hunt, A.P.; Frier, M.; Johnson, R.A.; Berezenko, S.; Perkins, A.C. Preparation of Tc-99m-macroaggregated albumin from recombinant human albumin for lung perfusion imaging. *Eur. J. Pharm. Biopharm.* **2006**, *62*, 26–31. [CrossRef]
21. Schembri, G.P.; Roach, P.J.; Bailey, D.L.; Freeman, L. Artifacts and Anatomical Variants Affecting Ventilation and Perfusion Lung Imaging. *Semin. Nucl. Med.* **2015**, *45*, 373–391. [CrossRef] [PubMed]
22. Shanehsazzadeh, S.; Jalilian, A.R.; Lahooti, A.; Geramifar, P.; Beiki, D.; Yousefnia, H.; Rabiee, A.; Mazidi, M.; Mirshojaei, S.F.; Maus, S. Preclinical Evaluation of ^{68}Ga-MAA from Commercial Available ^{99m}Tc-MAA Kit. *Iran. J. Pharm. Res. IJPR* **2017**, *16*, 1415–1423. [PubMed]
23. Malone, L.A.; Malone, J.F.; Ennis, J.T. Kinetics of technetium 99m labelled macroaggregated albumin in humans. *Br. J. Radiol.* **1983**, *56*, 109–112. [CrossRef] [PubMed]
24. Even, G.A.; Green, M.A. Gallium-68-labeled macroaggregated human serum albumin, ^{68}Ga-MAA. *Int. J. Radiat. Appl. Instrum. Part B Nucl. Med. Biol.* **1989**, *16*, 319–321. [CrossRef]
25. Watanabe, N.; Shirakami, Y.; Tomiyoshi, K.; Oriuchi, N.; Hirano, T.; Yukihiro, M.; Inoue, T.; Endo, K. Indirect labeling of macroaggregated albumin with indium-111 via diethylenetriaminepentaacetic acid. *Nucl. Med. Biol.* **1996**, *23*, 595–598. [CrossRef]
26. Hnatowich, D.J. Labeling of tin-soaked albumin microspheres with ^{68}Ga. *J. Nucl. Med. Off. Publ. Soc. Nucl. Med.* **1976**, *17*, 57–60.
27. Velikyan, I. Prospective of ^{68}Ga-radiopharmaceutical development. *Theranostics* **2014**, *4*, 34. [CrossRef] [PubMed]
28. Mathias, C.J.; Green, M.A. A convenient route to [^{68}Ga]Ga-MAA for use as a particulate PET perfusion tracer. *Appl. Radiat. Isot. Incl. Data Instrum. Methods Use Agric. Ind. Med.* **2008**, *66*, 1910–1912. [CrossRef] [PubMed]
29. Jain, A.; Subramanian, S.; Pandey, U.; Sarma, H.D.; Ram, R.; Dash, A. In-house preparation of macroaggregated albumin (MAA) for ^{68}Ga labeling and its comparison with comercially available MAA. *J. Radioanal. Nucl. Chem.* **2016**, *308*, 817–824. [CrossRef]
30. Hayes, R.L.; Carlton, J.E.; Kuniyasu, Y. A new method for labeling microspheres with ^{68}Ga. *Eur. J. Nucl. Med.* **1981**, *6*, 531–533. [CrossRef]
31. Maziere, B.; Loc'h, C.; Steinling, M.; Comar, D. Stable labelling of serum albumin microspheres with gallium-68. *Int. J. Radiat. Appl. Instrum. Part A Appl. Radiat. Isot.* **1986**, *37*, 360–361. [CrossRef]
32. Maus, S.; Buchholz, H.G.; Ament, S.; Brochhausen, C.; Bausbacher, N.; Schreckenberger, M. Labelling of commercially available human serum albumin kits with ^{68}Ga as surrogates for ^{99m}Tc-MAA microspheres. *Appl. Radiat. Isot. Incl. Data Instrum. Methods Use Agric. Ind. Med.* **2011**, *69*, 171–175. [CrossRef] [PubMed]
33. Ament, S.J.; Maus, S.; Reber, H.; Buchholz, H.G.; Bausbacher, N.; Brochhausen, C.; Graf, F.; Miederer, M.; Schreckenberger, M. PET lung ventilation/perfusion imaging using ^{68}Ga aerosol (Galligas) and ^{68}Ga-labeled macroaggregated albumin. *Recent Results Cancer Res.* **2013**, *194*, 395–423.
34. Amor-Coarasa, A.; Milera, A.; Carvajal, D.; Gulec, S.; McGoron, A.J. Lyophilized Kit for the Preparation of the PET Perfusion Agent [^{68}Ga]-MAA. *Int. J. Mol. Imaging* **2014**, *2014*, 269365. [CrossRef] [PubMed]
35. Mueller, D.; Kulkarni, H.; Baum, R.P.; Odparlik, A. Rapid Synthesis of ^{68}Ga-labeled macroaggregated human serum albumin (MAA) for routine application in perfusion imaging using PET/CT. *Appl. Radiat. Isot. Incl. Data Instrum. Methods Use Agric. Ind. Med.* **2017**, *122*, 72–77. [CrossRef] [PubMed]
36. Persico, M.G.; Marenco, M.; De Matteis, G.; Manfrinato, G.; Cavenaghi, G.; Sgarella, A.; Aprile, C.; Lodola, L. ^{99m}Tc-^{68}Ga-ICG-Labelled Macroaggregates and Nanocolloids of Human Serum Albumin: Synthesis Procedures of a Trimodal Imaging Agent Using Commercial Kits. *Contrast Media Mol. Imaging* **2020**, *2020*, 3629705. [CrossRef] [PubMed]
37. Gültekin, A.; Cayir, M.C.; Ugur, A.; Bir, F.; Yüksel, D. Detection of Pulmonary Embolism with Gallium-68 Macroaggregated Albumin Perfusion PET/CT: An Experimental Study in Rabbits. *Contrast Media Mol. Imaging* **2020**, *2020*, 5607951. [CrossRef]
38. Ayse, U.; Aziz, G.; Dogangun, Y. High-Efficiency Cationic Labeling Algorithm of Macroaggregated Albumin with ^{68}Gallium. *Nucl. Med. Mol. Imaging* **2021**, *55*, 79–85. [CrossRef] [PubMed]
39. Blanc-Béguin, F.; Masset, J.; Robin, P.; Tripier, R.; Hennebicq, H.; Guilloux, V.; Vriamont, C.; Wargnier, C.; Cogulet, V.; Eu, P.; et al. Fully automated ^{68}Ga-labeling and purification of macroaggregated albumin particles for lung perfusion PET imaging. *Front. Nucl. Med.* **2021**, *1*, 10. [CrossRef]
40. Velikyan, I. ^{68}Ga-Based radiopharmaceuticals: Production and application relationship. *Molecules* **2015**, *20*, 12913–12943. [CrossRef]
41. Zhernosekov, K.P.; Filosofov, D.V.; Baum, R.P.; Aschoff, P.; Bihl, H.; Razbash, A.A.; Jahn, M.; Jennewein, M.; Rosch, F. Processing of generator-produced ^{68}Ga for medical application. *J. Nucl. Med. Off. Publ. Soc. Nucl. Med.* **2007**, *48*, 1741–1748. [CrossRef] [PubMed]
42. Harris, W.R.; Pecoraro, V.L. Thermodynamic binding constants for gallium transferrin. *Biochemistry* **1983**, *22*, 292–299. [CrossRef]
43. Kotzerke, J.; Andreeff, M.; Wunderlich, G.; Wiggermann, P.; Zophel, K. Ventilation-perfusion-lungscintigraphy using PET and ^{68}Ga-labeled radiopharmaceuticals. *Nuklearmedizin. Nucl. Med.* **2010**, *49*, 203–208.

44. Senden, T.J.; Moock, K.H.; Gerald, J.F.; Burch, W.M.; Browitt, R.J.; Ling, C.D.; Heath, G.A. The physical and chemical nature of technegas. *J. Nucl. Med. Off. Publ. Soc. Nucl. Med.* **1997**, *38*, 1327–1333.
45. Lloyd, J.J.; Shields, R.A.; Taylor, C.J.; Lawson, R.S.; James, J.M.; Testra, H.J. Technegas and Pertechnegas particle size distribution. *Eur. J. Nucl. Med.* **1995**, *22*, 473–476. [CrossRef]
46. Mackey, D.W.; Burch, W.M.; Dance, I.G.; Fisher, K.J.; Willett, G.D. The observation of fullerenes in a Technegas lung ventilation unit. *Nucl. Med. Commun.* **1994**, *15*, 430–434. [CrossRef]
47. Burch, W.M.; Sullivan, P.J.; McLaren, C.J. Technegas—A new ventilation agent for lung scanning. *Nucl. Med. Commun.* **1986**, *7*, 865–871. [CrossRef]
48. Monaghan, P.; Provan, I.; Murray, C.; Mackey, D.W.; Van der Wall, H.; Walker, B.M.; Jones, P.D. An improved radionuclide technique for the detection of altered pulmonary permeability. *J. Nucl. Med. Off. Publ. Soc. Nucl. Med.* **1991**, *32*, 1945–1949.
49. Isawa, T.; Teshima, T.; Anazawa, Y.; Miki, M.; Motomiya, M. Technegas for inhalation lung imaging. *Nucl. Med. Commun.* **1991**, *12*, 47–55. [CrossRef]
50. Isawa, T.; Teshima, T.; Anazawa, Y.; Miki, M.; Soni, P.S. Technegas versus krypton-81m gas as an inhalation agent. Comparison of pulmonary distribution at total lung capacity. *Clin. Nucl. Med.* **1994**, *19*, 1085–1090.
51. Pourchez, J.; Albuquerque-Silva, I.M.D.; Cottier, M.; Clotagatide, A.; Vecellio, L.; Durand, M.; Dubois, F. Generation and characterization of radiolabelled nanosized carbonaceous aerosols for human inhalation studies. *J. Aerosol Sci.* **2013**, *55*, 1–11. [CrossRef]
52. Scalzetti, E.M.; Gagne, G.M. The transition from technegas to pertechnegas. *J. Nucl. Med. Off. Publ. Soc. Nucl. Med.* **1995**, *36*, 267–269.
53. Mackey, D.W.; Jackson, P.; Baker, R.J.; Dasaklis, C.; Fisher, K.J.; Magee, M.; Bush, V.; Burch, W.M.; Van der Wall, H.; Willett, G.D. Physical properties and use of pertechnegas as a ventilation agent. *J. Nucl. Med. Off. Publ. Soc. Nucl. Med.* **1997**, *38*, 163–167.
54. Fanti, S.; Compagnone, G.; Pancaldi, D.; Franchi, R.; Corbelli, C.; Marengo, M.; Onofri, C.; Galassi, R.; Levorato, M.; Monetti, N. Evaluation of lung clearance of inhaled pertechnegas. *Ann. Nucl. Med.* **1996**, *10*, 147–151. [CrossRef]
55. Sanchez-Crespo, A. Lung Scintigraphy in the Assessment of Aerosol Deposition and Clearance. *Semin. Nucl. Med.* **2019**, *49*, 47–57. [CrossRef]
56. Hartmann, I.J.; Hagen, P.J.; Stokkel, M.P.; Hoekstra, O.S.; Prins, M.H. Technegas Versus 81mKr Ventilation–Perfusion Scintigraphy: A Comparative Study in Patients with Suspected Acute Pulmonary Embolism. *J. Nucl. Med. Off. Publ. Soc. Nucl. Med.* **2001**, *42*, 8.
57. Möller, W.; Felten, K.; Seitz, J.; Sommerer, K.; Takenata, S.; Wiebert, P.; Philipson, K.; Svartengren, M.; Kreyling, W.G. A generator for the production of radiolabelled ultrafine carbonaceous particles for deposition and clearance studies in the respiratory tract. *J. Aerosol Sci.* **2006**, *37*, 631–644. [CrossRef]
58. Strong, J.C.; Agnew, J.E. The particle size distribution of technegas and its influence on regional lung deposition. *Nucl. Med. Commun.* **1989**, *10*, 425–430. [CrossRef]
59. Lemb, M.; Oei, T.H.; Eifert, H.; Gunther, B. Technegas: A study of particle structure, size and distribution. *Eur. J. Nucl. Med.* **1993**, *20*, 576–579. [CrossRef]
60. Blanc-Beguin, F.; Elies, P.; Robin, P.; Tripier, R.; Kervarec, N.; Lemarie, C.A.; Hennebicq, S.; Tromeur, C.; Cogulet, V.; Salaun, P.Y.; et al. ^{68}Ga-Labelled Carbon Nanoparticles for Ventilation PET/CT Imaging: Physical Properties Study and Comparison with Technegas®. *Mol. Imaging Biol.* **2021**, *23*, 62–69. [CrossRef]
61. Oehme, L.; Zophel, K.; Golgor, F.; Andreeff, M.; Wunderlich, G.; Brogsitter, C.; de Abreu, M.G.; Kotzerke, J. Quantitative analysis of regional lung ventilation and perfusion PET with ^{68}Ga-labelled tracers. *Nucl. Med. Commun.* **2014**, *35*, 501–510. [CrossRef]
62. Borges, J.B.; Velikyan, I.; Langstrom, B.; Sorensen, J.; Ulin, J.; Maripuu, E.; Sandstrom, M.; Widstrom, C.; Hedenstierna, G. Ventilation distribution studies comparing Technegas and "Gallgas" using [68]GaCl$_3$ as the label. *J. Nucl. Med. Off. Publ. Soc. Nucl. Med.* **2011**, *52*, 206–209. [CrossRef]
63. Bailey, D.L.; Eslick, E.M.; Schembri, G.P.; Roach, P.J. [68]Ga PET Ventilation and Perfusion Lung Imaging-Current Status and Future Challenges. *Semin. Nucl. Med.* **2016**, *46*, 428–435. [CrossRef]

MDPI

St. Alban-Anlage 66

4052 Basel

Switzerland

Tel. +41 61 683 77 34

Fax +41 61 302 89 18

www.mdpi.com

Pharmaceuticals Editorial Office

E-mail: pharmaceuticals@mdpi.com

www.mdpi.com/journal/pharmaceuticals

www.ingramcontent.com/pod-product-compliance
Lightning Source LLC
LaVergne TN
LVHW071242170726
843515LV00006BA/1306